Praise for *Feather Trails*

"I cannot put *Feather Trails* down. The book flies, and gracefully. It is timely, important, and applies widely. It reaches wide, and I hope it will be widely read. All Sophie says is true and can and should also be said."

—**Bernd Heinrich**, biologist; professor emeritus,
University of Vermont; author of *Mind of the Raven*

"*Feather Trails* takes you behind the front lines of conservation to experience the daily rigor and reward of a field biologist. Osborn guides you through forests and canyons to the haunts where our most endangered birds eat, sleep, and breed. As she works to understand and restore their populations, her splendid recounting explores the science of birds and their conservation, reveals the challenges a young technician must overcome, and inspires care for the rarest creatures among us."

—**John M. Marzluff**, professor of wildlife science,
University of Washington; author of *Gifts of the Crow*

"Field biologists are my heroes. Sophie Osborn is one of the main reasons why. Her fieldwork, like her writing, does more than draw attention to the plight of sacred species like the California Condor, the Peregrine Falcon, and the Hawaiian Crow. It also shows us how and why those creatures can be saved."

—**John Nielsen**, former National Public Radio environment
correspondent; author of *Condor: To the Brink and Back*

"Few biologists have had Sophie Osborn's front-row seat to the challenges and triumphs of saving critically endangered birds. In *Feather Trails*, Osborn does more than take us deep inside her decades of work to bring Peregrine Falcons, Hawaiian Crows, and California Condors back from the brink. By immersing us in the lives of these birds, we see them as she sees them—unique individuals whose successes and tragedies become our own. I can't recommend it highly enough."

—**Scott Weidensaul**, author of *A World on the Wing*

Feather Trails

Feather Trails

A Journey of Discovery Among Endangered Birds

Sophie A. H. Osborn

Foreword by Pete Dunne

Chelsea Green Publishing
White River Junction, Vermont
London, UK

Cover photograph *Peregrine, On Your Mark* was captured by Morris Finkelstein at State Line Lookout Park in New Jersey, overlooking the Hudson River. Morris Finkelstein's website is: morris-finkelstein.fineartamerica.com.
Back cover photograph of Condor 119 flying over the Grand Canyon is by Christie Van Cleve.
Back cover photograph of one of the last wild Hawaiian Crows (the Keālia male) is by Jack Jeffrey. More of Jack's photos can be found at https://jackjeffreyphoto.com.

Unless otherwise noted, all illustrations copyright © 2024 by Kira A. Cassidy.

The excerpt on page 194 is from "Do not go gentle into that good night" from *The Poems of Dylan Thomas* by Dylan Thomas. Copyright © The Dylan Thomas Trust. Reprinted with permission of New Directions Publishing.

The excerpt on page 304 is from "The Hill We Climb," copyright © 2021 by Amanda Gorman. Used by permission of the author.

Project Manager: Natalie Wallace
Developmental Editor: Matthew Derr
Copy Editor: Diane Durrett
Proofreader: Angela Boyle
Indexer: Kathy Berry
Designer: Melissa Jacobson
Page Layout: Abrah Griggs

Printed in Canada.
First printing May 2024.
10 9 8 7 6 5 4 3 2 1   24 25 26 27 28

Our Commitment to Green Publishing
Chelsea Green sees publishing as a tool for cultural change and ecological stewardship. We strive to align our book manufacturing practices with our editorial mission and to reduce the impact of our business enterprise in the environment. We print our books using vegetable-based inks whenever possible. This book may cost slightly more because it was printed on paper that contains recycled fiber, and we hope you'll agree that it's worth it. *Feather Trails* was printed on paper supplied by Marquis that is made of recycled materials and other controlled sources.

Library of Congress Cataloging-in-Publication Data
Names: Osborn, Sophie A. H., author. | Dunne, Pete, 1951- writer of foreword.
Title: Feather trails : a journey of discovery among endangered birds / Sophie A.H. Osborn ; foreword by
    Pete Dunne.
Description: White River Junction, Vermont : Chelsea Green Publishing, [2024] | Includes bibliographical
    references and index.
Identifiers: LCCN 2023059599 (print) | LCCN 2023059600 (ebook) | ISBN 9781645022428 (hardcover)
    | ISBN 9781645022435 (ebook) | ISBN 9781645022442 (audio)
Subjects: LCSH: Rare birds—United States. | Birds—Conservation—United States. | Osborn, Sophie A. H.
    | Women biologists—United States—Biography.
Classification: LCC QL68 .O83 2024 (print) | LCC QL68 (ebook) | DDC 598.1680973—dc23/eng/20240310
LC record available at https://lccn.loc.gov/2023059599
LC ebook record available at https://lccn.loc.gov/2023059600

Chelsea Green Publishing
White River Junction, Vermont, USA
London, UK

www.chelseagreen.com

FSC
www.fsc.org

MIX
Paper from
responsible sources
FSC® C103567

For the birds that inspired me
to give back to them;

For the unsung heroes who fought
so valiantly to keep them flying;

And for my sisters,
Lisa and Natasha, the bookends
who kept me upright in the dark
and soared with me in the light.

One of the penalties of an ecological education is that one lives alone in a world of wounds. Much of the damage inflicted on land is quite invisible to laymen. An ecologist must either harden his shell and make believe that the consequences of science are none of his business, or he must be the doctor who sees the marks of death in a community that believes itself well and does not want to be told otherwise.

—**Aldo Leopold**,
*A Sand County Almanac—
with Essays on Conservation
from Round River*, 1949

The beauty and genius of a work of art may be reconceived, though its first material expression be destroyed; a vanished harmony may yet again inspire the composer; but when the last individual of a race of living beings breathes no more, another heaven and another earth must pass before such a one can be again.

—**William Beebe**,
The Bird: Its Form and Function,
Volume 1, 1906

Contents

Foreword

I n Sophie Osborn's marvelous reflections about her life with birds, readers will learn much about the challenging, but gratifying, life of a wildlife biologist; readers will find much about their own lives reflected in the pages of *Feather Trails*, too.

As a postgrad, the author found herself searching for a purpose and meaning when she chanced to see the crossbow shape of a Peregrine Falcon soaring over a Vermont lake. This chance encounter with one of the planet's apex denizens gave her life wings, guiding her into a life of bird study and bird conservation. A few weeks later, working as a peregrine technician at a release site in the mountains of Wyoming, she would find herself gazing into the "fathomless" eyes of one of the young peregrines now in her charge. Looking past her reflected image in the bird's eyes, her focus passed into those dark guileless tunnels and into a world anchored in nature and guided by higher purpose.

Fieldwork is not glamorous but has its rewards, despite the biting gnats that descend right when your eye is fused to a spotting scope, rocks that seem always to find a path through your sleeping pad, and the bears (real and imagined) that are ever prowling just beyond the campfire's light. Jane Goodall was determined enough to take this path. George Schaller, too. Despite the challenges, Sophie embraced this life and so brings us a story of three endangered bird species.

In this book Sophie will guide you into a world of focused awareness where you, too, will be transformed by birds and ascend to a higher plane of appreciation. Sophie touched this elevated awareness on her first assignment as a peregrine hack-site technician, when her senses were captivated by the song of a Rock Wren—a new bird for her Eastern-attuned ears and a direct link to this exciting and soul-expanding immersion into nature. Birds, you see, are our ambassadors to the natural world, partners in our long ride into discovery and wonder.

Birds have been part of the human experience since before our ancestors vacated the limbs of trees, integrating themselves into every facet of our lives. From legends to literature, fables to flags, they have informed and influenced our religions, art, fashion, and everyday expression. Indeed, were birds excised from the human equation, we would scarcely recognize ourselves. Yet it is ironically human unmindfulness that has pushed many of our planetary partners to the brink of extinction, and now it is the job of people like Sophie to pull them back from that dark abyss. This focused ambition brings out the scientist and warrior in her. These pages uncloak the poet and the storyteller.

As I write these words, a troop of California Quail are just setting out on their daily rounds. In the oaks outside my office window a host of wintering warblers are dancing in the branches. I am brought to wonder how my own life might have fared had I not been introduced to Sophie's exquisite awareness. Would I have missed the quail or been oblivious to the warblers? I pray not.

But this book is not about me, it is about you and a gifted writer, because together you are a team, partners in a dance whose conjoined steps lead to discovery and wonder. Ma Nature calls the tune, but it's your choice, reader, whether to get out on the dance floor or sit this life out.

Congratulations to Sophie for crafting this insightful book, kudos to you for reading it, and may birds be the beneficiaries of this literary union. Just bear in mind, where *Feather Trails* ends, your journey into birds begins.

Pete Dunne, Student of Birds
Paso Robles, California

Preface

Like an arrow loosed from a celestial bow, the falcon shot into view over the cliff and imprinted itself forever in my mind. With rapid wingbeats, it flew over the Vermont lake, high above a hunting Osprey. Gliding leisurely over the wind-ruffled water, the Osprey peered into the clear depths for its fish prey, unaware of the falcon-shaped storm cloud gathering strength in the skies above. Hesitating for a fraction of an instant, the falcon pulled its wings in close to its body and dropped out of the sky with preternatural speed and a piercing scream. I could hardly follow the lightning velocity of its trajectory. Mere inches from the Osprey's back, the falcon pulled sharply out of its dive and headed skyward again as the Osprey rolled over and presented its talons in belated defense. Righting itself then flapping its wings with uncharacteristic haste, the Osprey fled to the safety of a dead tree.

The dark-helmeted falcon circled over the perched Osprey, reinforcing its territorial message to the perceived intruder, then returned to its cliffside home. I lowered my binoculars and exhaled. My first Peregrine Falcon. Only a few decades earlier, the species had been extirpated from the eastern United States, an unwitting victim of the pesticides that had been sprayed with enthusiastic and naive abandon over habitats countrywide. In the early 1990s, as I scanned the lakeshore cliffs, the peregrine was still in the early stages of recovery, slowly recolonizing its former haunts. Despite nearly two decades of reintroductions to former breeding territories in the eastern United States, only a few pairs nested in Vermont. Unbeknownst to me, the cliff overlooking Lake Willoughby, which this fierce peregrine had recolonized, was the site of perhaps the last credible observation (in 1970) of an eastern peregrine in the United States prior to the population's disappearance.[1]

As I stood by the lake, my mind's eye played over the falcon's purposeful flight, implausibly fast dive, and focused aggression toward the Osprey that had intruded on its territory. The images resonated with unknown portent. Even without the benefit of hindsight, the moment felt to me like an awakening—the falcon a foreshadowing, a messenger carrying a host of unknown possibilities that time would crystallize into feathered substance. I held the feeling close, delighting in the knowledge that in a few weeks' time I would regularly watch peregrines streaming across the sky.

I was new to the bird realm, having only recently been initiated to the dazzling diversity of their forms and plumages, the magic of their unique songs, and the thrill of their aerial exploits. At the time, I was a somewhat reluctant initiate, enthralled as I still was by mammals, which seemed less alien, less prehistoric, and more engaging than birds. But that fierce peregrine opened the gates for me to an unanticipated future, and its image returned to me repeatedly over the years. The falcon epitomized the soul-stirring beauty of birds and hinted at their surprising fragility. And it came to mark my entry into the endlessly captivating avian world—a journey that became a lifelong odyssey of deep appreciation, profound concern, and ongoing effort to protect and restore dwindling bird populations and their diminished habitats.

PART I

Peregrine Falcon
Recapturing the Skies

I remember in my childhood seeing or rather recognizing my first Peregrine as he glided through immeasurable space among the clouds, and never in all my life can I recall having witnessed anything in Wild Nature which left an impression so indelible and so full of romance as that small black cross against the sky.

—**H. Mortimer Batten**, *Inland Birds*, 1923

. . . what's past is prologue.

—**William Shakespeare**, *The Tempest*, 1611

CHAPTER ONE

Beginnings

Dropping onto the parched ground, I crawled into a patch of shade by a dusty ditch as the world tilted and spun around me. Waves of nausea blinded me to the grasshoppers that snapped and popped when I inadvertently flushed them from their desiccated perches. The searing sun and oppressive afternoon heat conspired with gravity to impale me on the stubbly grass. Closing my eyes, I could still see the contours of the deceptively green hills, the long sweep of the valley behind me, and the snowcapped mountains beyond. The landscape was dramatic, bold, full of beauty, full of possibilities. Up close, though, the insects hummed, the earth cracked, and the air throbbed with a dry heat.

"Not an auspicious beginning," I moaned quietly as I poured tepid water on my face in a vain attempt to cool myself.

My companion and I had left Vermont's lush greenness a week earlier and had been greeted only this morning by the vastness of Wyoming. And now, I was attempting to hike up to what would be our home for the next two months. I had always been prone to heat exhaustion and, in my excitement to begin my first field job, I had forgotten to hydrate and to refrain from overexerting until I was more habituated to the aridity and the powerful sun. I hadn't even considered the elevation when I'd heaved on my absurdly heavy pack, intent on minimizing the trips between car and campsite. Cody, the closest town, sat approximately 5,000 feet (1,500 m) above sea level. In comparison, St. Albans, Vermont, which I currently called home, was no higher than a piece of flotsam at less than 500 feet. The pounds of dried beans that had grown to boulderlike proportions in my pack had been my undoing as I hiked up the pitted dirt road under an unforgiving sun to our future campsite.

"Peregrine Falcons." The words danced in my mind. Visions of a clear blue lake and the only peregrine that I had ever seen drifted back to me, pushing away the heat and the prickle of the already yellowing grass. A blue-gray streak across a late-spring sky. The whistle of the wind over finely cut wings. And that was just going to be the beginning.

Several months before venturing west to Wyoming, on a Vermont day when every twig and branch was encased in ice after a March storm, I had idled away time between classes by skimming the chaos of flyers that fought for preeminence on the biology department noticeboard. I had almost missed the simple announcement: "Field assistants needed to help in release and monitoring of Peregrine Falcons. Contact The Peregrine Fund. Boise, Idaho." It sounded impossibly exotic. Falcons. The West. All so far removed from where I had been only a few years earlier as I finished my undergraduate degree at the University of Pennsylvania, double-majoring in international relations and French literature. Thinking I might become a diplomat, I had moved to Washington, DC, and worked temporary jobs while looking for a more permanent start to my career. The realization that diplomacy was more about politics than it was about the languages and travel with which I was enamored soon disabused me of the notion that I belonged in the diplomatic corps. I floundered. I couldn't envision an occupation that would satisfy my disparate interests: a deep love for animals, a desire to do good, a fascination with languages, a thirst for international travel.

The lifeline that finally provided a direction came from an unexpected source one spring morning, as my little rented townhouse vibrated with foot-stomping gospel music sung at top volume by the talented parishioners of next door's Tried Stone Church of Christ. As I leafed through the previous day's mail, a glossy catalogue caught my eye. It was from the School for Field Studies, an organization dedicated to offering study abroad experiences through field-based learning and research. My housemate, Lois, had taken one of their field courses our junior year, traveling to Australia to learn about tropical rainforest ecology.

Idly, I flipped through the picture-filled pages, stopping only when I came to photos of the African savannah and its resplendent mammals. "Course in Wildlife Biology and Management in Kenya," the catalogue proclaimed, and, as I read a description of what the classes entailed, I felt my pulse quicken. *This* was what I wanted to be doing. I had always loved animals but, while I had enjoyed working as a veterinary assistant the previous summer, I could not conceive of spending my life in a vet's office neutering and spaying untold numbers of pets. But learning about wild animals, their relationship to their environment, and how to evaluate and conserve their populations? That I could get excited about.

I had never heard of wildlife biology. I knew nothing about ecology. I had been daunted by science and math in high school and retreated to the safety of other disciplines. But introductory biology was a requirement for the School for Field Studies' course, so I signed up for a summer biology class and readied an application for the Kenya program.

Biology was a revelation. To my surprise, I enjoyed the class immensely, though part of that enjoyment was undoubtedly related to the joy of being in school again and the freedom of being out of high-heeled shoes and a cloistering office. I biked to my class each day, traveling from one side of Washington, DC, to the other, attended the all-morning session, then spent the afternoons working as a city bike messenger, relishing the intense physical activity, the adrenaline rush of navigating traffic, and my time outdoors.

As summer waned and cool nights brought Washingtonians relief from the stultifying August heat, I headed to Africa to try out my new direction. Though I was a few years older and far less knowledgeable about our chosen field of study than the other students, I put aside my discomfiture and embraced the rich experience of living what I was learning: basic ecology, wildlife biology, and wildlife management; concepts such as symbiosis, parasitism, and carrying capacity; processes such as photosynthesis, nutrient cycling, and energy flows.

I gloried in the sights of running giraffes that appeared to float across the savannah with mystifying grace, ponderous elephants kicking up trails of dust as they lumbered along in single file to a water source, shaggy-maned wildebeests filling the air with their strange moaning calls, the impenetrable golden eyes of a cheetah. I learned to use masking tape to remove the nearly invisible pepper ticks that blanketed my legs after a day in the field. I braved a small cobra that reared its head in protest when I almost stepped on it while out running on a ranch road, and I shrugged off a small but potentially deadly puff adder that found its way into our banda (the wood-and-thatch hut where four of us slept) and coiled itself at the foot of my roommate's bed. Having an intense phobia of spiders, I much less bravely punctuated my nights by searching out the large arachnids that crawled unimpeded into our banda through the many gaps between the wooden boards, shining my flashlight over the walls and hoping with quiet desperation that the terrifying intruders would decide against visiting my bed. (They didn't; on one occasion, one greeted me on my pillow.)

After four months in Kenya, I returned to Washington, DC, and did an internship with the World Wildlife Fund—a conservation organization dedicated to conserving nature and reducing threats to global biodiversity—before moving back to Vermont to begin classes. If I wanted to pursue a career in wildlife biology, I needed to take a host of biology, calculus, chemistry, statistics, and wildlife-related courses.

Now, in my second semester at the University of Vermont, I had chanced upon what sounded like a dream summer job. With trembling fingers, I jotted down the particulars from The Peregrine Fund's flyer, eager to share them with my friend Garrett, who would be as keen as I was to work with falcons.

A few days later, we received our application forms in the mail and scanned the inquiries about our experiences with birds and the outdoors. Noting with consternation such unexpected questions as "Have you used firearms?" and "Can you rock climb?" we worried that our acceptance might depend on our being able to answer *yes* to each question. To give our applications every chance at success, we began taking weekly target lessons at a local firearms range and signed up for a two-day rock-climbing course that provided an invaluable introduction to safe climbing techniques but also affirmed my terrible fear of heights. While the yeses we checked off on our application might be qualified yeses, at least our claim that we had some exposure to these outdoor pursuits would now be legitimate.

Five weeks after submitting our applications, we slid along the icy road to our rural mailbox on a bitterly cold, snowy night and discovered two thick envelopes from Idaho awaiting us. Eyes wide with anticipation, fingers a trifle shaky, we tore open our respective envelopes and each found a congratulatory letter and a cornucopia of information about releasing captive-raised peregrines to the wild.

"They've accepted us!" I shouted out gleefully to the quiet night.

We were heading to Wyoming and our first endeavor as field biologists!

———

"Are you okay?" A tinge of frustration laced the concern in Garrett's voice as he checked to see how I was faring. "We really need to get our things up to the campsite before the falcons arrive."

"The worst of the nausea's passed." Groggily, I checked my watch. "I should be good to go in a few minutes. We still have over an hour until The Peregrine Fund delivers our birds. What's our camp site like?" I asked, gaining strength with every passing moment, now that I wasn't expending energy beneath the sun's full glare.

Garrett, who had gone ahead to drop off his load while I recovered from my queasiness, laughed a trifle wryly. "It's not exactly what we envisioned. A patch of brushy ground by a nice little stream, but lots of cow pies all around."

"So not exactly pristine wilderness," I said, as I clambered to my feet and shouldered my pack. "I guess that's to be expected when your release site is on a ranch and not in a national forest." My gaze wandered to the distant peaks that thrust upward in a disorderly array on the horizon. "Oh, but what beautiful country," I sighed, already captivated by my surroundings.

"Here, let me take some of that food." Garrett pulled me back to the task at hand as he reached to transfer some bags of beans from my overloaded backpack to his now-empty pack.

"No. I'm fine," I protested, eager as always to pull my weight. "I just have to hydrate and move a little more slowly. I think I'll be okay now." Garrett shrugged, knowing all too well that my determination was matched only by my stubbornness.

"Okay. I'm going to head down to the jeep to get another load. I'll meet you up there. There's no camp or anything. Just a big, brown metal container by the edge of the road."

Feeling rejuvenated by the brief rest and lukewarm water, I shuffled slowly up the dirt track. Hands holding my pack straps to ease the weight on my back, I absorbed the western landscape as I walked, already recalibrating my metric for evaluating greenery, moisture, and wildness. A shallow stream ran alongside this part of the trail, and the patchy vegetation that traced its course offered a verdant contrast to the encompassing wind-seared hills. The narrow streambed I followed would be considered barren and exposed by Vermont's lush standards. But in this arid land, the unimposing stream was a ribbon of rich greenery where birds, insects, and hints of coolness coalesced.

Losing myself in the stream's quiet song, I soon reached the bear-proof container in which we would store our food to dissuade any ursine visitors from invading our camp. I'd never seen a bear box before and hadn't known what to expect. After dropping my pack and transferring its contents into the rugged metal container, I realized that it made a perfect bench and sat down to wait for the arrival of the young falcons that were going to consume our attention for the next few months.

After what felt like an interminable wait, Garrett joined me at the bear box. Another eternity later, we heard feet trudging toward us. Flashing quick smiles to each other, eyes alight with nervous anticipation, we stood to meet the newcomer. Seconds later, Bryan Kimsey came into view carrying a pack on his back and a large cardboard box—with circular breathing holes cut into the top of each side—in his arms. I felt a rush of adrenaline as I looked at the box. Despite its innocuous exterior, I knew it concealed five immature Peregrine Falcons that faced an unknown future—one that depended, at least in part, on my efforts and dedication. I tried to imagine what the youngsters might look like. How big were they? Were they fully feathered? My mind whirled with questions as I struggled to convey a calm composure and restrain my bubbling excitement so as not to overwhelm Bryan with my enthusiasm.

Dragging my gaze away from the cardboard box, I greeted Bryan, whose warm, brown eyes, shaded by a brimmed felt hat, were a reassuring contrast to his unsmiling demeanor. After brief introductions, Bryan told us he was anxious to get the birds settled. He had left The Peregrine Fund's captive-breeding facility in Boise, Idaho, with his charges early that morning and had driven over nine hours to reach us. To minimize the nestling falcons' stress and discomfort, release-site supervisors had to leave at whatever time of day or night was necessary to travel directly to each site and allow for time to transfer the birds to their new home. Garrett and I grabbed our smaller day packs and followed Bryan up to the site where our peregrines would be introduced to their new world.

Glancing away from the surrounding hills, which folded into each other like chaotic sea swells immobilized by a photograph, I looked down at my hiking boots. Each step I took brought me closer to those who had worked to save this imperiled species. And each step marked the beginning of my own journey to help this iconic falcon recover its depleted numbers and reclaim its former range.

Though relatively uncommon wherever it occurs, the Peregrine Falcon (*Falco peregrinus*) is the most widely distributed, naturally occurring bird species in the world and one of the most widely occurring vertebrate animals on Earth. A top predator, it is found on every continent except Antarctica and pursues its prey in skies over open habitats that range from windswept tundra to humid tropics, craggy mountains to lush wetlands, surf-pounded seacoasts to inland river canyons, wide-open deserts to sprawling forests, undeveloped maritime islands to densely inhabited cities.[1]

Peregrine, which originated from the Latin *peregrinus* meaning "coming from foreign parts," is defined as wandering, traveling, or migratory, and aptly captures the bird's expansive range and long-distance movements. Some North American Peregrine Falcons travel more than 15,500 miles (approximately 25,000 km) from their northerly breeding grounds to their winter range and back each year.[2]

Despite the peregrine's adaptability and widespread occurrence, the bird is sparsely distributed throughout its range. The availability of suitable nest sites (typically a ledge on a high cliff or tall structure) and the territoriality of breeding pairs, whose spacing is itself influenced by the abundance of food, limit its presence. Perhaps because of the scarcity of suitable nesting locations, peregrines show tremendous fidelity to quality nest sites. In the years before World War II, when peregrine populations were relatively stable, the birds used traditional aeries (nests located in high, inaccessible places) over periods of decades and even centuries in some places.[3]

Peregrine Falcons have never been abundant in North America. Indeed, with the exception of several neotropical species whose ranges reach their northern limit in the southern United States, the peregrine may be one of the least numerous raptors (birds of prey) north of Mexico.[4] Even before their numbers plummeted beginning in the 1950s, peregrines likely occupied only 7,000 to 10,000 territories in all of North America, ranging from Alaska and the Canadian Arctic to Baja California and the mountains of Mexico.[5] So unless one happens to live or work near a nest site, seeing a peregrine is usually rare, and most birders are thrilled to get even a glimpse of this elegant, swift-flying bird.

Human appreciation for the peregrine has ebbed and flowed over the course of history. Falcons were worshipped by the ancient Egyptians, who considered the bird to be a sky god and companion of the sun, as well as a symbol of transformation, rebirth, and eternal life.[6] Because of their surprisingly docile nature, falcons were captured and trained to hunt for their Egyptian and Chinese keepers as early as 4,000 years ago. Although falconry was widespread in the Orient, it was not practiced in Britain until the ninth century, and did not become widely prevalent until the early

Crusaders—who were introduced to Arab falconers and their ways of training and flying hawks—popularized the sport early in the thirteenth century.[7]

Despite being revered by the royal classes during falconry's heyday in the Middle Ages, by the eighteenth century—with the advent of firearms—Peregrine Falcons came to be viewed as competitors for game and were persecuted unrelentingly, along with hawks and eagles. Europeans dedicated themselves with singular passion to eliminating these predators in the hope that small game animals would thereby flourish. This malevolent attitude toward predators and the carnage that followed spread to the rest of the globe and continues in many areas to this day, though peregrines eventually received federal protection in North America in the 1970s.[8] Given their individual scarcity and their widespread distribution, peregrines nevertheless fared better in most areas under this onslaught than did many other raptors.

Although peregrines had long been killed opportunistically, their persecution reached unprecedented levels when Britain sanctioned their systematic extermination during World War II to protect carrier pigeons, which transported messages between military personnel and sometimes met their demise in the talons of hungry falcons.[9] More than 600 British peregrines were shot during that time.[10] Ever-resilient, though, peregrines in Britain began to recover soon after the wartime persecution ended. But their recovery proved to be short-lived.

Today, it is almost inconceivable that a bird with the peregrine's cachet and charisma could disappear virtually without notice. But in an era without mass communication and with far fewer birders combing the landscape, the falcon's sparse distribution and tendency to nest in remote and inaccessible areas meant that few noticed when the species disappeared from its former haunts. Ironically, it was the bird's most vocal detractors who drew the world's attention to the peregrine's population declines in the 1950s and who precipitated the eventual discovery of a far more profound threat to the falcon's well-being than those inflicted by generations of protective gamekeepers, fanatic egg collectors, unethical falconers, and trigger-happy shooters.

Because the peregrine preys on domestic pigeons—descendants of the wild Rock Pigeon—the falcon is not a particular favorite among "pigeon fanciers." For thousands of years, people worldwide have kept pigeons for food, sport, and entertainment. Some fanciers race homing pigeons, which are transported long distances then released to fly home to their lofts or coops. Others breed pigeons for unique flying characteristics, such as rolling, tumbling, or diving. (Such strange behaviors often attract raptors that have evolved to detect debilitated prey by focusing on birds that are not flying like their flockmates.) Pigeon fanciers include such notables as artist Pablo Picasso, deposed Panamanian dictator Manuel Noriega, father of evolution Charles Darwin, England's Queen Elizabeth II, and heavyweight boxing champion Mike Tyson, who threw his first punch at a thug who callously killed one of ten-year-old Tyson's beloved pigeons.[11]

In both Europe and the United States, extremist fanciers have occasionally expressed their frustration at raptors that target their cherished pigeons, by illegally killing their nemeses.[12] In 1960, Ivor George, a Welsh pigeon fancier, expressed his frustration more circumspectly, but with no less passion. In a television broadcast, he decried peregrines for the large number of homing pigeons they were killing each year in the mining valleys of South Wales—a stronghold for pigeon fanciers—disrupting the fanciers' sport and destroying the objects of their affection. Soon after, the fervent fanciers formalized these complaints; they petitioned the British government to remove newly established legal protections for peregrines, claiming that falcons were exacting an unacceptable toll.[13] Realizing that it knew little about actual peregrine numbers in Britain, the British Trust for Ornithology responded to the government's request for information about the bird's status by hiring biologist Derek Ratcliffe to organize a two-year survey of Britain's peregrines.[14]

The outcome of this survey, completed in 1962, was shockingly revelatory. Peregrine Falcons had declined precipitously throughout Britain during the 1950s, disappearing from many of their territories. In areas where birds remained, they often failed to raise young, and their nests sometimes contained broken eggs or eggshell fragments. Ratcliffe had first noted broken eggs in peregrine nests in 1951 and, in the intervening years, had even observed falcons eating their own eggs. Such baffling behavior and inexplicable egg breakages had never been recorded before. The pigeon fanciers' woes were superseded by the concerns of scientists and conservationists who scrambled to make sense of the unprecedented population declines of this apex predator.[15]

Disease and persecution were quickly ruled out as reasons for the widespread decline. But reports of large numbers of wild birds dying in fields recently sown with insecticide-treated seeds and the subsequent death of two otherwise healthy adult peregrines in Britain in 1961 raised the possibility that newly popular pesticides might be contributing to the falcon's disappearance.[16] Dichlorodiphenyltrichloroethane (DDT)—a synthetic organochlorine insecticide—was widely used for the first time during World War II to prevent the spread of insect-borne diseases. DDT was subsequently used in increasing amounts to control agricultural and forest insects. More toxic to mammals than DDT was, the related cyclodiene insecticides (such as aldrin and dieldrin) were introduced into British agriculture around 1955 and were also quickly embraced by farmers. These newest chemicals were soon found in the tissues of dead or convulsing birds collected from freshly sown fields. Birds that were experimentally dosed with these same pesticides in the lab died in a similarly convulsive manner.[17]

European animals that preyed on wild seed–eating birds—sparrowhawks, Common Kestrels, Tawny Owls, foxes, and even domestic cats—were also found dead on or near the same fields as their poisoned prey. That the Peregrine Falcon, which often nested far from agricultural lands, might similarly be poisoned by preying on sickened birds began to seem possible when an addled peregrine egg was analyzed for chemicals for the first time in 1961 and found to contain small amounts of dieldrin, dichlorodiphenyldichlorethylene (DDE—a breakdown product of DDT), and other organochlorine pesticides. It was the first direct evidence that peregrines could accumulate pesticide residues from eating contaminated prey.[18]

————

In 1963, Joe Hickey—a much-beloved University of Wisconsin professor—read Derek Ratcliffe's recently published paper on the peregrine's status in Britain. The report highlighted the bird's desertion of nesting territories and suggested that pesticides might be responsible for the falcon's precipitous decline. As he absorbed Ratcliffe's alarming analysis, Hickey suddenly recalled a rumor he had heard the previous year that not a single peregrine chick had fledged in the northeastern United States.

Hickey loved birds and had a particular fascination with peregrines, which he'd studied extensively in the 1930s.[19] As a child, his passion for birds had led him to membership in New York City's famed Bronx County Bird Club, where he became an early birding companion—and eventually lifelong friend—to famed artist, birder, and creator of the modern field guide, Roger Tory Peterson. En route to a career in ecology, Hickey was a student of renowned writer and conservationist Aldo Leopold, under whose tutelage he completed a master's degree in 1943. That same year, Hickey published *A Guide to Bird Watching*, the first American book to describe this increasingly popular pursuit.[20] Now, alarmed by hearsay about peregrines not producing young, Hickey mobilized a team to document the bird's status in the eastern United States by repeating surveys he and others had conducted in the region in the 1930s.

On April 1, 1964, peregrine surveyor and businessman Dan Berger and his assistant Charles Sindelar left Milwaukee, Wisconsin, in a Volkswagen microbus—specially outfitted to store 250 topographic maps and a summer's worth of field gear—and embarked on a survey of eastern peregrine aeries. Dan Berger was captivated by peregrines. The first time he'd ever searched for *Falco peregrinus* along the upper Mississippi River, he'd found a beautiful adult silhouetted against the sky, perched on a snag jutting from a cliff. That sight fostered his desire to monitor peregrines and precipitated annual surveys that he conducted for the next thirteen years—surveys that documented fewer and fewer birds.[21] Now, under Joe Hickey's directive, Berger was undertaking his most ambitious survey yet. Driving 14,000 miles (22,530 km) through fourteen states and one Canadian province, Berger and Sindelar checked 133 known peregrine territories between the Atlantic coast and the Mississippi River, from Alabama up to Nova Scotia. Arriving at each historic aerie, the duo eagerly scanned the cliffs, searching for peregrine nesting activity. Disappointment followed disappointment. And heavy hearts soon gave way to growing despair. Every territory they checked was deserted. The eastern Peregrine Falcon had disappeared.[22] Its former haunts echoed with silence as the inconceivable specter of extinction cast a somber pall over what had long been peregrine nesting cliffs.

Armed with the alarming news about eastern peregrines and reports of declines in many parts of Europe, Joe Hickey organized a conference in Madison, Wisconsin, in 1965, to discuss the unprecedented population crashes on both sides of the Atlantic. Scientists, falconers, and amateur naturalists spoke about the disappearance or decline of the iconic bird in their part of the world. Eventually, population declines were reported from thirty-six countries on five continents, leading conservationists to label the species' rapid and unprecedented losses an ecological disaster.[23] The Madison Conference also publicized the first tantalizing evidence that pesticides might be to blame, when British attendees Derek Ratcliffe and his colleague Ian Prestt superimposed maps depicting areas from which the peregrine had disappeared in Great Britain with the locations of pesticide-laden cereal farms. The map features overlapped perfectly. While the evidence thus far was circumstantial, it laid a foundation for a case scientists were beginning to make that organochlorine pesticides were responsible for the peregrine's global demise. And yet, no North American peregrine had ever been tested for pesticide residues. Conference participants left energized to further test hypotheses that organochlorine pesticides were either killing peregrines outright or were reducing their reproductive capabilities in unknown ways.

Daunted by the strength of the agrochemical industry and the force of the political winds against stricter pesticide regulations, another group of attendees left the conference committed to developing techniques to breed peregrines in captivity so that some falcons would remain even if wild peregrines disappeared altogether. More important, this group of biologists and falconers harbored the then-radical hope that if the chemical contamination of the environment *could* be reduced, the progeny of captive birds might one day be released to the wild to take back the skies from which their forebears had fallen. These were not idle longings. The origins of what became the four principal organizations dedicated to captive-breeding and reintroducing peregrines—The Peregrine Fund in Boise, Idaho; the Predatory Bird Research Group in Santa Cruz, California; The Raptor Center at the University of Minnesota; and the Canadian Wildlife Service—can be traced to this seminal meeting.[24]

Laying the groundwork for what would become one of the great conservation stories of modern times, the Madison Conference marked the beginning of a decades-long effort to restore the Peregrine Falcon to its prior abundance. The falcon's population declines were now recognized and publicized. The dangers of pesticides were now considered, and then intensively studied. And countless individuals embarked on the novel path of dedicating their careers and even their lives to ensuring that an elegant bird whose speed and grace had captured their imaginations and ignited their souls didn't disappear forevermore into the black hole of extinction.

CHAPTER TWO

Hacking

Thoughts about the efforts of so many peregrine devotees lent a certain gravitas to our proceedings as we followed Bryan across a shallow side stream and began climbing steadily toward the site of our falcons' new home. Though a trace headache reminded me to proceed with caution under the high-elevation sun, I was relieved that my heat exhaustion episode had preceded Bryan's arrival. "The silver lining to an overabundance of weighty beans," I thought wryly. Energized by the beauty of the surrounding landscape and the anticipation of seeing our falcons, I was more eager than ever to embark on my own effort to help peregrines return to the skies.

Sweating despite being buffeted by the wind, we paused to catch our breath as we prepared to hike up the last steep incline to a large plywood box. Resting atop a small set of cliffs, the "hack box" would serve as a sheltered nest until our falcons were ready to fly. Having gained significant elevation during our thirty-minute hike, we now had a spectacular 360-degree view of our surroundings. Far below us, to the east, lay the small stream along which we would make our camp. Panning clockwise, the South Fork of the Shoshone River wound its way through irrigated farm fields, tracing a green corridor between rolling hills clustered along the base of distant snowcapped peaks. And to the west and northwest, the horizon was dominated by a horizontal amphitheater of flat-topped, clifflike mountains whose distant crevices were highlighted with horizontal bands of snow, like icing in a layer cake.

Tick-ear, tick-ear, churree churree. Raising my binoculars, I searched for the source of the repetitive song emanating from a jumbled series of sandstone boulders.

"Rock Wren," Bryan said nonchalantly as he saw me searching out the sound. Scanning the boulders, I caught sight of the camouflaged songster. Its brown back and buffy front blended into the rocks over which it hopped, pausing every few seconds to issue its cheerful call.

"That's a new bird for me," I exclaimed, as I studied the bird's slightly drooped bill, cinnamon-colored flanks, and pale eyebrow stripe. The likelihood that I would recognize the wren's distinctive song upon hearing it again was still a source of wonder to me. Months earlier I'd read a book by a biologist who was studying Common Ravens in Maine's northern forests. As he watched a raven nest, he described the birds that he heard singing around him, identifying twenty-one different species in a ten-minute period.[1] It was the first time I'd ever heard of anyone identifying birds by their songs, and I read in amazement as the author listed the species with as much certainty as if each bird had landed next to him. Soon after, upon taking my first field ornithology course, I began to learn to identify birds by their songs as well, and I felt as if I'd acquired a brand-new sense. Where I'd previously heard background bird-chirping or noticed no noise at all, I now heard American Robins, Common Yellowthroats, and Warbling Vireos. It was as though I'd only ever used a fraction of my hearing before but was now enjoying the sound of a full orchestra, whose every instrument was identifiable, rich, and unique.

Tick-ear, tick-ear, the Rock Wren called again, before disappearing into a crack between boulders.

Heading up the last and steepest pitch, we hiked the grassy back side of the small cliffs that would be our peregrines' home base in the ensuing weeks, and finally reached the 500-pound hack box that was anchored with strong cables to the top of the cliffs.

"I'm just going to check on the condition of the hack box," Bryan announced, setting down the still-hidden falcons. Garrett and I removed our packs and watched as Bryan stepped out onto the cliff, then jumped across a large gap in the rock to reach the box. As he landed, he suddenly jumped sideways.

"Rattlesnake," he yelled back to us, his words almost disappearing into the wind that had gathered strength with every foot of elevation we'd gained. "A little one. About a foot long. Almost stepped on it. It dropped into the crack you have to jump over to get out here. This place is crawling with them."

Nervously, I looked around, expecting to see a mass of serpentine bodies, but only bunches of diminutive magenta, yellow, and white flowers greeted my wary eyes. After opening the side door of the hack box and glancing into the interior, Bryan gestured for us to hand over the cardboard box.

"Okay," he said, the snake already forgotten. "Sophie, why don't you climb onto the hack box so you can open and close the door from above when I put each falcon into the box."

I needed no more encouragement. Forgetting my newly discovered fear of rattlesnakes and squelching my long-standing fear of heights, I took a breath, jumped the crevice that split the top of the cliff, and scrambled up onto the four-foot-high hack box. Focusing on Bryan's cardboard box, I tried not to look at the cliff edge that dropped away into a steep, boulder-strewn slope. Finally, Bryan lifted one cardboard flap after another until, at last, I stared down at five puffs of down and feathers. And eyes. Oh, those eyes. Large, dark, unfathomable. Looking into those onyx eyes, I, like so many before me, was seduced by the mystical, strikingly beautiful Peregrine Falcon. The mesmerizing eyes drew me in, folding me into the peregrine's world and its long history—worshipped by the Egyptians; prized by emperors, kings, and sultans; revered by bird lovers on every continent.

Reaching into the box, Bryan placed a hand on either side of one of the youngsters and gently lifted it out, restraining its wings to protect the growing flight feathers. Still half covered in down, the five-week-old male clenched its feet, its astonishingly long toes—an adaptation for grasping avian prey—looking outsized and unwieldy. As the wind stirred its feathers, the nestling cast its eyes over a landscape that would soon become intimately familiar.

"I'm putting the males in first," Bryan announced as he placed the nestling with the green band on its right leg behind a vertical piece of plywood known as the hide. Positioned in a back corner of the hack box, the hide created a protective barrier behind which the birds could sleep, shelter from rain that might blow into the box, or conceal themselves if they felt threatened.[2]

"You have three males and two females. The females are the slightly larger ones," he continued, lifting out the yellow-banded male and placing him behind the hide. On its left leg, each nestling wore an aluminum US Fish and Wildlife Service band engraved with a number that would identify the bird if it was found dead or if someone viewed it closely enough to read the band with a spotting scope. On its right leg, each bird wore a colored aluminum band—known as an alphanumeric band—that was engraved with two letters or numbers.

"Believe it or not, this little guy is your oldest bird at thirty-six days old. He hatched on May 12." Bryan lifted out the smallest and least developed of the five birds. It wore a red band and still looked quite downy. "The two females are thirty-four days old and the other two males are thirty-five days old."

The eggs that had transformed into these nestlings had developed in an artificial incubator for thirty-two days. Once the eggs had hatched, the nestlings were fed first by human caretakers, then by adoptive peregrine parents for several weeks. Able to tear their own food from bird carcasses at about four weeks of age, the nestlings were subsequently transported to their respective release areas approximately a week later, when they could eat on their own.[3] Young peregrines take to the skies about six weeks after hatching, and it seemed inconceivable that these youngsters, which seemed more white down than sleek feather, would be ready to fly in a week's time. But their appearance was slightly deceiving, since the conspicuous patches of down on top of their heads, under their wings, and around their belly and legs overshadowed the tan-and-brown feathering that now covered much of their bodies.

Next, Bryan lifted out a white-banded female, followed by a dark, husky youngster with a black band that made a brief *kak kak kak* sound in protest as Bryan placed her in the box. The females were indeed noticeably larger than the males, particularly the little red-banded male, which looked like the very definition of scrawny. As adults, the female peregrines would be as much as 20 percent larger and 50 percent heavier than the males.[4] Such size disparities, while typical for birds of prey, are especially exaggerated in bird-eating raptors.[5] Whereas the smaller and more agile males—known as tiercels, from the French word for "a third," *tiers*, since the males are approximately a third smaller than females—typically catch passerines (songbirds) and other small birds, the larger female falcons are adept at catching waterfowl and larger birds, thereby minimizing competition for food between the sexes and allowing the pair to exploit a wide range of prey items on their breeding ground.[6] The female's larger size also might be related to egg production and nest defense, allowing her to better defend her nest and nestlings against intruders, predators, and would-be territory usurpers.[7]

Once Bryan had finished placing the birds in the box, I closed the door and he latched it. We then spent several minutes observing the falcons through peepholes in the side of the box. Two of the falcons, including the black-banded female, stayed out of sight, tucked behind the hide. Two others perched on large rocks, which had been placed at the front of the box, and looked out of the barred front of their temporary home. The fifth falcon stood by the door that we had just closed. Before leaving, Bryan poured water on seven of the dead quail that he had brought along to feed the falcons, then dropped them into a food chute on the side of the box. Eating wet quail would hydrate the peregrines and help to minimize the stress of their recent journey and new circumstances. Before stepping away from the box, I watched the red-banded male cacking (voicing his alarm with a harsh *kak-kak-kak* call) at another falcon that tried to interact with him. As we made our way back down to our camp, I fervently hoped the birds would adjust to their new surroundings and to their proximity to one another.

I fell asleep that night to the sound of flowing water and to the *peent* calls of a Common Nighthawk, which interspersed pursuit of its aerial insect prey with a booming courtship flight that sounded like a train going into a tunnel. It had taken me some time to get comfortable since my air mattress did little to soften the hard ground on which we had pitched our tent. I was relatively new to living out of a tent. And sleeping in a place where grizzly bears might be afoot was alien and discomfiting. Our dinner, too, had revealed our inexperience with outdoor living. Beans had seemed like a perfect camping food, but we hadn't taken into consideration the hours they needed to cook and the exorbitant amount of fuel our little gas stove required to render them edible. So we had gone to bed after eating peanut butter sandwiches, vowing to buy more suitable food after tomorrow's orientation session in town. Closing my eyes, I tried hard not to think about wandering bears and to focus instead on the world as seen through the large, dark eyes of a Peregrine Falcon.

———

Shortly after crossing the side stream the next morning on our way up to the hack site, I spotted a large falcon streaking across the sky, a posse of Violet-green Swallows tumbling in its wake. Raising my binoculars, I glimpsed a dusky head and boomerang-shaped wings as the falcon headed for distant hills. Unlike a Prairie Falcon, which would have dark feathering on the underside of its wings (in areas known as the axillaries and underwing coverts), this bird had pale, streaked underparts, confirming that it was an adult peregrine. Before I could exclaim in delight at the unexpected sight, a second peregrine appeared over a nearby ridge, flapping its wings as it traced a more leisurely path across the sky. I glimpsed a blue-gray back, distinguishing this bird from our brown-backed juveniles and the sandy-brown back of a Prairie Falcon.

We had been told that we might see adult peregrines in the area, though the chances were remote. Reintroduced peregrines were typically released at locations that contained suitable nest sites, in the hopes that some of the falcons would return to breed in later years. But our hack site, located near the South Fork of the Shoshone River on what had once been Buffalo Bill Cody's TE Ranch, had no nest sites. Its modest cliffs were too small and accessible to potential predators, and prey was too sparse to support breeding peregrines. Nevertheless, the South Fork hack site had proven to be a successful release area because of the high survival rate of introduced birds and its relative proximity to suitable nesting areas. Where these two peregrines had originated was anyone's guess, but their passage through the sky overhead infused my legs with energy, making the uphill trek to the hack site pass almost imperceptibly.

We soon settled into the observation site that we had chosen the evening before. Located on the opposite side of the drainage that carved its way below the band of cliffs harboring our peregrines, our lookout spot afforded us a clear view of the front of the hack box, but it was far enough away that it wouldn't disturb our birds. One lone limber pine provided the only shade around, though it did little to protect us from the wind, which periodically pelted us with dust, making our eyes tear as we strained to watch our peregrines through the spotting scope. Our first morning of observation would be brief since we needed to hike up to the hack box to feed the falcons and make sure they had experienced no mishaps before we headed into Cody for the orientation session. But first, we had to come to the all-important decision about what to name our birds.

"Let's name them after their color bands—Red, Green, and Yellow for the males, Black and White for the females," said the practical, emerging young scientist in me, striving for dispassionate objectivity though my heart soared with our peregrines' every move. But Garrett overruled me. Having a passion for mythology, he advocated naming the birds after Egyptian gods and other mythological characters.

Ra—Egyptian god of the sun, ruler of the heavens, and bringer of light—was an obvious choice, particularly since he was often portrayed with a human body and a falcon's head.

"We should give our little, red-banded male that name," I said, after Garrett enlightened me about this major god of the Egyptian pantheon. "He needs a powerful name since he's so small." And I'll remember that Ra equals red, I thought to myself.

Isis, preeminent Egyptian goddess, was the next name Garrett selected, and our beautiful, white-banded female seemed like the logical choice to endow with this regal name. Not realizing that Osiris—husband of Isis, son of Ra, lord of the dead in the underworld—was often portrayed as a green-faced deity, we bestowed his name on our yellow-banded male. The name Osiris seemed to have a yellowish ring to it, and Garrett, in a sudden departure from his allegiance to Egyptian gods, was adamant that our green-banded male should be named after the famed medieval knight, Sir Gawain, who accepted a chivalric challenge from a mysterious green knight in one of the more famous legends involving King Arthur's Knights of the Round Table.

But what to name our black-banded female who, thus far, had barely emerged from behind the hack box's hide. Garrett settled on Freya—a beautiful goddess of love, fertility, and war in Norse mythology—because she owned a cloak of bird feathers that allowed its wearer to change into a falcon. To me the name felt appropriate because we could make the alliteration: Freya seemed afraid.

When we peeked into the hack box to check on our birds, our falcons seemed to grow into their names, morphing mysteriously into unique individuals rather than just being five look-alike falcons with differently colored leg bands. Freya, of course, was out of sight behind the hide, along with Ra. Gawain and Isis were perched on rocks at the front of the box. Gawain clearly was aware of our presence and kept peering at the peepholes in the box, but Isis glanced toward us only once then focused her attention on the valley below and on the Violet-green Swallows and White-throated Swifts that zipped by in pursuit of flying insects. Osiris, meanwhile, appeared restless, alternating between standing near Gawain and Isis and tucking himself behind the hide with Ra and Freya. After feeding our birds five more quails, we headed down to our camp and into Cody for our orientation meeting.

———

Cody, Wyoming—the gateway to Yellowstone National Park's east entrance—endeared itself to me not only because of the grandeur of its surrounding mountains, Western-themed stores, and frontier history, but also because it became our link to civilization. Here we would shower off a week's worth of accumulated dust and sweat, indulge in hot breakfasts at a cozy restaurant, and use the pay phones that could connect us to our families. At a US Forest Service warehouse, we met Shoshone National Forest wildlife biologist Barb Franklin and one of the Wyoming Game and Fish Department's engaging grizzly bear biologists, Kirk Inberg (heartbreakingly, Kirk would be killed in a plane crash a few months later while searching for a wounded grizzly in the Absaroka Range). Barb's friendliness set us immediately at ease, as she showed us the freezer from which we would restock our supply of frozen quail, an industrial shower, and the phone we could use to update our supervisor at The Peregrine Fund on our birds' doings. But Kirk and Barb's warnings about grizzly bears were anything but reassuring to two Easterners who were spending their first summer camping out West. Although we were *unlikely* to see any bears, Barb told us, it was possible. Our food and quail had to be stored in our bear box. We had to carry a can of pepper spray on a belt holster at all times to defend ourselves against any overly curious or aggressive bruin that might investigate our camp or surprise us on our walk to and from the hack site. And we should call Kirk, aka Mr. Grizzly Bear, in case of any bear interactions—providing we could make it to town after such an encounter.

Despite these abundant cautions and our concern about grizzlies, our falcons were the only thing on our minds as we drove back to the ranch. We each would make a weekly trip into town, but once our birds were released a week from now, we would spend virtually every other waking minute focused on our charges. I couldn't wait to get started.

————

Little appeared to have changed when we peered into the hack box later that afternoon. Ra, Freya, and Osiris were tucked behind the hide. I fervently hoped they had concealed themselves upon hearing us approach and hadn't spent their entire day hiding. Isis and Gawain were again perched on the rocks at the front of the hack box. As he had before, Gawain looked toward us when we looked through the peepholes, head bobbing slightly as he searched out the source of the noises he heard. He seemed every bit the brave knight, staring down an unseen enemy. As afternoon slipped imperceptibly into evening, Gawain turned and watched the valley, though he still glanced periodically toward the peepholes. Meanwhile, once the swallows and swifts had gone to roost, Isis focused on the still life outside the hack box. Stretching her wings repeatedly, she yawned several times, finally impressing on us our own desire to settle in for the night. We left our birds and hiked back down to camp, moving languidly through day's end and Wyoming's fitful wind.

When we visited the hack box the next morning, we were delighted to see four falcons perched on the rocks at the front of the hack box. Gawain, Osiris, and Ra partially obscured the female who perched with them, though we guessed it was Isis, given her prior behavior. I relished the close-up view of our nestlings. Already they seemed more grown up than when Bryan had lifted them out of their transport box two days earlier. Their cream-colored fronts, infused with a faint chestnut wash, were streaked with vertical lines (the dark central stripe of each pale breast feather) that matched the brown of their backs and heads. Each dorsal feather was edged with a lighter brown that gave the falcons' backs a subtle scalloped appearance when viewed at close range. A bold, dark vertical patch extended below each eye. These malar stripes, which framed the birds' buffy throats and set off their paler cheeks, would help reduce the sun's glare when the peregrines began pursuing avian prey.[8] The soft skin around their eyes (the orbital rings), the fleshy area—known as a cere—around their beak, and their unfeathered legs were a soft blue-gray color that would turn a bright yellow in adulthood.

A piece of down that had been dislodged by one of the peregrine's growing feathers drifted out of the front of the hack box, carried on a breath of wind. Three pairs of bright swallow eyes watched its progress from a nearby branch. Flitting to the front of the box, one of the Violet-green Swallows snapped up the downy feather and carried it away to line its nest. I couldn't help noting the irony of swallows using the down of their future predators to create a cozy space in which to raise their own young. Another piece of down drifted out of the box and a second swallow quickly followed the first's example, flying away with a beakful of feathered comfort.

The falcons watched these activities assiduously, following the swallows' aerial paths with their alluring eyes. Peregrine eyes are enormous, taking up half of the bird's skull (our own eyes comprise less than 5 percent of our cranium).[9] In general, relative to body size, birds' eyes are almost twice the size of those of most mammals. And among birds, eagles, falcons, and owls have the largest eyes relative to body size. Larger eyes generally mean better vision since bigger eyes have larger retinas—a layer of tissue at the back of the eye onto which the image of an object that an animal is looking at is projected. Bigger retinas have more light receptors (light-sensitive cells or photoreceptors known as rods and cones), which allow for a better image.[10] Because raptors have a higher number and density of photoreceptors in their retinas than humans do, their visual acuity, which is so essential for capturing fast-moving prey, is slightly more than twice as great as ours.[11] Based on one falcon enthusiast's calculations, the peregrine may be able to spot a pigeon over 2 miles (3.2 km) away.[12]

Diurnal (active during the day) raptors also see particularly well because they have two small pits or depressions (foveae) in each retina—humans and most birds have one—where projected images are the clearest.[13] Raptor eyes are relatively fixed in their sockets, so perched peregrines make small, abrupt head movements and look from side to side to focus images on different parts of their eyes. The falcons view close-up objects using their shallow fovea—located near the back of the eye, like ours—which has lower visual acuity but provides three-dimensional binocular vision. The birds focus on distant objects using their deep fovea, which increases the depth of the eye and acts like a telephoto lens that magnifies images to provide high resolution, helping raptors judge the speed and distance of fast-moving prey. Because the deep fovea is located on the interior side of the eye, closer to the middle of the head, a perched falcon must turn its head sideways to see a distant object that is directly ahead.[14]

To maximize its speed, aerodynamic efficiency, and visual acuity when pursuing distant prey, a peregrine must follow a curved flight path that allows it to keep the line of sight of the deep fovea pointed sideways at its prey. If the bird flew directly at its target, it would need to turn its head to focus on the prey, thereby increasing its aerodynamic drag and reducing its flight efficiency. Once the falcon is close enough to see its prey with its less acute binocular vision, it straightens its path and flies directly at its target.[15]

Not yet familiar with the complexities of a falcon's eyes, I could only watch my birds' curious head motions with wonder, noting that no movement or passing bird escaped their notice.

Over the next few days, we were relieved to see all of our falcons at the front of the hack box. With each passing day, they became increasingly active. In addition to preening their feathers, the falcons focused ever more intently on the birds that frequented this seemingly desolate, high-elevation landscape. Violet-green Swallows and White-throated Swifts careened over the cliffs like winged torpedoes, snatching morsels of food from an aerial sea of planktonlike insects. Common Ravens chased each other, tumbling on invisible wind currents like exuberant acrobats. American Kestrels, Prairie Falcons, Red-tailed Hawks, and the occasional Golden Eagle cruised by with purpose.

Although Osiris appeared to be our boldest falcon, often running toward whichever peephole we were looking through, as though eager to confront an unseen threat, Gawain was the first to begin testing the boundary between being a nestling and a flighted bird. On his fourth full day in the hack box, he began running back and forth, poking his head between the round aluminum bars that offered a view of the valley below but prevented his escape. Osiris, Isis, and even Freya, who usually ducked behind the hide whenever we arrived at the hack box, soon followed suit, flapping their wings, running around the box, and hopping on and off the rocks that served as the bird's lookout perches. The lone exception in these activities was little Ra, who continued to look scruffy and underdeveloped. While the other birds stretched, preened, and flapped, Ra often lay on the pea gravel lining the bottom of the hack box and napped. I worried that the other birds were pecking him because his neck feathers looked sparse rather than sleek, but we never saw any of the other falcons act aggressively toward him. One morning, Ra clambered onto a rock in the middle of the hack box and perched next to Isis, who looked nearly twice his size. Gently, she reached out and preened his cheek. Not seeming to relish the attention, Ra jumped off the rock and went to a corner of the hack box to lie down. Fortunately, he did seem to be eating since his beak, like those of the other young peregrines, frequently showed traces of the quail on which he had fed.

On one occasion, I watched Osiris and Freya make repeated yawning and gagging motions for several minutes, while periodically scratching at their throat with a foot. At the time, I didn't realize that the birds were trying to regurgitate (or cast) a compact pellet of undigested feathers and small bones. A pellet's contents depend on the raptor's diet but can consist of fur, feathers, bones, insect exoskeletons, bird bills and feet, mammal teeth and claws, and even bird bands if the bird that was killed wore identifying leg bands. By examining regurgitated pellets, biologists can glean important information about a raptor's diet. In addition to allowing a bird to rid itself of difficult-to-digest prey parts, casting pellets also helps raptors maintain healthy digestive systems. Forming and expelling pellets scours the bird's digestive tract, particularly its crop—an extension of the esophagus that serves as a temporary food storage area, which allows a bird to eat large amounts quickly before flying off to digest in a safe place.

With our birds securely ensconced in their hack box, I had time to explore my surroundings and read The Peregrine Fund's hacking manual to better prepare for the upcoming release. Although finding and eliminating the threats that were responsible for the peregrine's global population decline was essential to the bird's ultimate recovery, the species' restoration was spurred by the development of captive-breeding and reintroduction programs that bolstered numbers where the falcons had declined or been extirpated.[16]

Successfully breeding peregrines in captivity was no foregone conclusion. Breeding techniques—spearheaded by biologist Tom Cade of Cornell University and later of The Peregrine Fund, which he cofounded—were developed over many years, starting in the early 1970s. The first fertile peregrine egg produced in captivity was at Cornell University in 1972. In 1973, the program raised twenty peregrines.[17] Since captive falcons often laid infertile eggs, one of the major captive-breeding breakthroughs followed the discovery that artificially inseminating females produced more peregrines. Falconers and biologists had learned that male falcons raised by people often became sexually imprinted on their human handlers, viewing them as potential mates. Those working to produce falcons in captivity must have been relieved when Les Boyd designed his renowned "copulation hat" in 1978, on which imprinted tiercels were trained to mount and deposit their semen, thereby saving countless cupped hands, shoulders, and heads from this rather unusual collecting task.[18]

Upon resolving additional challenges—such as how to successfully incubate falcon eggs and how to produce some chicks docile enough for captive breeding and others wild enough to be released—The Peregrine Fund and other entities were ready to begin bolstering wild peregrine populations. Reintroductions of captive-reared birds to the wild began in 1974.[19] After experimenting with techniques that included putting captive-reared young into the nests of wild peregrines (direct-fostering) and those of the closely related Prairie Falcons (cross-fostering), biologists found that the easiest, most successful way to return peregrines to the wild was through hacking—the controlled release of young falcons from an artificial aerie.[20]

Developed by long-ago falconers, hacking originally allowed falcon chicks that had been taken from the nest before they could fly to develop somewhat naturally. Hacking acquired its name in the second half of the sixteenth century when falconers transported nestlings to a hilltop in a wagon known as a hack. The falconer delivered food to the area each day so that the birds had a reliable food source as they learned to fly. Over the ensuing weeks, the falcons developed their flight skills and learned how to make their own kills. The falconer then recaptured the young falcons and trained them to hunt for their handler.[21]

Hacking was refined by modern-day falconers and biologists to release parentless young raptors to the wild. Juvenile falcons were placed in an enclosed wooden hack box—which protected them from predators and the elements—once they could eat on their own. Wide, round metal bars on the front of the box allowed the falcons to view their surroundings and provided a smooth surface that wouldn't damage the birds' developing feathers. Once the youngsters reached fledging age—the age at which they were fully flighted and ready to leave the nest—the front of the box was removed so the birds could fly free. Attendants monitored their activities and provided food until the young falcons learned to hunt for themselves and dispersed from the area. This reintroduction technique was predicated on the falcons' having an innate hunting drive. Although young peregrines benefit from watching their parents hunt and from having the adults bring live prey to the nest or aerially transfer food to them, juveniles are capable of learning to hunt successfully without parental influence.[22]

Since large numbers of peregrines could be released through hacking, this technique saturated an area with introduced falcons for several seasons, ensuring that some individuals ultimately returned to the area to breed.

Our falcons had lost most of their down during their week in the hack box and now looked sleek and beautiful, as though each of their feathers had been laundered and pressed. Once those feathers had sliced through windy skies and brushed against rocks and trees, they would become frayed and tattered, but would be replaced by new feathers during the bird's annual molt.

As we approached the hack box to prepare for the next day's release, a cliff-nesting Rock Pigeon flew past and disappeared behind a rocky outcropping. Though the pigeon would one day be ideal prey for our peregrines, it would be several weeks before our falcons threatened the cliff's current residents. At the top of the cliff, I set down my pack and, without any trepidation now, jumped the gap in the cliffs to reach the hack box, inadvertently flushing a male Mountain Bluebird from a nearby ledge. Looking like a celestial fragment that had fallen to Earth, he flew off with a soft *phew* call. Whenever I was treated to a view of this resplendent western thrush, the oft-used description of "the bird that carries the sky on his back" came to mind. Really, though, he might have been dipped in liquid sky since, aside from a gray wash on his belly and undertail, no part of him was untouched by cerulean hues. The more subdued female, cloaked in gray-brown with only her wings and tail hinting at her mate's vibrant color, remained hidden, likely incubating pale-blue eggs in a cliffside nest.

Reaching into the hack box, I removed three leftover quail then moved quietly away to give the agitated falcons a chance to settle down after the unprecedented intrusion of my arm. We would forego feeding them today so that they would be hungry and stay close to the food we would offer them tomorrow.

To my relief, when I finally looked through the peepholes, little Ra seemed as energetic and anxious to take his first flight as the other youngsters did. Flapping his wings, he ran from rock to rock then, with a burst of energy, landed on Osiris's back. Osiris squawked then ran, flapped, and twice jumped onto Isis's back. Like a chain of dominoes, each bird's energy seemed to set one of the others into motion. Taking up the torch, Gawain lunged repeatedly at the bars at the front of the box then clung to the chute through which we dropped the quail. The two females, though alert, were less active, watching the males' antics with what seemed like stoic forbearance. Finally, though, as if unable to remain indifferent to the surrounding frenzy of flapping wings, Isis flapped and jumped on top of Gawain and Ra, both of which had settled momentarily on a rock at the front of the box. With a quick motion of her head, Freya took one look at the ensuing chaos and retreated to the hide. As I made my way back down to camp, I felt like spreading my arms and running full tilt downhill, spurred by my falcons' unbridled energy and their instinctive desire to leap into the unknown.

CHAPTER THREE

Release

Release day—June 24, 1991—dawned clear, sunny, and, best of all, calm. The real work of hacking falcons was finally about to begin, and nervous anticipation thrummed through me as Bryan, his wife Georgia, Garrett, and I hiked up to the hack box to prepare for our birds' first flights. Only passing lightning storms had seemed ominous when our peregrines were securely ensconced in their hack box, but now threats seemed to lurk everywhere.

Golden Eagles, which sometimes flew over the area, posed a serious threat to inexperienced falcons. Though these fierce predators typically target rabbits and hares, they are opportunistic and will kill adult peregrines, given the chance. A juvenile falcon that had not yet mastered its flight skills and had no parents to defend it was particularly vulnerable. We would need to be vigilant and avail ourselves of The Peregrine Fund's special permit to shoot noisemaking cracker shells, if necessary, to scare off predatory eagles. In areas with eagles, hack-site attendants were tasked with being on guard during every minute of daylight for the first week post-release.[1]

A Great Horned Owl occasionally perched in a snag near the stream beside which we camped. Such sightings had previously been welcome. Now we would dread the presence of this large, yellow-eyed owl with feather tufts on its head, like cat ears. Fearsome nocturnal predators, Great Horned Owls had hampered peregrine reintroductions at several eastern sites because they readily killed young falcons. Coyotes and red foxes occasionally passed by our peregrines' cliff and might pounce on an unwary peregrine that perched in an unsafe location. Even the adult peregrines, whose presence had so thrilled us only days before, were now a concern because territorial adults might attack the youngsters, driving them away from the hack site and their only food source. After a week of flying, our peregrines would be safe from most raptors and mammalian predators, but Golden Eagles and Great Horned Owls would be an ongoing threat.

Arriving at the hack box, the four of us quietly dropped our packs. Opening his, Bryan pulled out a bag of dead quail and motioned for Garrett and me to approach him so he could share the bounty. We would lay out forty recently thawed quail on and around the hack box to provide the soon-to-be-released falcons with several days' worth of food. In the process, Bryan would show us how and where to feed our peregrines post-release. Whenever I fed quail to our birds, I felt a twinge of sadness about the role captive-reared Japanese Quail (also known as Coturnix Quail) played in restoring the Peregrine Falcon. Native to East Asia, this small game bird has been domesticated and farmed for its eggs and meat since at least the twelfth century.[2] Easily managed, requiring little space, and reproducing quickly, Japanese Quail proved to be an ideal food source for the peregrine recovery effort. The Peregrine Fund raised its own quail and supplied dead ones to its release sites.

More habituated than I was to dealing with the reality of working with a predator that required meat from other animals to survive, Bryan cavalierly tore some of the quail open to tempt the falcons to stay close to the easily accessible food. He also broke some of the dead quails' wings so that they would flap in the wind and catch the falcons' eyes, drawing the young birds back to the hack site if they ventured too far on their initial flights. In natural situations, parent peregrines find and feed their vocalizing fledglings wherever their offspring land after taking their first flights. But our immature falcons, which had never seen the outside of the hack box and had yet to get an aerial view of the cliff on which the box rested, would need to learn that this was where they must return to for food until they could hunt for themselves.[3] Inwardly cringing and trying not to reveal my unwillingness to inflict harm on an animal even when it was in no condition to notice, I followed Bryan's example. Staying away from the front of the box, we worked quietly so that the falcons would remain calm. Frightened peregrines might bolt from the release site once given the opportunity to fly, thereby endangering themselves during this most critical part of the hacking process—the release.

After Bryan put quail on the cliff rim and on the hack box, he then placed additional quail on a board affixed to a long metal pole that was attached to the top of the box. Known as a marten board, this elevated platform was inaccessible to hungry mammals such as pine martens (for which the board was named, though they did not occur at our site), raccoons, weasels, and foxes that might otherwise abscond with our falcons' food. The quail on the marten board would also serve as a beacon to the young falcons as they flew over the release site.

Once we had laid out the quail, Garrett and I headed down to our observation point under the limber pine to observe the falcons' activities when Bryan opened the hack box. Meanwhile, Bryan extracted a spray bottle filled with water, a roll of duct tape, and a large piece of cardboard from his pack. Following his unspoken directions, Georgia positioned herself on the side of the hack box to block any falcons that might try to escape while Bryan cautiously opened the door. The falcons rushed to conceal themselves behind the hide when they caught sight of him. Pulling out the spray bottle, Bryan spritzed water over the youngsters. The water would encourage them to preen their damp feathers and, like the superabundance of available food, would help dissuade the birds from bolting.

Next, Bryan blocked the entrance to the hide with the cardboard, securing it with duct tape so that the birds remained out of sight in the calming darkness during the most critical part of his preparations. Balanced precariously on the edge of the cliff, he carefully unscrewed the heavy wood, wire, and metal panel that formed the front of the hack box and removed it. After storing it nearby, he motioned to Georgia, who was already backing away from the top of the cliff. Finally, Bryan pulled away the cardboard covering the hide, swung the side door closed, then moved away from the hack box as speedily and inconspicuously as possible.

In the stillness that followed, I realized I had been holding my breath. Releasing it shakily, I focused intently on the hack box, watching through my binoculars and our spotting scope for any sign of movement.

"8:57," I scribbled in my field notebook, unconsciously effecting the essential technique of recording the time and taking notes while watching the birds' activities. "Blind (covering hide) open." Our peregrines were free to fly.

A mere three minutes later, Osiris and Gawain left the security of the hide and hopped out of the hack box and onto the cliff. Isis quickly followed. I had read that the young falcons might come out from behind the hide within a few minutes or might wait until the following day. In the wild, the smaller tiercels typically left the nest several days earlier than the larger, slower-to-develop females.[4] So it was with mixed pride and trepidation that I watched Isis, Osiris, and Gawain flap their wings and hold them out to dry. By 9:03 a.m., Ra emerged from the hide and approached the edge of the hack box, but he remained within the security of its three walls. For once, none of the falcons paid any attention to the Violet-green Swallows that streamed past on their erratic, insect-seeking missions. Instead, Osiris grabbed (or footed, in technical parlance) a quail with his talons and began plucking and eating it. Despite the surrounding bounty, he appeared nervous that the other falcons might pirate his meal. Flapping his wings, he dragged his prey away from the watchful eyes of the other falcons.

We struggled to identify our birds since their activities sometimes obscured our view of their colored leg bands and masked the subtle differences in the falcons' sizes and coloration. Isis soon mimicked Osiris's example, and the two birds fed. Gawain, who had gone back into the hack box, where Ra was now perched on a rock, also began eating a quail he'd dragged in with him. But when Osiris ran by with a quail in his beak, Gawain jumped back out and joined Isis and Osiris in a bout of frenzied wing flapping. Despite, being absorbed in their activities, I began to relax, realizing with relief that our falcons seemed more inclined to eat and flap their wings than to bolt from the release site. Even so, their fates might change in an instant. At one of the hack sites Bryan had supervised in Rainier, Washington, a tiercel had flown out of the hack box for the first time and had been swept skyward by a thermal of rising air. He was never seen again.[5]

A Clark's Nutcracker landed in our limber pine, lulled into proximity by our immobility. I spared it a quick glance, admiring the way its bold black-and-white wings and tail contrasted with the warm grayish-brown of its body, as the bird uttered its harsh, grating call. It seemed inconceivable to me that, each spring, this high-elevation member of the crow, raven, and jay family remembered where it had hidden the thousands of pine seeds it had cached the fall before, whereas I found it challenging to remember where I had put my wallet at any given moment. All thoughts of the nutcracker fled as I glimpsed movement through the scope. Ra was flapping his wings! At last, at 10:51 a.m. he hopped out of the hack box. Still flapping, he walked awkwardly over to some nearby rocks. True to character, Freya remained behind the blind.

Suddenly, pandemonium. A raven swooped over the cliff, descended toward the falcons in a sharp glide, and grabbed a quail in its bill. All four peregrines cacked madly, their cacophonous voices sounding loud even to our distant ears. Since ravens could quickly decimate our peregrines' food supply, Bryan, who had joined us at our viewing site, picked up the rifle he'd brought with him and fired a cracker shell. Dropping its prize, the raven hastily retreated, issuing a protesting croak as it dropped behind the cliff. Our peregrines, having no negative association with a loud noise that emanated from an unknown source, remained unperturbed and began watching passing swallows. Unwilling to give up on an easy meal, the raven returned minutes later. The peregrines immediately cacked and Bryan fired off another shot. Falling away in a glide, the raven disappeared behind the ridge.

True to its reputation as an opportunistic scavenger, the raven—or one of its consorts—soon returned. Our peregrines, naive as they were about the world around them, usually noticed the raven before we did, cacking in unison the moment they spotted the intruder. Twice more the corvid flew over the hack box then dropped out of view. We watched intently, waiting for more decisive action by the raven before we fired again. Given its legendary intelligence, the bird might quickly realize that our noisemakers were an idle threat, so Bryan wanted to use them judiciously.

As though inspired by the corvid's flyovers, Gawain tested his wings with a modest first flight, launching himself from a rock outcrop and landing by a quail, a demure 3 feet away. While he fed, the raven passed by yet again. Absorbed as we were in its persistent forays over the release site, Osiris and Gawain blindsided us by leaping into flight simultaneously, two hours and thirty-six minutes after leaving the hack box. We scrambled to follow the birds' trajectories with our binoculars. And beautiful, timid, dark-plumaged Freya chose that moment to emerge from the hide and begin flapping her wings.

To our relief, Osiris circled and landed back at the hack cliff. Gawain, however, cut across the cliff, losing elevation before coming to an awkward rest on a boulder. Clearly, he had not yet grasped how to use his wings and patches of upward-rising air to maintain his altitude. Moreover, at this stage of development, his flight feathers were only three-quarters grown—the base of the feathers were still encased in waxy keratin sheaths that protected them during their development—limiting their efficacy.[6] As Gawain landed, a male American Kestrel flushed from the cliff. North America's smallest and most colorful falcon, the diminutive kestrel weighs between 3 to 6 ounces (about 80 to 165 grams), less than the average apple. Flashing his finely tapered, smoke-gray wings and black-and-white-tipped chestnut tail, the departing kestrel captured Gawain's attention. Fortunately, this was one raptor that our inexperienced peregrines didn't need to fear.

Not to be outdone by Gawain, Osiris took off again and disappeared around the north side of the cliff. After telling us to hike around and locate Osiris if he didn't return within a few minutes, Bryan left us for a few hours. In the meantime, Freya, with what seemed like remarkable nonchalance given the hours she had remained behind the hide, hopped out of the hack box, a mere ten minutes after stepping into view. Flapping her wings and making short hops, she joined Osiris, who had reappeared on the south side of the box.

Two ravens flew over the cliff, and Garrett stood up and raised the gun to his shoulder. To our relief, the wily ravens quickly dropped out of view. Readying ourselves for their return, we were surprised by the sudden appearance of an adult peregrine, which materialized out of the firmament and glided with deadly purpose toward our hack box. Heart in my throat, I stood up, hoping that doing so might dissuade the peregrine from attacking our vulnerable youngsters. I needn't have worried. Swooping over the hack box, the adult footed a quail, and continued on its flight path, leaving five cacking juveniles and two wide-eyed hack-site attendants in its wake.

Quick to follow the peregrine's example, the two ravens again cruised over the hack box. As one dropped down to grab a quail, Garrett fired a cracker shell, foiling the determined scavenger. Seconds later, our young falcons cacked again and we readied ourselves for another raven assault. Instead, an adult Prairie Falcon shot into view over the valley below the cliff, turning its head to register the activity at the hack box but never changing its trajectory before disappearing from view. We could no longer see any of our peregrines, but we were reassured to hear cacking near the hack box when the ravens returned minutes later. Before we could decide if one of us should go and try to locate our charges, an adult peregrine rocketed over the cliff, hovered for a moment over the marten board, then footed a quail and took off with it. We weren't sure if this was the same adult that had robbed a quail earlier or if it was its mate. Regardless, we couldn't fire noisemakers at endangered peregrines, though we felt our anxiety mount with each disappearing quail—each quail that would no longer be available to our inexperienced youngsters.

It was as though we had posted a metaphorical neon sign over the hack site, proclaiming, "Free food—for one and all." Caught up in the urgency of keeping track of our falcons' every move and documenting the chaotic activity in my field notebook, I was nevertheless struck by how attuned every bird in the neighborhood was to changes in its environment. Even peregrines, the most predatory of birds, were eager to capitalize on an unexpected free meal.

Our concern about our falcons' unknown locations was relieved when Ra walked into view from behind the hack box. Hopping onto a small log in front of the box, he preened and rested, his relaxed demeanor easing my anxiety. Garrett set off to search for the other peregrines, while I kept a close watch over Ra and Gawain, who was still perched on his boulder at the base of the cliff. After ten blissful minutes of inactivity, a third peregrine appeared from behind the hack box. I watched it closely, trying to make out the color of its leg band, while keeping an eye on Gawain. But my identification efforts were foiled by another passing raven. Standing, I clapped loudly and watched with satisfaction as it swerved away from the hack box and dropped behind the cliff. My smile quickly faded as I glanced toward Gawain and saw only the boulder on which he had been perched. Walking around, I scanned the cliff and the jumbled boulders scattered on the slope below it, trying to spot Gawain on rocky outcrops and in shadowed recesses. I had no luck, but managed to identify the mystery falcon as Isis. Minutes later, Freya walked into the hack box through the side door, which had blown open in the strengthening wind. Now, only Osiris and Gawain were missing.

During the next hour, ravens and a Prairie Falcon flew repeatedly over the hack cliff. Fortunately, the ravens fled as readily when I clapped—the sound echoing loudly off the surrounding rocks and cliff—as they did when we fired cracker shells. Freya flapped energetically in the hack box, while Isis preened Ra's face. Ra responded by lifting a wing, and Isis obligingly preened his underwing. Soon after, Garrett returned to our observation site for lunch while I hiked to a nearby ridge to look for our missing birds. To my delight, I finally located Gawain perched midway up the cliff southwest of the hack box. I watched as he rested, closing his eyes periodically. Idly, he pecked at a tuft of grass decorating his cliff ledge then made a short, hopping flight to a higher ledge as the wind gathered force and rain began to fall.

Shielding my field notebook and sweeping my windblown hair from my eyes, I headed back to our observation site to retrieve my raincoat but was stopped short by Garrett's arrival. Breathlessly, he announced that disaster had struck. The side door of the hack box, which had not been latched and had blown open earlier in the afternoon, had slammed shut with a bang. Ra, Isis, and Freya had leaped into the air and scattered in every direction, as each tried inexpertly to maneuver in the violent wind that now buffeted the cliff. Garrett had lost the birds from view and had hoped that I'd spotted them. With a helpless shrug, I shook my head and we rushed back toward our observation site to begin another search. Minutes later, we saw Bryan heading toward us, and we hurried to tell him the bad news.

Grim-faced, and with no alternative but to hike up and latch the door, Bryan trudged up the hill. An open door would flush the birds each time it banged shut in the wind, and if it did so at night, the results could be fatal to a young peregrine that was unlikely to be able to fly to a safe roost spot in the dark. But by closing the door now, Bryan would violate the sacred dictum of not going anywhere near the flighty young falcons during this critically sensitive period. Seeing his furrowed brow, my heart went out to him as he did what he knew he must though it ran counter to release protocols. I could only hope that any birds he might flush would have sufficient time to move to safety before darkness fell in a few hours' time.

As Bryan moved uphill, Gawain flushed from his perch and flapped to a nearby ledge. Bryan tried to give the youngster as much space as he could, but Gawain flew straight out from the cliff and then circled back toward his earlier boulder perch. His strong wing strokes belied his lack of expertise, but he tired quickly, even in the diminishing wind, and lost altitude. Dropping rapidly, he landed on the ground midway down the boulder-strewn slope below the cliff. Having accounted for Gawain, Garrett and I continued on to our observation point to look for the females.

Reassuringly, we soon spotted Isis and Freya on the cliff. The sight of two Golden Eagles flying high overhead was less welcome. I was sure their powerful eyes had spotted the scattered quail and several of our vulnerable peregrines, but fortunately the eagles didn't waver in the steady arcs they traced across the clearing sky, and soon disappeared. Shadows lengthened. Freya and Isis preened and napped, as though wholly accustomed to this new existence of cliff ledges without walls, of a world with no ceiling. The omnipresent ravens made half-hearted attempts to pirate quail, and we stood tiredly and clapped them away. Bryan had gone to search for Osiris and Ra, after securing the hack box door.

Succumbing to the inherent tranquility of day's end, I was startled by a falcon's piercing *kak kak kak*. A peregrine was in flight, flapping along a nearby ridge, then gliding expertly toward the hack cliff. I nearly convinced myself that the silhouetted bird was an adult, but as it landed on a bulge in the cliff, the falcon fell forward, flapping awkwardly to avoid hitting the rock with its beak. It was Gawain, showing us that we no longer needed to worry about him. He'd instinctively known to seek out the safety of the cliff as the sun approached the horizon. A pesky raven flew over to our intrepid knight, threatening to knock him off his perch, but Gawain stood his ground, cacking as the raven soared over the site.

Minutes later, Bryan, who had returned after a fruitless search for Ra and Osiris, spotted a peregrine taking off from a ridgetop to our north. Skimming along the ridge, the bird approached the hack cliff and, flaring its wings, landed almost on top of two ravens that were quietly preening each other. With a squawk the ravens jumped apart and lunged at the intruder—Osiris. Launching back into flight, Osiris flapped toward the hack box, circled, and made his best landing yet. Grabbing a quail, he indulged in an end-of-the-day snack. The ravens, though, as if avenging their disrupted evening, started harassing Gawain, diving at him unrelentingly. Spreading his wings, Gawain stepped into the wind, made a wobbly aerial circle, then tried repeatedly to land on the cliff. The ravens would have none of it, diving at him again and again. Finally, Gawain flew from the cliff and landed on a distant outcropping.

Brushed in the stillness that often permeates those last moments before the alchemy of the gloaming turns day's light into night's shadows, we struggled to keep our birds in view, discomfited that little Ra had eluded our best efforts at finding him. And now we could no longer see Gawain in the fading light. Osiris cleaned his beak on the cliff then flew over to join the females, as though flying to roost with his nestmates was a routine endeavor. Isis had the last word, cacking at a raven that refused to relinquish its nearby clifftop perch. Exhausted, apprehensive, exhilarated, we stumbled back to camp for a warm meal and a much-needed rest.

"Soph, Soph!" Garrett's urgent but hushed whispers dragged me out of the deep well of sleep into which I'd quickly sunk after crawling into the feathered comfort of my sleeping bag. "I think there's a bear out there."

I sat up abruptly, fingers of fear slithering up my back and conspiring with the cold night air to make me shiver. I leaned forward to see what Garrett was looking at through the tent door. A waxing gibbous moon cast silver light onto the sparse grasses and shrubs that surrounded us and, there, protruding over the crest of the closest hill, was a humped form. Mesmerized, trapped in our stampeding fear and careening thoughts, we stared at the animal's largely obscured body, waiting in an agony of apprehension for it to notice us, to turn toward us. Finally, it lifted its head. And I sat back on my heels, exhaling a huff of breath.

"Cow," I said, as relief flooded through me.

"Cow," Garrett confirmed with a wry, apologetic grin, turning away from the tent door. It was the first we had seen near our camping area, which we hadn't expected to be used by cattle during our stay on the ranch. In Wyoming, peregrines were hacked in two national parks, three national forests, a national wildlife refuge, a national monument, four Bureau of Land Management resource areas, and three private ranches.[7] This first cow was a harbinger of the challenges we would face while tending a hack site located on private grazing land: trampled tents, bent tent poles, broken lanterns, an excess of fresh cow patties, and the pungent smell of too many overly curious bovines.

I lay back down, wide awake now as the adrenaline that had coursed through me slowly dissipated. And I thought about Ra. I thought about his small, slim, 1-pound, feather-clad body. He seemed vulnerable in daylight. But how much more so was he now, perched in some unknown location, every element of his world new and unfamiliar. I pictured him with his head tucked under his wing, but wakeful, watchful, looking around every now and then. I imagined a small noise, the soft footfall of a canid, Ra's head turning sharply to view the intruder. Repositioning my pillow with an impatient fist, I burrowed into my sleeping bag and tried to corral my wayward thoughts. If we had been unable to find him, perhaps he was safe from any night creatures that might do him harm. I willed him to be high on the cliff, tucked safely behind a rock, as unthreatened by nocturnal predators as we had been by our fearsome "bear."

CHAPTER FOUR

Sky Birds

As we hiked up to our observation area just after sunrise the next morning, my mind churned, wondering whether the day would bring peace of mind about Ra or heartbreak. We spotted three peregrines near the hack box upon reaching our observation point. But then, as if in portent, two separate groups of coyotes raised their voices in an exuberant yipping, yapping ode to morning—or perhaps it was a celebration of last night's hunt. Despite my all-consuming concern about Ra, I couldn't help but feel exhilarated by the inherent wildness of the coyotes' song. It spoke of untamed country, the sociability of wild canines, the timeless relationship between hunter and hunted.

We quickly identified Freya, Gawain, and Osiris. Then, after a fruitless search for Isis and Ra, we settled into our camp chairs. Watching the birds through my binoculars and our spotting scope, I took notes on their activities while Garrett made coffee on our small camp stove. Precocial Osiris launched into flight when a pair of ravens glided over the hack box. On the ravens' second pass, Osiris fell in behind them, chasing the unruly corvids from the area with a confidence that seemed disproportionate to his experience. He soon returned to feast on quail with Gawain. Moments later, Isis stepped into view from behind the hack box. I tried to enjoy my coffee and the sight of four peregrines, but Ra's absence was disquieting.

Aside from the antics of the ravens that cruised over the hack site, the relatively tranquil morning was enlivened only by our juveniles' variable reactions to the visiting adult peregrine. Materializing out of a breath of wind, she landed close to the hack box and cacked several times. Gawain and Osiris immediately flattened their bodies and lowered their heads, as the adult hopped up onto the box.

Eechip, eechip, eechip, they called out, as if in timid greeting. Male and female peregrines typically emit the *eechip* call—one of four main vocalizations—during food transfers, courtship, or nest-site displays when one bird bows low to appease its mate.[1] In apparent contradiction to its appeasement function, the call may also be given around a nest during aerial encounters with an intruding peregrine.[2] New as I was to the bird world, the messages transmitted by Gawain and Osiris's lowered body postures were a mystery to me, but the youngsters appeared to be acting submissively and seemed intent on diffusing potential conflict with the intruding female. Ignoring the groveling youngsters, she grabbed a quail and departed.

The adult peregrine's second morning visit was more protracted. Flying over the hack site, cacking, she touched down briefly on the cliff rim then circled again. In contrast to the males, Isis responded by puffing out her feathers and squatting on the clifftop like an impassive feathered Buddha for several minutes before coming to life with a flap of her wings. When the adult landed nearby, Isis clambered awkwardly toward her, cacking querulously. But before the adult could react to Isis's advance, a passing raven distracted the falcons. With a flick of her wings, the adult launched into the air, then traced the raven's flight path with the precision of a figure skater before diving at the corvid with desultory grace. As the two birds disappeared from view, I couldn't help but think that our parentless youngsters could only profit from watching the aerial mastery of the intrepid adult, particularly since she seemed disinclined to harm them.

A minute later, the falcon returned, nonchalantly footed a quail, and flew off with it while Osiris and Gawain, in seeming solidarity with the voluble Isis, cacked vociferously.

Distracted as we were by the peregrines' activities, we scarcely noticed the rapidly gathering storm clouds and fading light, but a sudden lightning bolt commanded our immediate attention and the deluge that followed had us scrambling to protect our gear and take cover. Though unaccustomed to the often-abrupt ferocity of Wyoming's weather, we knew enough not to stand under our lone limber pine during a lightning storm. Slipping on the suddenly slick soil, which couldn't absorb the sheet of water connecting the tumultuous heavens with the wind-scoured earth, we tucked binoculars and spotting scope under hastily donned raincoats, hoisted already-drenched packs onto our backs, and stumbled toward an eroded gully we had discovered several days earlier. Moments later, we huddled under the inadequate shelter of an overhanging rock, watching a celestial chiaroscuro of livid sky slashed by lightning, as rivulets of water poured over our feet and dripped from our baseball caps. An avalanche of enormous hailstones followed the lashing rain, covering the ground in a layer of white that gave the illusion of snow.

In Vermont, we took pride in saying: "If you don't like the weather, wait a minute." Wyoming made a mockery of that adage, subjecting those who worked and recreated below the ever-changing tapestry of its big skies to a quick succession of every feasible type of extreme weather. We often began the day shivering under layers of clothes only to swelter soon afterward under a fierce, desiccating sun. Sunshine and calm regularly turned, with astonishing rapidity, into thunderous skies, battering storms, perilous lightning, violent winds, blinding rain, and painful hail. Yet weathering these often-inclement conditions was the insignificant price we paid for awaking each morning to an exuberant dawn chorus, for being surrounded by mountains suffused in alpenglow, and for relishing our peregrines' hesitant early dances with a limitless, if not always accommodating, sky.

We returned to our observation post at 1345. (As befitted a budding biologist, I had set my watch to military time—a twenty-four-hour clock—soon after starting our peregrine work, and I was now recording the timing of my observations like a "real" scientist.) The morning's endeavors had been swept away with the rain and we began, once again, to relocate our birds. The very real threat the storm had posed to our inexperienced falcons left me feeling unsettled. Large hailstones could kill birds of all kinds. Lightning had struck hack boxes, and peregrines had perished in storms soon after being released. We could only hope that our birds had found shelter.

To our relief, twenty minutes later, Osiris and Gawain came into view from behind the hack box hide, where they had taken refuge from the sky's assault. My eyes tearing in the wind that followed in the storm's wake, my wet gloves doing little to protect my stiff fingers from the cold, I recorded our observations and kept searching for missing peregrines.

Oh, where was Ra? My anxiety over his well-being had only escalated as the storm unleashed its fury. Was he tucked securely into a cozy crevice (and if so, why hadn't we seen him all day?) or had the rain washed away the sad remains of his scavenged body? After several hours of scanning the cliff and its surroundings, we finally sighted another peregrine—Freya—standing on a boulder near the base of the cliff. Flapping her wings, she flew up to a low cliff ledge then worked her way upward.

Ravens continuously attempted to steal our falcons' food—and were successful at least once—as the roar of the wind obscured our peregrines' cacking protests. With clouds dissipating, the ravens' unrelenting piratical maneuvers were forestalled by an immature Golden Eagle that streaked over the clifftop. The motley crew of clownish ravens quickly coalesced into a coordinated acrobatic air defense, mobbing the eagle and chasing it away from the enticing food and from the peregrines that the inconstant corvids were only too eager to harass. I felt a surge of gratitude for the black-feathered gang that heretofore had been subtle but unrelenting thorns digging into the vulnerabilities of our hack site.

Two hours later, as day's end crept closer with subtle changes in light and a lessening of the wind under now-sunny skies, Garrett spotted Isis flapping her wings mid-cliff. And, suddenly, Gawain and Osiris were speeding across the sky in their first chase. The lead bird's feint to the left was shadowed by the follower, and we watched their dramatically improved flight skills in surprised awe until the falcons disappeared behind the cliff. Two peregrines were perched on the cliff rim and we turned our eyes to them, noting that the light-brown chest of one bird stood out next to the other bird—Freya's—dark plumage. Ra. It was Ra! His red leg band seemed to flash like a beacon, igniting smiles that split our weathered faces.

"He's alive!" we exclaimed, exchanging high fives. Little Ra had made it through his first night. He had made it through a ferocious storm. He was a survivor after all.

I could hardly tear my eyes away from Ra's diminutive form, as he busily preened his feathers, unaware of the impact his every move had on two distant forms in the valley below the cliff. But the need to account for every bird was deeply ingrained in me, and soon I was searching out Ra's companions. Nearby, Freya watched the Violet-green Swallows and flapped her wings. Isis preened nonchalantly on her mid-cliff perch. Then, Osiris dropped out of the sky and landed, with far less grace than he showed in flight, on the clifftop. Minutes later, Gawain walked into view next to Ra, after landing behind the hack box. Five falcons. We had all five falcons. I sat down in my camp chair beaming, the damp chill forgotten.

"It's been a good day's work," I smiled to myself. But my satisfaction at having all of our birds in view was short-lived. Like a kernel of dried corn on high heat, Osiris popped into the air when four ravens flew over. He circled and returned. Then Gawain took off, disappeared momentarily to the west, and returned to perch again next to Ra. For a brief moment, in the rapidly fading light, we had all five in view once again. Then one of the clifftop perchers lifted off and, flapping vigorously, disappeared westward into darkness. Night's curtain descended on the cliff, and Garrett and I turned toward camp, wondering if our westbound bird would roost safely, and ready to resume searching for our increasingly mobile falcons the next day.

————

Although first Freya and then Gawain eluded us until late afternoon over the next two days, we managed to locate all five of our peregrines. Now, on the fifth day post-release, I reflected on the prior days' highlights while hiking down to camp after two hours of early-morning observation. I was heading into town for supplies, but even though I relished the thought of a shower and a hot breakfast, I was already wondering what peregrine adventures I was missing. Yesterday, I had hiked up to the hack box in the morning to deliver the first batch of quail since our birds' release. I'd had no alternative but to flush the falcons as I approached, but their hasty departure from the cliff when they spotted me was reassuring evidence of their wildness. All flew to cliff ledges or boulders, out of sight of the box, so I distributed the quail without the youngsters' associating me with food. Within a few hours, they returned to the clifftop and fed.

In addition to being filled with tempestuous weather, the last two days had revealed subtle changes in our falcons' flying behavior. Whereas their first flights had consisted mainly of moving from one perch to another, our falcons were now flying more extensively, soaring for minutes at a time over the release site, and even starting to dive at nonexistent prey. Once again, Osiris had exhibited a confidence that exceeded his already impressive flying abilities. After circling above the hack site the day before, he had dived at an invisible target, then reversed course and sped skyward again, cacking loudly. When the Golden Eagle he'd targeted suddenly surged over the cliff, Osiris pursued the much larger and eminently more dangerous raptor without hesitation. To our relief, when the eagle disappeared over a ridge, Osiris broke away, circled, then alighted on the hack box. Despite his daring flight, his wobbly landing showed he was still a novice. For several minutes he watched the skies, tilting his head from side to side, as if daring another intruder to cross the patch of blue that constituted his perceived domain.

Despite Osiris's vigilance, the ever-present ravens, adult peregrines, and even a Prairie Falcon continued to abscond with our falcons' food. But after seeing how many leftover quail remained despite the incessant robberies, I felt less anxious about our falcons going hungry when freeloaders made off with a meal. And I couldn't wait to tell The Peregrine Fund that the adult peregrine had finally lingered long enough on the cliff for us to determine that she was banded. The white band on her right leg and black band on her left revealed that she had been raised in captivity and released to the wild by The Peregrine Fund and its cooperators.

As they did elsewhere in the United States, breeding populations of peregrines disappeared from Wyoming between 1950 and 1973.[3] Intensive surveys between 1978 and 1980 confirmed that no breeding peregrines remained in the Cowboy State. Reintroductions started in 1980, with the release of eleven peregrines at three different sites in Grand Teton National Park. In the 1980s, the first nesting pairs in Wyoming all wore leg bands, meaning that the population originated entirely from captive-raised birds that had been reintroduced to the wild at hack sites like ours rather than coming from wild birds that had moved into the state from other areas.[4]

Arriving back at camp, I hastily gathered up dirty laundry, water containers, and other items for my trip to town. I could almost taste the hash browns at my favorite restaurant. But a squeaky little mewing sound suddenly disrupted my work. I sat back on my heels and listened.

Mew, I heard again. Surely a kitten couldn't have made it as far as our camp from the ranch house. Cattle were now abundant, but we had seen no cats of any kind. Again, the plaintive feline voice called. Maybe this was the progeny of a feral cat. Grabbing my binoculars, I scanned the ground under the tangled bushes along the stream but saw no movement. For good measure, I searched the bushes, as well. I couldn't leave a helpless kitten alone in the semi-wild of Wyoming. I searched and searched to no avail, though the intermittent mewing continued. Minutes passed while I waited for the shy creature to call again so I could pinpoint its location.

Then a sudden flutter of wings caught my eye and I swung my binoculars toward the bird that had distracted me from my search. I expected a small brown sparrow or a cryptic wren, but the beautiful bird that filled my binocular field was unlike any I had seen and had me momentarily forgetting the defenseless kitten. About the size of a tanager, the bird had a jaunty rufous cap that sat above a gray face and dark-red eyes. Elegant white mustache stripes and a patch of white under its bill contrasted boldly with the bird's cinereous front. And its back, wings, and tail were suffused with a lovely olive green. I didn't remember seeing a bird like this in my field guide. And I'd certainly never seen this species in the East. As I stared at it, transfixed, it opened its beak and issued a small *mew*. My kitten. *This* was my kitten! This beautiful, surprising mystery bird. I shook my head, wryly admonishing my ignorance.

Jumping to my feet, I reached into my backpack, grabbed my western bird guide, and rifled through its already worn pages. Most guidebooks arrange species taxonomically, based on different birds' relationships to one another. I didn't even know where to begin looking for this bird, but its bold, unmistakable colors soon jumped from the page. A Green-tailed Towhee. A denizen of shrubby western habitats, the towhee is a secretive bird that resides mainly in the undergrowth. As it flitted away, I rushed to finish gathering up my paraphernalia for town, glad that no one had witnessed my misdirected search as I fumbled to become familiar with new geographies, new habitats, and the often-unexpected inhabitants that filled them.

Returning from my trip into Cody early that evening, I headed up to our observation site to relieve Garrett of the last few hours of falcon watching. Our peregrines were now spending more time airborne, particularly early and late in the day. And they appeared to be increasingly possessive of their turf. When an adult soared over the clifftop, Osiris leaped into the air and intercepted the peregrine's flight with a direct attack. The female instantly flipped sideways to defend herself and the two falcons grabbed each other's talons—an aggressive behavior sometimes referred to as crabbing. Talons locked, the birds fell for a millisecond before breaking apart. Both birds disappeared behind the cliffs, but half an hour later, the adult flew over again, as if to reassert her ownership of the skies. She passed unchallenged. A short while later Osiris reappeared and dived repeatedly at one of the other young peregrines, as though redirecting his aggression to a more manageable target. The unidentified juvenile cacked and fell away repeatedly, rolling to its side to present its talons each time Osiris attacked. I marveled at how quickly our youngsters had mastered the evasive maneuver of tucking their wings and falling away, how instinctively they presented their talons in defense, how able they already seemed to be at defying gravity.

For the moment, Osiris was the fiercest about defending our young falcons' realm, but once our peregrines had matured and were raising their own young, it would be the larger females that actively defended the aeries, while the males served as virtually the sole providers for their mate and chicks until their offspring were half-grown.[5]

Clean, relaxed, I watched our falcons' flights over cliffs suffused in golden light as the sun descended toward the horizon. A crisp wind rustled the pages of my field notebook. Cirrus clouds streamed across the sky like the blur of wild mustangs running across open prairie. Just before dark, the falcons settled on the cliff near the hack box. And, finally, I had no choice but to tear myself away from their silhouetted forms. "Left 2115," I wrote in my notes before heading down to camp.

———

As if in defiance of their reputation for being celestial creatures, our peregrines took a step back from their growing aerial exploits on the sixth day after their release and embraced the terrestrial realm. To my relief all five were visible, perched on the hack box or a nearby cliff ledge, when I arrived at our observation point. All of them launched into flight, cacking loudly in alarm, when Garrett approached the hack box to put out more quail. Despite sometimes falling awkwardly onto their chests upon landing or having to make multiple attempts before coming to rest on the cliff, our falcons continued to impress us with their rapidly improving flight skills.

After their brief flurry of activity, I worked to relocate the birds and confirm their identities. The color bands were invisible on flying birds and often difficult to see when the falcons were perched. Isis's white band and Osiris's yellow band were virtually indistinguishable under certain light conditions. And even though the males were smaller than the females, a lone bird could look like either sex. As a result, it often took time to confirm identifications based on a combination of size, plumage characteristics, and color band.

I finally identified the falcon that had been making repeated short flights as Gawain when he landed on a small crag adjacent to the main cliff. Hopping down to a grassy ledge, he drew my attention to another falcon that was resting there: Isis. While Gawain rested with drooped wings, Isis pecked desultorily at the grass then clambered onto a rock. And then I noticed a third bird. Freya preened and pecked at the grass, then, as though exhausted by her own inactivity, lay down for a few seconds, before getting up to tug on a nearby twig. Soon, Osiris joined the group.

Succumbing to the intense midday heat, the birds preened or lay down. Heat waves shimmered and danced, obscuring my view through the spotting scope. But I almost laughed aloud at the sight of four immature peregrines lying together like lumpy feathered pancakes on a sizzling ledge. Steve Sherrod, who spent years studying the behavior of fledgling peregrines, remarked on their tendency to lie down when resting—up until they reach independence or learn to hunt for themselves—often congregating in a manner reminiscent of their huddled days as nestlings. Since the youngsters will lie down in such varied locations as on top of boulders, tree branches, fence posts, and even flat ground, I was relieved that our seemingly oblivious falcons had little to fear on the safety of their grassy ledge.

Gawain, appearing indecisive about indulging in a prolonged nap-fest, stood up and preened while the others slumbered on. After nearly half an hour of that productive activity, and with no response from his nestmates, Gawain lay down once more. Within minutes, though, he stood up. Then, seeming to lose his battle against the soporific effect of the sun and the lounging falcons, Gawain again started to lie down before deciding instead to clamber onto a rock next to his slumbering compadres. Standing on his slightly elevated perch, Gawain dragged his foot across the closest prone peregrine three times then determinedly walked across its back. To no effect. As though giving up, Gawain started to lie down once again, but then flew to a higher ledge.

Mimicking Gawain's restlessness, another falcon stood up and dragged its foot across its neighbor's back. When that induced no response, the youngster briefly preened its sleeping companion then walked over to the third bird and stepped on its back. The trodden-on falcon jumped to its feet. Mission accomplished, the attacker lay back down. After preening and tugging grass, the lone wakeful falcon now took its turn at trying to wake up another falcon. Walking over to its flattened neighbor, it preened it briefly, then dragged a foot down its back, then tugged on its feathers, before preening it again. Awake now, the harassed falcon preened its awakener's face and grabbed its beak, before the whole crew went back to sleep once more.

Absorbed in our falcons' activities, I scarcely noticed the fading light and coalescing clouds until they brought a welcome respite from the blazing sun and shed droplets of rain. And still the falcons slept on. Fifteen minutes of light drizzle finally roused the heat-drugged birds and soon they were behaving like the falcons I'd come to know in the last week—taking short flights, clambering over rocks, eating quail, and evading a Prairie Falcon's half-hearted attack.

Hours later, after an afternoon of intermittent rain, rumbling thunder, and fierce wind, the falcons took to the skies in a series of aerial chases. Ra flew alone, seeming to play with the wind, letting it hurl him through the skies at supreme speeds before he turned into the powerful force and climbed skyward again. Eventually he tired and tried to land, repeatedly floating over the clifftop with his legs extended. But he couldn't slow down sufficiently, stall, or control his descent. Unable to land in the gusty conditions, he was forced to remain aloft. Bringing his tail forward to steady his flight whenever the wind grabbed at one of his wings and threw him off balance, Ra stitched a wobbly line across the sky before disappearing over a ridge. As I watched another of our young peregrines smoothly pursue a passing adult peregrine, I could only hope that Ra would win his battle with the elements and land without suffering any mishaps.

———————

The days began to settle into a pattern. Garrett and I took turns arriving early to begin the day's observations while the other lagged about an hour behind. Whoever started the day first sometimes headed down to camp slightly earlier. We took turns feeding the falcons and keeping track of those that flushed when we approached. Our falcons' days consisted of long periods of inactivity—when the birds rested, preened, or watched the doings of the area's other avian residents—punctuated by almost continual movement. When our birds were inactive, one of us kept an eye on them while the other sometimes passed the time with a book. But, avid reader though I was, my eyes rarely strayed from the falcons' activities for more than a few minutes at a time. We could have divided up our peregrine monitoring, as other hack-site attendants did, rather than sitting at our observation spot for thirteen to fifteen hours a day. But neither of us wanted to miss out on our falcons' activities or risk being absent if our birds needed help. Hating to have any blank hours in my field notebook, I assiduously recorded our falcons' every move.

After the first week post-release, we unconsciously relaxed. Our peregrines' most vulnerable period had passed. Our greatest responsibilities now were to make sure that our birds were well-fed and to monitor their behavior. Experimentation had shown that withholding food at hack sites did not stimulate an earlier pursuit and capture of prey by young peregrines. Hungry juveniles were likely to scream plaintively for food at the release site whereas well-fed youngsters were more apt to develop their flight skills and begin chasing prey more quickly.[6]

As many have noted, peregrines are first and foremost birds of the air. With each passing day, we noticed minor changes in our birds' behavior that spoke of their increasing comfort in this medium, and of their growing strength and developing skills. Rather than walking awkwardly from one spot to another to join a fellow peregrine or feed on a quail, the falcons now spread their wings and fluttered or glided to their chosen destination. And the birds increasingly made longer flights over the release area, tracing the length of the cliff by gliding along on currents of air that rose up and over it. As their flight feathers finished growing out and their pectoral muscles developed, the falcons' quick shallow wing beats deepened and strengthened. Ra, in particular, was transformed by these physical and behavioral changes. His narrow chest broadened and his pencil-thin physique grew noticeably huskier. Nine days post-release, he looked stronger and more confident, though he still sported a few endearing tufts of down on his head that belied his maturing build.

The birds also spent more time soaring on updrafts of air, joining other raptors in tracing effortless circular paths on these invisible sky elevators. Most entertaining for us, our peregrines began fine-tuning their flight skills and developing their nascent predatory tendencies with endless pursuits and mock battles. Just as playing kittens develop the agility and strength to catch fleet and elusive prey, so too did the falcons' tireless aerial games fine-tune their bodies to respond to the evasive flights of the shorebirds, songbirds, and waterfowl that evolved over the eons to dodge the speed-made predators that hunted them. Though the wild chases had a playful air, these flights, this practice, would determine whether our peregrines became successful hunters and survivors or whether they went hungry and perished.

Evenings and mornings were now filled with the falcons' screams and aerial exploits as one bird streamed after another like a small fighter plane locked onto a fast-moving target. The pursued bird dodged and rolled, tucked its wings, and fell earthward in ever-more sophisticated evasive maneuvers. Becoming one with the sky, the peregrines flew with ineffable grace, dazzling speed, and breathtaking agility. As I watched their transcendent flights, I felt my own heart quicken. Transported by the falcons' inimitable aerial prowess, I could almost feel the wind stream behind me, feel every subtle movement of wing or tail that determined my trajectory, my speed, my effortless engagement with the wind. All the while, earthbound, I dispassionately scribbled my field notes, recording our falcons' flights while never taking my eyes off the birds as they traced invisible loops across the firmament, imprinting their movements, their exuberance, their vitality forever in my mind's eye, and lifting my spirits in ways that only such epiphanic sights in nature can. Small wonder that legendary raptor expert Leslie Brown called "the unmatched and unmatchable peregrine, the prince of all flying birds."[7]

Now Ra launched from the clifftop and, with strong wingbeats, chased after Isis, who had glided over him trailing an invisible wake that seemed to pull the young male into motion. Climbing above her, he screamed a challenge, then drew in his wings and dropped toward her at a steep angle. Cacking in protest, Isis tucked a wing, fell away sideways, and rolled over, presenting her ferocious talons. Undaunted, Ra extended his legs and grabbed Isis's talons in his own. Briefly the two cartwheeled through the sky, their screams echoing off the cliffs and ricocheting over the valley. Seconds later, they released each other and Isis circled over the cliff before extending her wings, flaring her tail, and landing with an almost imperceptible wobble.

And then another tiercel materialized out of the clouds and the pursuer became the pursued. Drawing his wings close to his body, Osiris dropped toward Ra with breathtaking speed and an agility that the larger-bodied, proportionately shorter-winged Isis could never match. Undaunted, Ra dropped to his side, cacking fiercely as Osiris shot harmlessly past him. Leveling out, Ra flapped his wings, found an updraft and shot skyward. Circling briefly, he dived at Osiris and the two sped like nano-jets out of view behind the cliffs.

Ever-watchful and attuned to the world around him, Gawain tilted his head to the right then left, keeping a close eye on Ra and Osiris's antics from the clifftop as Isis preened assiduously beside him. His large eyes seemed to take in the entirety of the heavens, and he watched his fellow tiercels long after I had lost them from view. Although the open skies were now a significant component of our falcons' world, our view of the birds was all too often blocked by cliffs, rocks, trees, and hills. But even when I could no longer see our falcons, the trails they had blazed across the sky still lingered in my mind, and the peregrines' surreal speed and arresting acrobatics—even at this young age—left me awed.

The Peregrine Falcon is often referred to as the world's fastest animal, eclipsing such iconic speedsters as the cheetah and the pronghorn. Nevertheless, such claims come with caveats. In level flight, the peregrine's cruising speed is comparable to speeds attained by many other birds, and ranges from about 40 to 60 miles per hour (64 to 97 km per hour).[8] Sharp-shinned Hawks, for example, may pursue their feathered prey through dense forests at speeds of approximately 40 mph while Golden Eagles can attain speeds of 60 mph when chasing fleet-footed jackrabbits.[9] Bird flight—even fast flapping flight—so often looks effortless and relatively languid from our earthbound perspective, that it seems surprising that the fastest race horse ever recorded—a blur of galloping legs, straining muscles, and streaming mane—only reached a speed of 44 mph (71 kph) in 2008.[10] The peregrine's maximum speed in horizontal flight—70 mph (113 kph)—was measured by a helicopter, as the bird hunted over the arctic tundra, and is similar to the top running speed of the golden-eyed cheetah.[11]

It is in its stoop, or aerial dive, in pursuit of prey that the peregrine attains its famed, record-breaking speed, though here, too, circumstances all too often prevent the bird from attaining its maximum velocity. To fly over 200 mph (322 kph)—a speed at which a peregrine was clocked flying with its skydiving human trainer—it must drop from altitudes at which it rarely hunts.[12] Typical stooping speeds usually range from about 80 to 150 mph (129 to 241 kph)—more than sufficient to overtake any winged prey a peregrine pursues.[13] With its long, pointed wings; relatively stiff flight feathers; powerful, well-developed breast muscles; streamlined, aerodynamic body; and oversized feet and toes, the Peregrine Falcon is perfectly designed to chase, strike, and carry away its fast-fleeing avian prey.

The peregrine also has a number of more subtle anatomical features that contribute to its record-breaking speed and predatory prowess. Were it not for certain ocular adaptations, the rush of wind streaming past a stooping falcon's head would evaporate the delicate tear film—a thin, fluid layer that protects, lubricates, and nourishes the eye, ensuring clear vision. Like other birds, the falcon has a nictitating membrane—a transparent third eyelid—that sweeps horizontally across the eye to protect, moisten, or clear it of debris without obstructing the bird's vision. As with human eyes, the surface of a bird's eye is lubricated by the lacrimal gland, which maintains the tear film. In addition, though, animals with nictitating membranes also have a second secretory gland—the Harderian gland—that's located at the base of the membrane. In falcons, the Harderian gland produces a viscous solution that provides additional moisture to the cornea during the bird's dives and is less prone to rapid evaporation than the more dilute tear film.[14]

Whether or not the small keratinous cone or tubercle that nearly fills each of a peregrine's nostrils—located in the cere above the bird's beak—also helps the falcon during its high-velocity dives is less clear. Although the tubercles look like they might impede the bird's breathing, some have hypothesized that they serve as baffles that either relieve pressure as the peregrine stoops or modify air flow at high speeds, preventing wind from ballooning the falcon's lungs. However, such tubercles are not found in other high-speed fliers such as eagles and do occur in birds such as South America's Savanna Hawk, which doesn't fly at such high speeds. Others have therefore proposed that these tubercles may instead serve as air-speed indicators that are sensitive to pressure and temperature during the peregrine's high-speed attacks.[15]

As I watched Gawain and Osiris speed over the clifftop in yet another chase, I wasn't thinking of the birds' adaptations for high-velocity flight. Instead, I was relishing the beauty of powerful birds defying gravity and moving with dizzying speed through a transparent medium whose currents and complexities the falcons navigated with aplomb. Stacked cirrus clouds glowed on the horizon, their cotton-candy-colored layers interspersed with the dusky blue of juniper berries. The falcons' dark forms finally came to rest on the cliff, and each bird preened its feathers, conditioning and realigning them for the next day's flights.

A sneezelike snort disrupted the evening's stillness. Scanning the jumble of inclines and declivities surrounding the hack cliff, I searched for the source of the distinctive sound. The snorting exhalation sounded again, and my eyes finally found the elegant silhouette of a pronghorn buck standing on a ridge looking fixedly into the distance. Captivated by its graceful form and intrigued by its chuffing alarm call, I didn't at first notice what had drawn its attention. But as it suddenly spun on delicate legs and high-trotted out of view over the ridge, I noticed Garrett walking toward our observation point. I felt a contradictory twinge of regret, as I prepared to relinquish the day's cherished solitude, and anticipation, as I looked forward to sharing our respective adventures. Returning from his trip to town, Garrett carried a large, flat object in his hands. As he approached, a smile splitting his face, eyes bright with the joy of a long-held surprise, he announced, "Pizza delivery for the South Fork hack site."

CHAPTER FIVE

Osiris

When Osiris first went missing, we weren't unduly alarmed. His absence from the hack site revealed itself to us slowly, the way a silhouetted shape gradually acquires color and detail with the progression of dawn. We were becoming accustomed to the increasingly elusive habits and ephemeral absences of our independent youngsters. With the passing of each long summer day, our peregrines ranged more widely and their daily activities became less known to us. If we didn't see all of our falcons on any given day, we hiked from one vantage point to another, checking their favorite perches and searching out new ones until, inevitably, the missing bird reappeared, feeding by the hack box or perched on a cliff ledge.

We didn't see Osiris at all on the seventh day after our falcons' release. Nor did we see him on the eleventh or thirteenth days after his first flight. But though the increased frequency and length of our falcons' absences no longer alarmed us, we were always reassured to see all five birds roost around the hack box, as they did on July 9, sixteen days post-release. I had returned near day's end from what should have been a quick morning trip into town to buy a new lantern. Ours had been destroyed by the cattle that seemed all too eager to share our camp with us. Half jogging up to our observation point, I wasn't sure whether to curse or be grateful for the learning experience of having to deal with my first-ever flat tire. With the help of the vehicle's manual, I had managed to put on the spare, but then I had to get the flat fixed at a garage. Seeing all of our falcons that evening was now unusual enough that I exclaimed, "Five peregrines total!" in my field notes before heading down to camp at 2100 hours. The observation affirmed that, though our birds were traveling more widely, they were still tied to the hack site's bounty.

We weren't concerned when we didn't see Osiris the next day or the day after that. I tried to squelch my anxiety as we searched for him to no avail on the third day of his absence. Of all our falcons, he was the most independent, the most capable, I reassured myself. And our days were filled with the many small dramas and distractions that all too often obscure larger, more significant patterns or events.

Brian Mutch, a long-term Peregrine Fund hack-site supervisor, whose field abilities and falcon knowledge were legendary, visited our site after taking over its supervision from Bryan Kimsey, who was overseeing several sites outside Wyoming. In addition to checking on our birds and our monitoring work, Brian wanted to determine where our visiting adult peregrine might be nesting. Given the rarity of breeding peregrines in the West at the time, it was critical to document any reproduction in the wild. Unfortunately, though, the distances the adult traveled and the rugged topography around our hack site precluded following her when she left our site with a quail clutched in her talons. Instead, to discover where she was taking her stolen bounty, Brian affixed tiny wildlife radio transmitters to the legs of dead quail and placed the birds on the marten board. If the adult carried a transmittered quail to her aerie, Brian might be able to locate the nest by tracking the radio signals emitted by the transmitter, using a special receiver and antenna. Because rocks and topography could block radio signals and the transmitters had limited ranges, success wasn't guaranteed, but the effort might illuminate where the adult was traveling when she left our site, heading northwest, as she invariably did.

While we waited for the adult to appear, Brian regaled us with stories about his peregrine work, Montana childhood, and experiences as a falconer. Athletic and rangy, with a boundless energy that had us jogging to keep up, Brian epitomized what I came to view as the ideal field biologist: indefatigable, resourceful, and able to thrive under any conditions. I listened agog as he laughingly told us about being bitten by rattlesnakes (twice!) and about an arc of blue light jumping between his finger and a lightning rod that hummed atop a hack box while he was transferring peregrine nestlings from a disintegrating cardboard transport box during a thunderstorm.

We were also joined by Dale Mutch—a blond version of his younger brother—who worked as a peregrine biologist for the Wyoming Game and Fish Department. Amidst animated conversations, we watched transfixed as two of our falcons chased a passing Red-tailed Hawk. Stooping a short distance, the peregrines drove the buteo toward the ground before nonchalantly resuming their aerial chases and talon grappling.

Several hours later, Isis cacked a warning from the cliff, and the sky suddenly reverberated with a penetrating, low-frequency sound.

"Did you hear that?" Dale exclaimed, pointing to an adult Golden Eagle that was just leveling out over the hack cliff, about 20 feet from one of our flying peregrines. "I think that eagle was going after one of your falcons!"

"That sound was an eagle?" I asked in disbelief, astonished that what had sounded like a passing jet was actually an eagle diving at high velocity and with deadly purpose. A pair of Golden Eagles had flown high over the hack site minutes earlier but, distracted by our conversation, we had missed seeing one of them stoop.

"Yeah, that was an eagle." Dale confirmed grimly as we watched the powerful predator fly off to the northeast. Our peregrine, seeming unaffected by its close call, dropped out of view behind the cliff. I shivered despite the warm sun, struck by how easily our bird might have been killed despite our assiduous watches over the last few weeks and our best efforts to protect our still-vulnerable youngsters. Brian and Dale were quicker to shrug off the near miss since the danger had passed and our falcons were safe. But I couldn't help feeling that I had failed my charges by not giving them my undivided attention. I vowed to be ever more vigilant.

And the day's ongoing activities commanded my complete attention. Not long after the eagles disappeared, we heard loud screeching as three peregrines sped over the cliff. The highest-flying bird—the adult female—dove at one of our falcons, driving it almost to the ground before one of the young males chased her out of view. Once again, the pursuer had become the pursued in one of the endless raptor interactions in the sky-blue realm above our heads. Soon the adult was back, unfazed by the young tiercel that had "escorted" her from the area. Swooping over the marten board, she footed a transmittered quail and headed northwest. Beaming at their success, Dale and Brian hastily gathered their gear and left to pursue the quail's radio signal. Meanwhile, I settled into my camp chair and relaxed into day's end, enjoying the quiet moments when Ra and Freya perched on the cliff, bathed in the evening's golden light, almost as much as the dramatic moments when each of them took turns skirmishing with an always voluble Isis.

Despite her apparent confidence when flying with the other falcons, Isis cacked in alarm at every large bird that flew over the area. She appeared to be especially intimidated by the ravens that regularly perched on cliff ledges and patrolled the airspace over the hack site. Seeming to sense their advantage, the ravens harassed her unrelentingly. Just over two weeks after the falcons' release, a raven had flushed Isis from the clifftop by the hack box, where she had been feeding, and driven her almost to the ground, jabbing at her with its large, powerful bill. The world's largest songbird was attacking the world's fastest predator. Twenty minutes after being chased out of sight, Isis returned to the cliff only to be driven away once more. Again she returned, warily footed a quail, and attempted to eat. But the moment another raven flew over the hack box, she flushed from the cliff, dropping the quail in her apparent haste to avoid yet another dark-winged tormentor. The opportunistic raven immediately changed course and gave chase. Rolling over repeatedly and presenting her talons, Isis finally dissuaded her pursuer, which abandoned its impromptu harassment and continued its journey. After that challenging afternoon, though, Isis became even more reactive to ravens, often leaving the cliff the moment one appeared in the area, which inevitably led to her being chased by the unrelenting corvids.

Unlike Isis, Gawain seemed only too eager to pursue any corvid that came close to him. Freya, meanwhile, appeared to be largely indifferent to the black-feathered denizens that congregated in small parties on the hack cliff and engaged in aerial acrobatics in the skies overhead. But she was quick to join the flights of the other young peregrines and practice her own dazzling maneuvers. Although Ra often joined the aerial chases and skirmishes, he still spent a disproportionate amount of time lying on cliff ledges. We learned from Brian that Ra had had an illness as a nestling. It explained his scruffy scrawniness when he arrived at our hack site and his slower development, which tracked that of the larger females and lagged several days behind the fleeter, more agile males.

Despite our falcons' mixed reactions to the ravens that frequented the hack site, they seemed virtually indifferent to the local Prairie Falcons, which were far more amicable though no less opportunistic. Discovering the quail bounty at the hack site shortly after fledging from a nearby aerie, a juvenile Prairie Falcon and one of its parents became frequent visitors, often feeding and roosting near our peregrines. As much of a novice as our captive-reared birds, the immature prairie almost landed on a feeding peregrine on one of its first visits to the hack cliff. Undaunted by the prairie's lack of manners and coordination, our peregrine watched it grab a nearby quail and the two birds ate peaceably, side by side.

———

On an evening when exuberant falcons seemed powerless to resist a wind that teased with its fitful gusts and powerful currents, three of our peregrines took to the skies and engaged in tumultuous mock battles. Chasing and darting, falling and rolling, they streamed across the sky, pursuing each other with extended legs, as though ready to snatch their companions from the invisible airflows that each of them navigated with such finesse. Their screams rivaled the howl of the wind as their streamlined bodies sped over the cliff, skimmed over boulders and trees, and dived alarmingly close to a barbed wire fence that marched its way over a nearby hill.

As one of the females raced across the sky, I noticed that she was holding a small branch in her feet. Reaching down to it with her beak, she plucked off bits of leaf as easily as if the branch were tucked under a taloned foot on the clifftop. I had read that immature falcons that were developing their hunting skills would snatch at tree branches, rocks, grass, floating sticks, or any other object that passed within grabbing distance of their long-toed feet while the birds were in flight.[1] Seeing our young female carrying her branch brought home to me how deep-seated our falcons' predatory urges were and how instinctively such traits developed in these parentless birds. Flying for just over two weeks, this youngster was the first of our birds to seize an object and cavalierly pluck it while hurtling through the air, the way she might one day do with a shorebird. Grabbing objects helped our peregrines develop their footing skills and their ability to grasp and hold on to prey, as well as their general coordination and agility in flight. Soon, no treetop or shrub was safe as our falcons repeatedly tagged trees, bushes, and other targets with their feet while buzzing over in pursuit of their nestmates or winged insects.

The neighborhood birds soon fared no better than the site's vegetation in avoiding harassment by our youngsters. The falcons increasingly chased every feathered form that came close to them. Young peregrines typically begin pursuing whichever birds (or mammals) are most abundant and conspicuous, regardless of whether they are suitable prey.[2] Our falcons honed their pursuit skills on the ubiquitous Violet-green Swallows and the slightly less abundant White-throated Swifts. What had begun as slight, investigatory digressions when our falcons' flight paths crossed those of these swift-flying birds evolved into half-hearted chases and then increasingly, into more prolonged and focused pursuits. Despite their efforts, though, none of our peregrines had come close to making contact with these tiny, agile aerialists—at least as far as we had seen. For the time being, the hack site's avian denizens seemed safe, though eventually most of the nonraptors might be hunted by the masterful falcons.

Worldwide, the Peregrine Falcon has preyed on an estimated 1,500 to 2,000—or as many as one-fifth—of the world's bird species.[3] Matching or exceeding the speed and agility of the fastest and most acrobatic fliers, peregrines are among the few raptors known to regularly kill swifts.[4] Unbeknownst to them, our falcons were cutting their teeth on some of the world's best fliers.

Seeing our falcons make their first kills was unlikely since our young birds quickly disappeared amidst the hack site's rolling topography. But the peregrines would gradually be able to keep up with the area's small birds, and the falcons' pursuits would then become more serious. These smaller birds typically comprised a young peregrine's first kills.[5]

The innate instinct of immature peregrines to chase anything that moved not only made it possible to hack these parentless birds, but it also ensured that they would develop the predatory skills essential to their survival. Our falcons would likely pursue hundreds of targets, with increasing intensity, before learning how to kill successfully. Although we never saw our falcons chase anything but birds, others have witnessed humorous examples of this hyper-motivation to pursue any moving object—regardless of its suitability as prey. Aside from watching juvenile peregrines chase a staggering variety of birds, researcher Steve Sherrod also observed them flying after voles, rabbits, woodchucks, arctic foxes, and even white-tailed deer![6]

We had no such comic relief at our hack site. Instead, as they began their fourth week as flying birds, our peregrines chased every raptor that passed over the release area. Whether pursuing other birds of prey is part of their instinctive reaction to fly after anything that moves or whether it is an agonistic response to perceived enemies isn't known. But despite the potentially deadly consequences, our peregrines were intent on the chase, regardless of the caliber of the opponent.

After a rare partial day away from our hack site to observe a Golden Eagle nest with Brian Mutch, we settled down for an evening of peregrine watching. Osiris was still worryingly absent. We hadn't seen him in five days. Five days that had felt like an eternity. The skies were overcast and thunder rumbled, but the usual afternoon gale had been replaced by a refreshing breeze, and when the sun peered through the clouds, it cast a warm glow over our observation site. We immediately picked out Isis perched on a favorite ledge below the cinereous bulge in the cliff that we'd dubbed

Pregnant Rock. No sooner had we spotted Gawain on the hack box than he took off. Flapping hard, he caught up to an adult Red-tailed Hawk that had drifted over the site seconds earlier. The buteo went into a steep glide and the two birds disappeared from view. A minute later, having escorted the red-tail on its way, Gawain landed with a flourish on the marter board then hopped down to the top of the hack box and began eating next to Freya.

As soon as the two falcons finished feeding, they spread their wings and disappeared and reappeared from our view as they engaged in mock battles in the strengthening wind. The sudden sight of an immature peregrine close on the tail of a juvenile Golden Eagle had me catching my breath and remembering the earlier attack on one of our falcons. But I need not have worried. Soon, the eagle had been abandoned and three of our youngsters were chasing ravens, a far safer pursuit, from which they soon digressed to chase one another. Their screams filled the air. Legs dangled threateningly. Dominion over the skies was on full display. One falcon landed, and I watched another circle over the hack cliff, gaining altitude on an invisible column of rising air. As it tucked its wings and started to dive, I belatedly noticed its target. Another juvenile Golden Eagle. I cringed, my face grimacing as I unconsciously shrank from an incipient disaster. Pulling out of its stoop just before hitting the eagle, our foolhardy peregrine swooped upward then circled around and settled in the eagle's wake, chasing the larger bird behind the cliff. Not to be outdone by its nestmate, another young peregrine joined in the fun, chasing its fellow peregrine and the fleeing eagle, which seemed unbothered by the diminutive falcons. To my relief, our two peregrines soon reappeared and as one landed, I identified Ra. Another day with only four falcons.

As though tiring of their benevolence, the weather gods suddenly stirred the wind into a frenzy. I scrambled to cover my notebook and belongings from the dust storm that engulfed us. Squinting through gritty eyes, we struggled to watch the cliff and keep our birds in view. As a hard rain began to fall, none of our falcons was visible. But we had faith now that they could outsmart the elements and find safe roosts. As long as they were in the vicinity of the hack cliff, we felt sure we would see them the next day. When—and if—we might see Osiris again, on the other hand, was a disquieting question that accompanied us back to our camp and settled around us for the night with the chill of a damp and penetrating fog.

———

Halfway through our hacking period, the days were infinitely quieter. We read more and saw our falcons less, as their universe expanded into unknown realms. Occasionally we had hints of their increasingly distant movements, spotting them, for instance, perched in trees in the meadow adjacent to the stream we crossed on our hike up to the hack box. My field notes were now filled with repeated notations of "No activity," and "No falcons visible." We might see only two or three of our falcons on any given day. Inevitably, though, over the course of a few days we saw Isis, Freya, Gawain, and Ra. But still no Osiris.

Our Idaho supervisor had not been optimistic about his fate when I called during my weekly trip into town and told him how long Osiris had been absent after disappearing just over two weeks post-release. It was highly unlikely that he could hunt successfully at this early stage of his development, and fasting for the length of time he'd been missing could leave him severely compromised, at best. It was possible that a newly released immature peregrine that didn't return to roost at the hack site had made a kill, gorged, and spent the night near its kill site rather than returning to the hack cliff. But a juvenile that happened to make a kill during those first two weeks could not be considered independent and would require supplementary food until its hunting skills had developed enough for it to catch prey regularly.[7] Young falcons that inadvertently made a kill usually returned to the hack cliff within a day or two. Osiris had now been absent for nine days.[8]

As I watched the elegant pair of adult peregrines fly over our site, *eechipping* and stooping nonchalantly at one of our perched youngsters, I thought about their legendary predatory abilities and the adaptations that made them possible. And I wondered for the thousandth time about Osiris's chances of catching prey if he managed to dodge Golden Eagles and avoid other threats. Despite our falcons' innate hunting drive, their nascent skills had not yet been honed by experience.

The day before, I had hiked up to the base of the cliff while looking for our birds since none of them had been visible from our observation site. Eager to learn about the area's other avian denizens, I decided to clamber a short distance up the cliff to peer into a crevice I had spotted from afar that held a Violet-green Swallow's nest. As I began to climb, I inadvertently flushed two Rock Pigeons from a hidden perch. Materializing out of nowhere, an immature peregrine dropped out of the sky, closing the distance to one of the pigeons with remarkable speed. The wily pigeon, whose evasive maneuvers still outmatched our young falcon's amateurish skills, easily dodged its attacker. The falcon then briefly soared over the cliff before disappearing again. Pigeons and doves are among the peregrine's preferred prey and are often the most readily available. Some scientists have speculated that these columbids evolved their characteristic soft plumage (some of which may be shed when the birds are grabbed) to facilitate their escape from the clutches of the apex predator that has pursued them through the ages.[9] Awed by our peregrine's keen eyesight and astonished at the rapidity of the falcon's response to the sudden movement of potential prey, I retreated, chastened that I had drawn its attention and disrupted the natural order of life on the cliff.

It seemed inconceivable that our young falcons would soon be able to catch such swift prey, and it was highly unlikely that Osiris had managed to do so. Even recently fledged falcons whose parents have "placed" still-live birds in the sky in front of them have low capture-success rates.[10] Experienced adults often have difficulty catching prey, too. Migrating peregrines may make as few as one successful kill for every thirteen hunting attempts. But hunting success rates are extremely variable. One remarkable tiercel in New Jersey had a reported hunting success rate of 93 percent, or more than nine successful kills for every ten attempts. Peregrines may require only a few minutes to make a kill, or it may take them several hours. Such variability depends on a falcon's age, experience, and hunger level; the proximity, density, and vulnerability of prey; the time of day; and the weather conditions.[11] Raptor expert Clayton White followed a tiercel by helicopter from its remote breeding site in Alaska and mapped twenty-one of its hunting sorties. On thirteen observed hunting forays, the falcon averaged about ten stoops before making a kill (the minimum number was two stoops at a single prey item and the maximum number was twenty-four stoops at nine prey items).[12]

Although Osiris likely lacked the experience to make more than a chance kill, he was anatomically equipped to do so even as a young fledgling. Aside from the adaptations that allowed him to fly with exceptional speed and maneuverability, he had short, thick lower legs, or tarsi (singular, tarsus)— the bone, technically called the tarsometatarsus, between a bird's toes and its heel, or the first bend in its leg—which would allow him to strike prey with tremendous force. His exceptionally long toes were almost the length of his tarsus, and were the reason John James Audubon called the peregrine the Great-footed Hawk. Whereas mammal- and insect-catching birds have shorter, fleshier toes, bird-killing peregrines have especially elongated middle and hind toes, the better to wrap around their prey. At high speeds, the falcon's contact with prey often looks like a closed-foot strike, but slow-motion videos have shown that a peregrine spreads its toes open to grab flying prey and often rakes it with its hind talon, which has been dubbed the killer talon.[13]

Even though peregrines can kill prey with their feet, they typically use their beaks to dispatch birds in the air—while holding them in their talons—or on the ground. All falcons have a special adaptation known as the tomial tooth (the *tomium* refers to the cutting edges of a bird's bill), which is a protrusion on the outer edges of the upper part of the bird's bill that corresponds with notches in the lower part. A peregrine dispatches its prey by biting its neck with its tomial tooth to disarticulate cervical vertebrae and sever the nerve cord. A female falcon can kill larger prey with her longer, huskier beak than a tiercel can with his smaller beak.

Despite my hopes that Osiris had capitalized on his bird-hunting arsenal, that possibility seemed increasingly remote. It had been nearly twelve days now since we had seen bold Osiris, who showed no qualms about attacking fearsome foes like Golden Eagles and adult peregrines. Although I tried to resign myself to having four out of five of our falcons reach independence, I kept wondering what we could have done differently to keep our intrepid, yellow-banded falcon from straying. As I watched the hack cliff's rocky protuberances and crevices sharpen and fade in the changeable evening light, I agonized endlessly over his whereabouts.

Brian and Dale weren't having any better luck finding the haunts of the adult peregrines that frequented our hack site. Although the female had taken several transmitter-bearing quail, the Mutches had been unable to track her to her aerie. Either she had fed on the quail and dropped them in transit or the transmitters' signals were obstructed by rocks and topography.

Determined to get more information, the Mutch brothers had even upped the ante by trying to trap the female so that they could outfit *her* with a transmitter and track her movements directly. Wrapping a dead quail in a little harness covered with small nooses made out of monofilament fishing line, they had placed the "trap" on the marten board. If the adult grabbed the quail, she was likely to get her long toes caught in a noose and, as she tried to pull away to free herself, the loop would tighten, ensnaring her foot. The quail was carefully weighted so that the peregrine couldn't fly away, with her feet bound up in fishing line, if she was caught.

Midmorning on an unusually calm day, the adult peregrine had streaked over the site then slowed, dropped, and reached out to grab a quail. But, as we held our collective breath in anticipation, the wily female somehow evaded capture. Her talons had not been ensnared and, perhaps feeling the resistance of the weighted quail, she had hesitated and released the prey. Circling over the site, she looked fixedly at the quail, then flew off. Something was awry, and her instincts seemed to tell her that the easy meal she had become used to taking was not worth the risk on this day.

Now, on July 21, the Mutches were trying once again. Just before noon, the adult peregrines soared over the site. Hovering high over the marten board, her head turning to and fro, the female eyed the bait. Her mate had shown little interest in the abundant quail in the preceding weeks and stooped repeatedly at the female. The two circled, *eechipping* to each other. But the female would not be deterred. Hovering over the marten board, she suddenly swooped down and attempted to snatch the transmittered quail. Again, she eluded the monofilament loops, though her strike knocked the quail off the marten board. Jumping to his feet, Brian jogged up the hill to replace the trap. But after circling high over the valley, the peregrine pair glided out of view.

Later that afternoon, as I settled into my camp chair for a quiet afternoon of peregrine watching—or cliff watching as was increasingly the case—I replayed the female's subsequent visits to the noosed quail. Although she'd hovered repeatedly over the bait, staring at it intently, she had not tried to grab it again, and had eventually lost interest. As much as I'd wished to see her in the hand, I couldn't help but admire a captive-raised bird that was as wily and cautious as one that had been raised on a remote cliff by wild parents. I could only hope that our juveniles would be equally wary if ever confronted by one of the untold hazards of a world so heavily impacted by humanity.

Repositioning the scope, I scanned the clifftop and spotted Osiris eating a quail next to the hack box. Osiris!

"I think I'm looking at Osiris," I yelped. Fiddling frantically with the focus knob on the scope, I stared and stared at the band on our falcon's leg, wondering if I could possibly be mistaking it for Isis's white band. But the yellow glowed like a flare, proclaiming our great adventurer's unexpected homecoming.

"It is! It's Osiris! It's Osiris!" I exclaimed, jumping out of the way so Garrett could look.

A huge smile spread across his face as he peered intently into the scope.

"It's Osiris!" he whooped in turn. And then we were jumping up and high-fiving, and I felt a knot in my throat as unexpected tears filled my eyes.

"He's back! He's back! He's alive!" we exclaimed as our young traveler gorged on quail. Almost an hour later, Osiris finally stopped eating, his crop so full it looked like a feathered baseball protruding from his chest. Ten minutes later he fed again. Elated by his presence, I stared at him through the spotting scope as if the intensity with which I watched him could unveil the mysteries surrounding his absence. Finally, he dragged what appeared to be his third quail out of view behind the hack box, and I sat back in my chair with a huff of breath and a huge smile. Five falcons. We had all five falcons again. Osiris had come home.

————

As I sat in camp that night, relishing the warmth of my chicken-flavored pasta, my sweatshirt and windbreaker pulled tightly around me to ward off the evening chill that had crept stealthily down the surrounding hills to wrap itself around our sparse cooking area, I thought about Osiris and where he might have been during the twelve days he had been missing. Surely, he must have made a kill. But what had kept him away so long? It seemed a miracle that he had returned, hungry but otherwise apparently healthy. If he had been injured, there was no trace of it. I thought about the myriad fates that could have befallen him. He could have been killed by an eagle or a Great Horned Owl or even another peregrine. He could have broken a wing on a barbed-wire fence or utility wire or been electrocuted by a transmission line. He could have been shot. Or hit by a car. Or poisoned. Songbirds that ingested poisons in agricultural areas were often targeted by raptors because the afflicted birds flew in unnatural or compromised ways, making them vulnerable to predators that keyed into signs of weakness in their prey. If Osiris had made it as far as the river valley, his first kill might have been a poisoned blackbird, a manageable target for an immature falcon.

My rampant speculations about the hazards he'd faced led me no closer to unveiling the mystery of our falcon's long absence, but they did start me thinking about poisons, about DDT, and about the risks such chemicals posed not only to unintentional targets like our inexperienced but incomparable Osiris but also to local bird populations, and even entire species. In the early 1960s, it was nearly unfathomable that a well-regarded pesticide like DDT could harm anything but its intended targets. But as the world was soon to learn, it was pesticides, after all, that had launched the globally distributed Peregrine Falcon on its improbable journey from relative abundance to near extinction.

CHAPTER SIX

A World Full of Poisons

When British peregrine researcher Derek Ratcliffe asked fellow scientists to review one of his manuscripts a year after the 1965 Madison Conference that electrified falcon aficionados around the world, the comments that two of them jotted in the margins precipitated an astonishing breakthrough in the ongoing mystery of the bird's widespread decline. Ratcliffe had highlighted the bizarre egg breakages and disappearances that he and others had documented in peregrine nests since 1949 and speculated on the causes of this unprecedented ornithological phenomenon. One possible cause, he wrote, might be "defects of the egg." "If for some reason, Peregrine eggshells had recently become thinner," he continued with visionary boldness, "an increase in breakage might be expected." But he conceded that he had not measured the thicknesses of broken eggshells and, given the normal thinning of shells during incubation, none of the eggs he'd examined seemed unusual.[1]

Reading these speculations, University of Wisconsin professor Joe Hickey noted that Ratcliffe's hypothesis could be objectively tested. "I am told poultry science people . . . routinely measure shell thickness," he wrote on Ratcliffe's manuscript, recommending that the British scientist measure peregrine eggs collected before and after 1950 and compare them. Another of Ratcliffe's colleagues, Desmond Nethersole-Thompson—an authority on birds of the Scottish Highlands—commented that eggshell thickness was related to eggshell weight, so Ratcliffe could evaluate shell thickness by weighing older eggs and comparing their weights to those of more recently laid eggs.[2]

Ratcliffe was quick to follow his colleagues' advice. Weighing and measuring peregrine eggs (from which the contents had been removed) from the British Museum and private collections—including many illegal ones to which he adroitly gained access—he found that eggs taken from aeries all over Britain after 1947 were almost 20 percent thinner-shelled than eggs taken before that year. Given this widespread change in eggshell thicknesses, Ratcliffe felt certain that the onset of thin-shelled eggs and their subsequent breakages could only have been caused by a pervasive environmental change that occurred in Britain around 1946–47. Suddenly, it hardly seemed coincidental that the timing of thin-shelled peregrine eggs—as well as the reduced shell thicknesses he also documented in Eurasian Sparrowhawk and Golden Eagle eggs—coincided with the introduction of organochlorine pesticides into general use to control agricultural, horticultural, and livestock pests.[3] Before publishing his breakthrough thin-eggshell finding, Ratcliffe gave Hickey permission to relay the remarkable story to the US Fish and Wildlife Service and the Canadian Wildlife Service, and to initiate a similar eggshell-measuring study in the United States.[4]

When Hickey and his affable graduate student, Daniel Anderson, decided to compare a variety of raptor eggs from before and during the modern-pesticide era, they expected to find eggs in public museums and private collections. However, they were certain that eggs from two species—Bald Eagles, which had been given federal protection years earlier, and Peregrine Falcons—would be unavailable for the latter period.[5] To their surprise, though, the researchers discovered that many private egg collectors were still illegally procuring raptor eggs. Fortunately, Anderson proved to be adept at gaining the confidence of those who engaged in these illicit activities, and he measured 73 Bald Eagle eggs taken by collectors *after* such activities had been strictly prohibited. The intrepid Anderson also found 123 peregrine eggs from the pesticide era. Like Ratcliffe, he found that peregrine eggs laid after the introduction of DDT were as much as 26 percent lighter than those laid before the modern insecticide era.[6]

Even though the eggshell-thickness findings in Britain and the United States were similar, the coincidental onset of thinner eggshells and widespread DDT use didn't irrefutably implicate the pesticide. Female birds, whose physical condition is reflected in the quality and composition of their eggs, might have been affected by any number of environmental perturbations, including another class of toxic chemicals—PCBs (polychlorinated biphenyls)—which had long been used as coolants, adhesives, and flame retardants, among other industrial applications. PCBs had only recently been recognized as a global environmental pollutant found in people, animals, and even in the air.[7]

Despite the many possible explanations for thinner eggshells, the likelihood that DDT—long considered a relatively innocuous pollutant—was responsible received scientific (rather than just circumstantial) support in 1968, when Hickey and Anderson compared eggshell thickness in five Herring Gull colonies in the north-central United States and New England. Variation in the shells' thicknesses was closely related to the levels of DDE—the breakdown product of DDT—found in the egg remnants.[8] Several years later, peregrine biologist Tom Cade and his colleagues at Cornell University bolstered these findings by showing that Alaskan peregrine eggs that contained higher levels of DDE had thinner eggshells, whereas those with lower levels of DDE (meaning the females that had laid them had lower exposure to DDT) were thicker shelled.[9]

Over time, study after study added to a growing body of evidence that DDE wasn't just *correlated* with eggshell thinning but had actually *caused* it.[10] When scientists fed DDE to Mallards in a laboratory, the ducks subsequently laid thin-shelled eggs.[11] Captive American Kestrels that were fed DDE in different doses showed a related range of eggshell thicknesses.[12] And British birdwatching chemist David Peakall showed conclusively that DDE was *not* present in eggs before the DDT era but *was* present in eggs from 1947 on. Furthermore, the levels of DDE in shell membranes and traces of the original egg contents were directly related to the thickness of the eggshells that enclosed them.[13] Skeptics' claims that eggshell thinning had occurred before DDT was widely used were finally laid to rest.

Though additional research and experimental refinements continued for many years, scientists had unraveled the mystery of the now-familiar symptoms of DDT poisoning in peregrines: abnormally thin-shelled eggs, high DDE residues in eggs and body tissues, and a reduced ability to successfully produce young. Whenever and wherever peregrines laid thin-shelled eggs, the birds' numbers declined. As a top predator, peregrines killed shorebirds, ducks, and songbirds that had accumulated DDE in their bodies after feeding on contaminated prey, vegetation, or seeds. Since the falcons consumed many of these birds over the course of a year, they often accumulated large amounts of the pesticide, which was subsequently stored in their fatty tissues and persisted in their bodies. Because of this biomagnification process, bird- and fish-eating predators at the top of the food chain were most vulnerable to DDT's pernicious effects.

Ultimately, researchers revealed slightly different poisoning patterns between the United States and Britain. In the United States, peregrines were harmed primarily by DDT, whose lower toxicity didn't necessarily kill them but reduced their ability to reproduce successfully. In Britain, on the other hand, the heavily used, highly toxic cyclodiene insecticides—aldrin, dieldrin, and heptachlor—killed adult peregrines outright, contributing more significantly to that country's precipitous population crash. Some poisoned falcons in Britain were seen falling out of the sky or off elevated perches. Meanwhile, DDT in Britain reduced the peregrines' reproductive output before and after cyclodiene insecticides were widely used, and slowed the birds' recovery when voluntary restrictions were placed on these more toxic chemicals in Britain, starting in 1962.[14]

———

Despite growing scientific evidence that DDT was harmful to wildlife, powerful agrochemical and farming interests fought mightily to preserve the use of the wonder product that had protected millions of soldiers and civilians from insect-borne diseases in World War II, spared acres of crops from the ravages of insect pests, and promised to help eliminate diseases such as malaria. In doing so, influential parties—backed by the US Department of Agriculture, which was responsible for the development and registration of pesticides until the founding of the Environmental Protection Agency (EPA) in 1970—confronted a growing tide of public concern about the indiscriminate use of pesticides and their potential effects on wildlife and human health.

In 1962, the publication of Rachel Carson's seminal book *Silent Spring* fostered a growing awareness of the deleterious effects of pesticides, converting "what had been a quiet, scientific discussion" over the unintended consequences of environmental contaminants "into a noisy, public debate."[15] Highlighting the persistence of pesticides like DDT in water, soils, and vegetation, Carson decried the use of such chemicals without a proper understanding of their ecological effects, and warned of potential connections between their indiscriminate use and escalating threats to human health. She also wrote poignantly about the death and disappearance of wildlife in the wake of broadcast insecticide spraying. Raising the specter of a world where no birds remained to cheer us with their beautiful songs and bright colors, Carson's words struck a sensitive chord with those who lamented wildlife losses.

In the Midwest, successive waves of American Robins—those much-loved harbingers of spring—had convulsed and died in shocking numbers on college campuses and private lawns after feeding on earthworms contaminated by DDT in the spring of 1955. The worms had ingested leaves coated with DDT following intensive spraying to control Dutch elm disease, which was striking down the elegant shade trees that graced so many of America's main streets.

A woman in Bloomfield Hills, Michigan, reported seeing twelve dead robins on her lawn at one time. Others decried the loss of cherished backyard birds—chickadees, nuthatches, and woodpeckers—that had been poisoned after gleaning contaminated insects from tree bark; tiny resplendent warblers that had dropped from the sky after foraging among contaminated tree leaves; raptors that had succumbed after feeding on birds and small mammals; aerial swallows and flycatchers, whose insect prey was toxic or no longer available. All told, birds representing about ninety different species were reported killed following spraying for Dutch elm disease in Michigan, and this was only one of an astronomical number of uses to which DDT was routinely (though not necessarily successfully) being applied. One Wisconsin naturalist echoed the sentiments of many when he wrote, "Summer mornings are without bird song . . . It is tragic and I can't bear it."[16]

Rachel Carson also described Bald Eagle declines in Florida, and East Coast eagles that had disappeared from former haunts or were unable to raise young. The timing and nature of these reproductive failures were eerily similar to those that researchers would soon uncover for Peregrine Falcons. And Carson's speculation that insecticides like DDT were inhibiting the eagles' ability to raise young would be confirmed in the coming decade.

Although *Silent Spring* evoked raw emotions and raised awareness and alarm about the pesticides that were cavalierly being used to combat an endless parade of insect "enemies," the widespread use of DDT and related chemicals persisted. Entomologists, chemical companies, and agricultural and forestry interests, unwilling to give up the perceived benefits of their poisoning practices, worked vigorously to bolster DDT's reputation and discredit Carson's work. Despite their efforts, a Science Advisory Committee convened by President Kennedy supported many of Carson's findings—particularly those related to DDT's impact on the environment—in 1963, and recommended the phaseout of persistent, toxic pesticides.[17] Yet the widespread use of DDT continued unabated.

In 1966, frustrated by their inability to protect cherished habitats from ongoing chemical assaults through traditional avenues such as public education and lobbying, a group of scientists and conservationists filed a lawsuit against a Long Island, New York, mosquito commission in an attempt to stop the use of DDT on local marshes. Using litigation based on scientific testimony to influence policy was a novel approach to environmental protection and proved to be highly successful. The plaintiffs won a temporary injunction to stop the spraying and, even though the judge refused to ban the use of DDT outright, the commission ultimately opted to use other forms of mosquito control. DDT was never again used on Long Island.[18]

Empowered by their success, which helped spawn the field of environmental law, ten of the advocates—most of whom were scientists—formed the Environmental Defense Fund (EDF), a national advocacy organization, in 1967. Led by Charles Wurster—a scrappy, passionate birder and environmental science professor with a PhD in organic chemistry from Stanford University, whose own research had revealed DDT's dire impact on northeastern songbirds—the group dedicated itself to combining science and law to achieve environmental protection. Within two weeks of incorporating their environmental advocacy group, the fledgling organization, armed with the fiery slogan "Sue the bastards," filed suit in federal court in Michigan against the use of DDT and dieldrin and, despite early setbacks, eventually succeeded in obtaining court orders banning the use of DDT in more than fifty of the state's cities.[19]

But the most significant and precedent-setting legal challenges to the widespread use of DDT came in Wisconsin in 1968–69, and then in the 1971–72 EPA hearings. By the time the EPA hearings were held in Washington, DC, the National Cancer Institute had shown that chronic low-level exposure to DDT in mice led to an increased incidence of liver tumors, and public concern with the pesticide had grown exponentially. Pitting pesticide manufacturers and the US Department of Agriculture against conservation organizations and the newly formed EPA (which had adopted responsibility for pesticide registration and regulation), the EPA hearings were the most extensive public discussion to date on the pros and cons of DDT.[20]

Of all the scientific research presented at the Wisconsin and EPA hearings, perhaps none was as effective for the case against DDT as the studies on the singularly captivating Peregrine Falcon. Joe Hickey, Tom Cade, David Peakall, toxicologist Robert Risbrough, and other scientists helped weave together interdisciplinary research that painted a dramatic picture of the falcon's precipitous decline on multiple continents, its inability to reproduce in many areas, and its thin eggshells and broken eggs. Their testimony highlighted the properties that made DDT so dangerous and disruptive to ecosystems—its chemical stability and persistence, its mobility through air and water, and its presence and biomagnification throughout food chains—all of which meant that it posed a serious threat to organisms far from where it was applied and for many years after it was initially used. Dramatic testimony on shell-less Brown Pelican eggs, plummeting Bald Eagle and Osprey populations, and the decreased eggshell thickness of virtually all North American fish-eating birds in the DDT era further undermined claims about the insecticide's safety and harmlessness.

With the close of the EPA hearing, administrator William Ruckelshaus shrugged off significant political pressure from agrochemical and farming interests, and effectively banned DDT on June 14, 1972, withdrawing its use for nearly all purposes nationwide. With a few temporary exceptions, DDT would no longer be used on crops, but could be used for quarantine and public health uses, and would still be produced and exported.[21]

Additional EDF litigation led to national bans on aldrin and dieldrin in 1974, but "emergency" use of DDT to control selected insect infestations continued for years. A worldwide DDT ban was eventually formalized under the 2001 Stockholm Convention on Persistent Organic Pollutants, which restricted DDT use to vector control—the killing of disease-carrying insects. Although 185 countries ultimately became parties to the convention, DDT is still used in malaria-prone countries, where the inside walls of homes are sprayed with the pesticide to kill or repel mosquitoes.[22]

———

Lest anyone wonder whether DDT supporters were right in asserting that the environmental consequences of the pesticide's use were negligible or would be ephemeral, the chemical lives up to its legendary persistence. More than half a century after its heyday, it periodically reasserts itself into the public consciousness. The Montrose Chemical Corporation manufactured, packaged, and distributed DDT from 1947 to 1983, discharging its toxic waste into the Los Angeles sewage system until it was forced to stop doing so in 1971. Flowing into the Pacific Ocean via the Los Angeles harbor, the

DDT-laced wastewater was likely responsible for the dramatic eggshell thinning and lowered reproductive success of the area's Brown Pelicans and Double-crested Cormorants.[23] The pelicans and cormorants, which fed on contaminated fish that foraged close to the water's surface, recovered fairly quickly once the toxic discharge was halted (though Bald Eagles laid thin-shelled eggs for decades). But more than 1,700 tons of DDT settled onto the ocean floor in an area known as the Palos Verdes Shelf.[24] Today, the area's DDT-laced fish pose an ongoing health threat to human consumers. The toxic fish also threaten the California sea lions that feed on them. After breeding, these sea lions migrate north along the California coast, stopping at a picturesque beach near Big Sur.

When biologist Joe Burnett scaled a California redwood, in 2006, and peered into the abandoned nest cavity of central California's first breeding pair of reintroduced condors, he expected merely to confirm the young birds' failed nesting attempt. Instead, he was confronted by abnormally thin eggshell fragments and the specter of an insidious threat that was all too familiar to avian biologists working decades earlier.[25] Follow-up research revealed that Big Sur condors, which scavenge the carcasses of sea lions and other marine mammals that feed on the Palos Verdes Shelf and along the California coast, had significantly thinner eggshells than those of inland populations of California Condors, which don't feed on marine mammals.[26] Even those who warned of DDT's persistence and widespread impacts likely would have been surprised that toxic wastes spewed by a Los Angeles chemical plant half a century earlier continued for decades to contaminate fish that fed sea lions, which were later consumed by imperiled condors nesting in California's iconic redwoods.

Other reminders of the DDT legacy are more direct, and their toxic pathways are clearer. In the spring of 2014, residents of St. Louis, Michigan, tested some of the dead songbirds they had been finding for years to determine what was killing them. It wasn't wholly surprising when researchers at Michigan State University announced that the birds were being poisoned by DDT, given that St. Louis was once home to a company that manufactured pesticides, including DDT, until 1963. What was surprising, however, was that the twenty-nine birds that were tested had some of the highest-ever recorded levels of DDE seen in wild birds. The birds appeared to have been poisoned after feeding on DDT-laced worms found in contaminated soils on the fifty-four-acre site that once supported the company's main production plant and is now a designated Superfund site.[27]

The EPA began testing St. Louis yards in 2006 and found approximately sixty that were highly contaminated with DDT either from residues that had drifted over from the chemical plant or from the free fill dirt that the company once offered to townspeople decades ago.[28] As cleanup efforts were undertaken, St. Louis's residents watched their local birds die year after year, and lived a version of Rachel Carson's "silent spring" more than fifty years after the banning of the pesticide that so many claimed was safe.[29]

DDT is not the last pesticide with countless adherents—who highlighted its benefits and supposed safety—to be used indiscriminately and without a full understanding of its ecological or human-health effects. In the years after DDT was banned, the use of carbamate and organophosphorus insecticides that were highly toxic to birds led to frequent large-scale poisoning events. As many as 20,000 Swainson's Hawks—a common denizen of North America's western grasslands—died in the mid-1990s after feeding on grasshoppers and caterpillars in agricultural fields that had been treated with monocrotophos, an organophosphate insecticide, in the hawk's wintering grounds in Argentina.[30] In the early 1990s, David Pimentel of Cornell University and his colleagues conservatively estimated that approximately sixty-seven million birds were killed each year as a result of pesticides.[31]

Birds that breed in grasslands—among the most threatened group of birds—have declined precipitously in recent decades (nearly 40 percent between 1968 and 2014).[32] The Horned Lark, whose ability to withstand the fiercest winter weather brightens many a birder's drear day; the Mountain Plover, which evolved to nest in shortgrass prairies grazed by American bison; the Western Meadowlark, so beloved as a sign of spring that six states claim it as their state bird; and the Ferruginous Hawk, whose fierce predatory mien is tempered by the bird's penchant for hunting from the ground, have all captured the attention of conservationists who have counted fewer and fewer of these birds during yearly surveys.

Many grassland bird species are believed to have declined because of habitat loss associated with the intensification or industrialization of agriculture, including the shift to larger fields, the expansion of monocultures, the widespread use of herbicides and pesticides, and the conversion of native pastures and other seminatural habitats. But a study, led by Canadian toxicologist Pierre Mineau, that sought to examine the individual impact of these stressors on North America's grassland birds found that acute pesticide toxicity was the best predictor of bird declines. Lethal pesticides were nearly four times as likely to be associated with bird declines as was the next

73

most likely cause. Although the scientists stressed that habitat loss remains a key factor in explaining the loss of grassland birds, their research findings suggest that the widespread use of highly toxic agricultural pesticides poses a significant threat to these birds.[33]

The acute lethal risk of agricultural pesticides to birds seemed to decline in the 1990s because of the withdrawal of particularly toxic products like carbofuran—a carbamate insecticide that is especially deadly to birds—but a new class of pesticides known as neonicotinoids, or neonics, was introduced during this time period. Initially welcomed because of their lower acute toxicity to vertebrates, neonics—now the world's most widely used insecticides—have provoked growing concern because of their persistence in the environment, their high water-solubility, and the ease with which they are transported by runoff from terrestrial to aquatic environments, where they contaminate surface and groundwater.[34] Despite concerns and warnings expressed by its own toxicologists about the potential environmental effects of neonicotinoids, the EPA had approved approximately 600 of these products by 2013. California alone has registered around 300.[35]

Because they are water-soluble, neonicotinoids are absorbed by leaves and roots, making entire plants toxic. They are also found in the nectar, pollen, and fruit of plants grown from neonic-coated seeds. These potent chemicals have come under increasing scrutiny because of their high toxicity to insect pollinators—studies have linked neonics to the widespread collapse of honeybee colonies—and to aquatic life.[36] Despite DDT's infamy, it takes 10,000 times more DDT to kill half a sample of honeybees than it takes for two common neonics—imidacloprid and clothianidin—to do so.[37]

In addition to being widely used in agriculture (seeds of more than 140 crops are coated prophylactically with neonicotinoids), investigations in 2013 and 2014 revealed that these pesticides were also found in about half of the bee-attractive nursery plants sold at large retail garden centers. The plants contained neonics at levels that could kill pollinating bees by attacking their nervous system, or could impair the bees' foraging and navigation abilities, disrupt their learning and communication, reduce their fertility, and suppress their immune systems.[38] When I learned of these investigations in 2015, I thought with dismay about the pollinator gardens I'd delighted in creating almost everywhere I'd lived, trying to leave each place better for wildlife than I'd found it. In trying to help the bees and butterflies that visited my salvias and penstemons, my bee balm and delphiniums, had I instead hastened the insects' demise?[39]

Although neonics are deadliest to insects, birds are susceptible to these poisons, too. A single corn seed treated with imidacloprid—the oldest and most widely used of the neonics—can kill a seed-eating bird the size of a Blue Jay. Daily consumption of one-tenth of a treated seed during the breeding season can disrupt a songbird's ability to reproduce.[40] White-crowned Sparrows that were experimentally dosed with imidacloprid during their spring migration subsequently fed less, accumulated less fat, and delayed their departure from a refueling area during migration. Such delays can lead to lower survival during migration, late arrival on the breeding grounds, and the raising of fewer young.[41]

As is often the case with environmental poisons, it is their unintended effects that are most insidious and become the overarching cause of concern. Countless birds either feed on insects year-round or provide insects to their fast-growing nestlings as a protein source during the breeding season. But the widespread use of neonics is increasingly believed to be a major contributor to the alarming disappearance of insects—a phenomenon serious enough to be dubbed an insect apocalypse—that is being documented by scientists in the United States, Europe, and other parts of the world.[42] The rest of us are taking notice, too. Car windshields used to be smeared with dead insects during night drives in our childhood summers, but windshields are now disturbingly clear.[43] And while I've always been chagrined to find so many drowned insects in my hummingbird feeders at day's end, I was alarmed to find almost none in my feeders in 2023.

The loss of insects inevitably leads to a loss of birds. Populations of common, insectivorous farmland birds in the Netherlands declined an average of 3.5 percent annually, from 2003 to 2009, wherever imidacloprid concentrations exceeded levels equivalent to 0.02 parts per billion (parts pesticide for every billion parts water) in the surface waters of lakes and ponds adjacent to agricultural areas.[44] Neonics are increasingly found in US waterways as well, at levels that harm aquatic ecosystems—including sensitive organisms like mayflies, upon which birds and other organisms feed.[45] Recent research in North America showed that the increased use of neonicotinoids between 2008 and 2014 contributed to significant bird population declines, particularly among grassland and insectivorous birds.[46]

In the last two decades, it has become increasingly clear that the toxicity of neonics to the earthworms that enrich our soils, to the aquatic invertebrates that feed our fish, to the bees and butterflies that pollinate our plants, and to the insects that sustain our birds and bats threaten global biodiversity and critical ecosystem services such as pollination, pest control, decomposition, nutrient cycling, and the regulation of soil and water quality—all of which impact humanity.[47]

More than half a century ago, Rachel Carson challenged the widespread use of pesticides whose ecological and human-health effects were poorly understood. A growing body of research on the dangers of neonicotinoids suggests that her cautions have largely gone unheeded and the lessons learned in the wake of *Silent Spring* have largely been forgotten or ignored. The parallels between DDT and the neonicotinoid insecticides are deeply troubling, with the cascading effects of both types of poisons promising to have ecosystem-wide effects that were never anticipated, were difficult to forestall, and would have repercussions for years to come.

Because of growing concerns about the potential danger neonicotinoids pose to bees, the European Union permanently banned the use of three neonics (clothianidin, imidacloprid, and thiamethoxam) on all field crops in 2018.[48] More conservative in its response to the steady accumulation of scientific studies linking the harmful effects of neonicotinoids to ongoing pollinator population declines, the EPA has yet to complete its years-long review of these pesticides (as of 2023) but has allowed their continued use.[49] Legislation such as the Saving America's Pollinators Act, which would force the EPA to suspend the use of neonicotinoids pending more extensive review of their ecological impacts, has repeatedly failed since its 2013 introduction.[50] Nevertheless, Eugene, Oregon, became the first town in the United States to ban the use of neonicotinoids on city property in 2014, and other cities have since followed suit.[51]

Also in 2014, the US Fish and Wildlife Service (or "the Service") announced a phaseout of neonicotinoid use on National Wildlife Refuges by January 2016. (Farmers and ranchers operating on these lands used over half a million pounds of toxic pesticides, including neonicotinoids, in 2014 alone.)[52] In a world still contaminated with DDT, the organophosphates that followed in its wake, and innumerable toxic products like flame retardants, rodenticides, and other harmful chemicals, the Service's neonic ban presaged a less toxic future for US wildlife. However, the Trump administration reversed the ban on neonicotinoid use in wildlife refuges in 2018.[53]

The state of the world's bird populations is widely viewed as an indicator of the health of our environment. And while the reasons for current bird declines are as varied as rays of sunlight deflected through a crystal prism, the dark-eyed, smoke-backed peregrine and its DDT-suffering compatriots taught us in the 1960s and beyond that pesticides are a threat that merits our utmost attention. Today's birds seem to be inadvertently sharing their wisdom again, and the severity of the existing pesticide threat should not be underestimated, either for the birds' sake or for our own, because, as biologist Tom Cade noted insightfully decades ago:

> Down through the centuries, not all the falcon trappers, egg collectors, war ministries concerned for their messenger pigeons, or misguided gunmen [were] able to effect a significant reduction in the numbers of breeding [peregrine] falcons. But the simple laboratory trick of adding a few chlorine molecules to a hydrocarbon and the massive application of this unnatural class of chemicals to the environment [did] what none of these other grosser, seemingly more harmful, agents could do.[54]

CHAPTER SEVEN

Delisted

We never saw Osiris again. But seeing any of our falcons was rare during our final two weeks as hack-site attendants. Our surroundings seemed unnaturally quiet now that the skies were no longer filled with screaming, skywriting peregrines. I sat in the tenuous shade of our limber pine under a sweltering late-July sun, torn between my desire for another sighting of our now-elusive falcons and my awareness that their absences revealed the birds' arrival at a critical threshold in their development—independence. Despite their earlier social nature, our falcons would become increasingly solitary as they branched out on their own. And, as they made their early forays into becoming self-sustaining predators, we could do little for our birds but offer them our best hopes for their successful future and a meal for them to return to should they need it.

Given their infrequent visits in our fifth week post-release, I was overjoyed to glimpse Ra and Gawain feeding on quail near the hack box. Two days later, Isis and Freya graced us with their elegant presence. And though each female now looked like a capable immature Peregrine Falcon, I smiled with tender affection when a raven flushed Isis and she flew from the cliff, cacking. Had I known that the next two days' brief sightings of little Ra napping, watching swallows, and eating quail on the cliff would be the last time I ever identified one of our birds, I would have savored the moments in the same way that I had done when looking at Osiris after his return. But I still had one more week of hope that one of our peregrines would materialize out of the sky that had long been my ceiling, or would appear on the cliff that had been the ever-changing canvas adorning the wall of my summer home.

We began our sixth and final week post-release by *not* feeding our falcons. For the last week of the hacking period, we would put food out every other day instead of daily, to help foster our peregrines' independence. On average, hacked peregrines dispersed thirty to thirty-three days after the birds' first flights, with males typically leaving the hack site a few days before females. Our six-week hacking period mimicked the dependency period of juvenile peregrines from migratory populations of wild peregrines. The offspring of nonmigratory pairs might remain with their parents for up to ten weeks after leaving the nest.[1]

I felt negligent when I headed to our observation site rather than trekking up to the hack box to deliver quail. So it was with a measure of relief that I climbed the familiar route to the top of the cliff the following day to deliver a supplemental food stash. As I hiked upward, focused on the dust that I kicked up with each step, an immature peregrine sped past in a fast glide. I fumbled with my binoculars, but before I could focus them, our bird had disappeared into an unknown future.

Although an occasional American Kestrel flew past the hack site and the adult peregrines made one last flight over the cliff in our final days, falcons of all sorts were scarce now, even though quail were still available. But Clark's Nutcrackers abounded, their grating call—so characteristic of the West's high country—revealing their presence long before I spied their flashy wings. Several American Robins that perched for brief intervals in our limber pine also captured my attention. Used as I was to seeing robins searching for worms or berries in backyards across the country, they seemed out of place amidst this vast, high-elevation landscape of jumbled rock and stunted sagebrush. But their familiarity belied their extraordinary adaptability. I was as likely to see the rust-breasted robin at tree line on mountain peaks as I was in a lowland forest or on a suburban lawn.

As the sun climbed to its zenith during one of our last observation days, a Mountain Chickadee hoarsely voiced its namesake call from an overhead branch. The black bandit stripe through each of the bird's eyes gave it a cranky look that its sweet-faced cousin—the Black-capped Chickadee—wholly lacked. Minutes later, a strident trill disrupted our cliff watching as a tiny iridescent fireball—our first hummingbird of the summer—appeared in front of us. Hovering in midair, its wings beating so fast that their movements were indiscernible, the hummingbird turned its rufescent head to the left and right, mere inches from Garrett's red-socked foot, which must have looked like a tropical bloom amidst the dry landscape. Not daring to move, we stared wide-eyed at the diminutive pollinator, entranced by its fiery throat, which seemed to ignite like the setting sun reflected in a glass skyscraper whenever it turned its head in our direction.

Better known for its aggressive defense of nectar sources than for its extraordinary migration, the pugnacious but sparkling Rufous Hummingbird breeds farther north than any other hummingbird and makes the longest migratory journey in the world relative to its small body length. With wings beating up to 3,700 times a minute, it leaves its wintering areas in southern Mexico's scrubby mountain forests in early spring and heads northward along the Pacific Coast—where early-blooming flowers abound—flying at an average speed of 25 miles per hour (40 km per hour). Some Rufous Hummingbirds migrate as far as southern Alaska; others stop to raise their young in the Pacific Northwest. Rather than retrace its spring flight along the coast once its young have fledged, the rufous completes a circuit. It travels southward through the Rocky Mountains following "nectar trails" of mountain flowers whose blooms coincide with the hummingbird's arrival.[2]

The hovering rufous suddenly sped away with a high-pitched buzz. Although it seemed an unlikely messenger, the hummingbird was one of the earliest signs that the yearly fall bird migration was now underway. The end of July still meant midsummer to me, but across North America, shorebirds and hummingbirds, songbirds and raptors were preparing for or beginning the long southward journeys that would highlight the changing seasons, linking still-warm summer days to lengthening fall nights. It seemed inconceivable that in as little as a month or two our young peregrines would join the billions of birds that streamed through the skies like feathered celestial rivers, heading for warmer climates and more accommodating habitats in which to spend their winters.

The adult peregrine that had frequented our hack site all summer likely would begin migrating while our immature falcons were still honing their predatory skills. Adult peregrines usually appear at staging areas—where raptors coalesce to feed on abundant prey or to rest before crossing large bodies of water—earlier in the fall than do juveniles. In places like Chincoteague and Assateague Islands on the Eastern Seaboard, and Padre Island on the Gulf Coast, peregrines temporarily forsake their southward flights to prey on flocking shorebirds and songbirds before resuming their journeys. Peregrines may also break up their water crossings by landing on oil rigs and ships in the Gulf of Mexico, feeding on the hapless birds that drop in exhaustion onto these floating refugia to avoid watery graves.[3]

Although they could end up almost anywhere in Latin America if they migrated south of the US border, our falcons were likely to spend their first winters in Mexico. Meanwhile, many of the northernmost tundra-breeding peregrines would leapfrog over wintering Rocky Mountain peregrines and fly as far as Patagonia for the austral summer. Transmitter-wearing peregrines—once tracked by aircraft and now recorded by satellites that transmit location coordinates to a biologist's computer—have provided fascinating insights into some of these birds' lengthy journeys. An adult female that was trapped and outfitted with a transmitter on Assateague Island, Maryland, in 1993, subsequently flew to Guatemala, Honduras, Nicaragua, Ecuador, and Bolivia, before crossing over the Andes Mountains to the headwaters of the Amazon River and ending up at a saline playa in northern Argentina's Altiplano. Another female peregrine—trapped on Padre Island, Texas, while on her northward migration in 1994—left her Alaska summer home on August 31, flew south on the east side of the Rocky Mountains, spent over a week on the Yucatán Peninsula before leaving on October 3 for a two-week stint in Panama, then crossed the Andes, and arrived in Argentina on November 21. An immature peregrine that left her summer haunts on King William Island in the Canadian Arctic, on October 12, passed through Churchill, Manitoba, before heading south through the eastern Midwest. She initiated her Gulf of Mexico crossing east of New Orleans and wintered in the Yucatán Peninsula.[4]

As I watched the cliffs that our peregrines had so recently animated, I could only dream about where the falcons might spend their winters, their future journeys as unknowable to me as their current haunts. Henceforth, I would have to be content with the peregrine figments that resided in my imagination and relinquish their flesh-and-blood counterparts to the unknown. And perhaps this was for the best. Since only about half of immature peregrines survive their first year, I was likely to imagine better outcomes for our falcons than fate—and its myriad obstacles—might dictate for them.

Between 2001 and 2010, The Center for Conservation Biology—a Virginia-based research organization that studies issues related to land management and birds of conservation concern—placed solar-powered satellite transmitters on immature peregrines (prior to fledging) in the mid-Atlantic region to study their movements and survival. Of the twenty-four falcons for which they were able to determine cause of death, eleven were lost to predation (mainly by Great Horned Owls and adult peregrines), seven flew into human-made structures (including transmission lines, towers, a high-rise building, and the side of a barn), three were killed in storms, one was hit by a truck, one drowned, and one was lost at sea near Bermuda.[5] In a larger sample of 455 recorded fatalities of midwestern peregrines, collisions with buildings and vehicles killed significantly more falcons than any other factors.[6]

The tranquil evening, settling like a veil over the familiarity of the hack cliff, was a welcome counterpoint to thoughts about potential threats and uncertain futures. I pulled on a sweatshirt to guard against the chill that enveloped our observation site once the sun set. Scanning the cliff, I checked our peregrines' favorite perches, trying to imprint their images and the memories each of them held for me in my mind as much as to look for any lingering falcons.

A flash of movement by the hack box grabbed my attention. I didn't need binoculars to identify the burnished-orange form that circled around the box, pointed muzzle sniffing for leftover quail. A red fox—our first of the summer—boldly seeking the food that had long attracted the neighborhood raptors. A contradictory surge of belated concern for our falcons' safety waged with the delight of watching a beautiful animal that wasn't watching me. Surely the fox had patrolled the area during the summer, I thought, as it chewed a piece of quail. But we'd never seen any sign of it on the clifftop. Its presence surely would have led to sleepless nights as I worried about our inexperienced peregrines. Now, though, I had every confidence that our falcons would perch out of harm's way. Relieved that the amber-eyed fox no longer posed a threat to our absentee charges, I packed up in near darkness and headed down to camp, leaving the fox to its investigations, still unconcerned by my distant presence.

When we delivered the next-to-last batch of food to our peregrines, the absence of any quail remains on the cliff was as much an indication of the fox's recent presence as the scat we found on top of the hack box. The clifftop looked as if it had been swept of quail debris by a cleaning crew that had left an abundance of calling cards in its wake. Clambering up onto the hack box with none of the agility I was sure the fox had displayed, I placed five quail on the marten board where they would be inaccessible to roving canids. We would take no chances. Should our falcons still need a supplemental meal at this late stage, we wanted to be sure that it wasn't stolen by an opportunistic fox.

———

August 5. Day forty-three post-release. Our last morning at the hack site. Despite the freshness of the day, my worn boots felt heavy as I trudged up to the hack box for the last falcon feeding. Though I was looking forward to visiting nearby Yellowstone National Park for my first time before beginning the long journey home, I felt like something precious was slipping through my fingers. Eager as I was to embark on more wildlife work the following summer, I already knew how profoundly I would always treasure—and miss—those halcyon summer days of watching peregrines light up the sky with their incandescent, gravity-defying aerial maneuvers.

Borne of memory—or perhaps longing—one of our juvenile peregrines suddenly materialized in the sky beside me, sped past, glided over the hack site, and disappeared. My heart leaped in response to the unexpected gift. And inevitably, on the heels of the small lump that formed in my throat, came the worries: Were we leaving too soon? Were our falcons really independent? Would they make it without us? Finely tapered falcon wings again sped past—a kestrel this time—unwittingly offering reassurance that life in this windswept place would continue as it had done over the last few weeks. Our falcons were on their own. We could do nothing more for them. As I turned away from the hack box, I glimpsed the fox trotting down the indistinct path we'd worn up the hill. Minutes later a Prairie Falcon flew by and disappeared to the southeast. I followed, heart lighter, hopes fuller, life immeasurably richer.

The Peregrine Fund released 140 peregrines to the wild at twenty-seven hack sites in the Rocky Mountains and Pacific Northwest during the summer of 1991. Of the twenty-five peregrines released at Wyoming's five hack sites, twenty-three—including all five from our South Fork hack site—reached independence, the highest success rate of any of the five states where peregrines were hacked that summer. In addition, biologists monitored fourteen wild peregrine pairs in Wyoming, which produced at least twenty-two young.[7] The recovery of the Peregrine Falcon was well underway in western North America. I had only played a small role in the grand endeavor, but I would feel the glow of that participation every time I saw a peregrine in the years to come.

I had come to Wyoming looking for a novel experience, meaningful work with animals, the opportunity to do something "of worth." My summer was all those things, but what I really found was birds. Each arc our falcons traced through the sky during that watershed summer stitched me more and more firmly to their world. For six weeks, I had been captivated. Now

I was committed. I wanted to observe birds, to work with them. I wanted to learn about their habits and their haunts. Above all, I wanted to protect and conserve them. I had found my direction. I had discovered my passion.

And more. There had been so much to see, to internalize beyond the daily observations, beyond the diligent cataloguing of peregrine sightings, feeding activities, and interactions with other birds. Like the sediments laid down by rivers over geologic time, a palimpsest of images and experiences had transformed me and would forevermore define me, sustain me—a Great Horned Owl, silent and watchful against the trunk of a cottonwood; a pronghorn silhouetted on a ridge at sunset; the dazzling blue of a Mountain Bluebird reflecting the firmament; Peregrine Falcons writing an awakening across summer skies.

———

On August 20, 1999, amidst a gathering of more than 1,000 celebrants, the American Peregrine Falcon (*Falco peregrinus* subsp. *anatum*) was officially delisted, or removed, from the federal list of endangered species.[8] The move followed thirty-five years of work by numerous organizations and thousands of people.[9]

A variety of biological factors contributed to the peregrine's successful recovery in North America. The falcon's relatively small population made it easier to establish and reach manageable restoration goals. The peregrine's docility and long relationship with humanity made breeding it in captivity relatively easy once successful techniques had been developed. The bird's longevity and generally high survival rate allowed for a greater return on captive breeding investments compared to organisms with low survival rates or shorter lifespans. The tendency of peregrines to return to their natal areas allowed vacated regions to be targeted and repopulated. And the falcon's flexibility in adopting atypical nest sites in areas where food was abundant helped bolster overall peregrine numbers.[10] But the banning of DDT in 1972 was, indisputably, the most important factor—and the most important action taken—in recovering North America's peregrines. The number of nesting pairs increased from the late 1970s onward, first in the north where DDT has been used more sparingly, and then in the south.[11]

Perhaps the most fundamental takeaway from the peregrine's conservation success story is that the primary threat to a declining species must be dramatically curtailed—if not eliminated—for sustainable population recovery to be achieved. Ultimately, every species in North America that experienced DDT-induced eggshell thinning, reproductive failure, and population declines recovered.[12] In the meantime, it became clear that birds—conspicuous, easily viewed, and intensively studied—could be successful indicators for detecting and monitoring the effects of contaminants and other environmental perturbations.[13]

Captive breeding and the reintroduction of nearly 7,000 peregrines in the US and Canada also contributed to restoring the falcon in North America, expediting a recovery that likely was inevitable following the DDT ban but otherwise might have been slower or more localized. The Peregrine Fund reared and released 3,318 falcons to the wild in the eastern United States, the Rocky Mountains, and the Pacific Northwest. The Santa Cruz Predatory Bird Research Group reintroduced peregrines to California and Nevada; The Raptor Center at the University of Minnesota released peregrines in Midwestern states; and the Canadian Wildlife Service managed reintroductions in Canada. Ultimately, the impact of captive rearing and reintroducing peregrines went far beyond restoring a much-loved falcon, since the techniques that were developed were later used for recovering other endangered birds, including the Mauritius Kestrel, Bald Eagle, and California Condor.

The legal framework that protected peregrines and structured recovery efforts also played an important role in conveying the falcon on its journey from beleaguered raptor to iconic conservation success story. In 1970, troubled by the falcon's rapid population decline and its association with the prevalent use of organochlorine pesticides, the US Fish and Wildlife Service included both the American and the Arctic Peregrine Falcons—two of the three North American subspecies—on a list of endangered species under the Endangered Species Conservation Act (ESCA) of 1969, which prohibited the killing or trade of listed foreign and domestic animals.[14]

Emerging from an era of growing public concern about the degradation of the nation's environment—including excessive pesticide use, widespread pollution, and disappearing species—the subsequent passage of the more comprehensive and powerful Endangered Species Act (ESA), in 1973, represented a historic, unprecedented, and enlightened commitment to preserving the nation's biological diversity. The peregrine was listed as endangered under the ESA the year the law was passed. Since peregrines were still routinely shot and their eggs targeted by collectors in the early days of the recovery program, legal protections provided by the ESCA and ESA were critical.[15]

When it considered delisting the *anatum* peregrine subspecies in 1998, the US Fish and Wildlife Service noted that residual pesticide contamination still caused eggshell thinning in certain areas—such as coastal California and western Texas—over twenty-five years after DDT was banned. But these effects were localized, and robust peregrine numbers and positive growth trends in almost all areas underscored the bird's extensive recovery.[16] The United States had established a breeding population goal of 456 peregrine pairs, with an additional objective of 60 pairs in Canada. In 1998, biologists documented 1,650 pairs in the two countries. The Peregrine Falcon had recaptured North America's skies, and cliffs once again reverberated with the birds' trademark calls.

As the Service worked through the delisting process—holding public hearings, considering public comments, and soliciting a scientific review of the delisting proposal—some peregrine supporters expressed lingering concerns about the falcon's future well-being. In midwestern and eastern states, where so many peregrines now nested on artificial structures, would building managers protect nesting peregrines when they were no longer required to do so by ESA strictures? Would rock climbers disturb peregrines at their nest cliffs, disrupting their breeding attempts? Would falconers take too many nestlings or migrating peregrines into captivity, depressing wild population numbers? Would ongoing DDT use in Latin America lead to continued eggshell thinning, disrupting breeding efforts by migratory peregrines?

The Service offered reassurances. The peregrine—along with hundreds of other native birds—would still be protected under the Migratory Bird Treaty Act—North America's signature avian conservation law. Building managers reported their continued enthusiasm for hosting nesting peregrines. Seasonal protection plans for nesting raptors had been established for popular rock-climbing areas. Falconers would be issued capture permits to ensure that their use of wild falcons for falconry wouldn't harm peregrine numbers. And the United States would work with its neighbors to reduce the ongoing legal (particularly for malaria control) and illegal uses of DDT in Latin America. Meanwhile, although blood samples collected from migratory peregrines showed that the birds continued to accumulate pesticides while wintering in Latin America, DDE residue levels in females captured during spring migration at Padre Island, Texas, decreased between the years 1978 and 1994, and were below levels that would harm reproduction.[17]

The Peregrine Falcon's recovery was a triumph for the entities and individuals who joined forces and worked for years to save a captivating bird. Falconers contributed their unique knowledge; scientists lent their hard-won expertise; donors invested critical funds; federal and state agencies provided habitat, oversight, funding, and the framework for the bird's protection; and hundreds of hack-site attendants (more than 170 in Wyoming alone) volunteered their time and poured their hearts into helping young birds grow into future generations of breeding peregrines.[18]

Many had at first been fatalistic about the peregrine's dismal recovery chances, believing that if we could allow the Passenger Pigeon—once the most abundant bird in North America, numbering in the billions—to go extinct, then the fast-disappearing falcon was doomed to follow its former prey into nonexistence. But each successful step in restoring peregrines, along with the bird's charisma and growing appeal, helped change attitudes, providing hope not only for the falcon's survival but also for the future of other seemingly doomed creatures, like the California Condor.[19]

T-shirts produced in support of The Peregrine Fund by the outdoor-clothing company Patagonia pictured a stooping peregrine and proclaimed: "The Peregrine Falcon: Back by popular demand." The sentiment, though hyperbolic, was almost as true as it was unprecedented. Henceforth, though, the correlation between an imperiled species' public appeal and the level of effort invested in saving it would repeat itself, with recovery efforts for species such as gray wolves getting a disproportionate level of support compared to little-known but no less deserving animals like the Hawaiian Crow, which would languish in comparative obscurity.

The Peregrine Falcon's August 1999 delisting celebration in Boise, Idaho, was a two-day extravaganza of marching bands, line dancers, live acts, speeches, exhibits, and captive peregrines. Participants ranged from senators to scientists, funders to falconers, celebrities to small children, bureaucrats to hack-site attendants. As drinks flowed and glasses clinked, attendees shared stories and reminisced, basking in the pervasive feeling that history had been and was being made.

Unable to join the festivities, my own delisting celebration was a quiet one. I thought about my first peregrine, diving at preternatural speed over a Vermont lake. I thought about the many subsequent peregrines that had graced my life at unexpected moments—the one that flew over me at sunset by a favorite lake in Idaho; the one that sped past, right below me, as I stood on a ridgetop in Tanzania; the pair I found perched by a willow-lined oasis in the Arizona desert; the female that eyed ducks in a National Wildlife Refuge in Montana where one day I would get married. And, of course, I thought fondly—and achingly—about my falcons, the five young birds I'd watched over with trepidation, care, and fierce, if naive, protectiveness.

I never discovered the fates of Gawain, Osiris, Ra, Isis, or Freya. For years, I yearned to know what had befallen the falcons that had shared my summer and inspired me, through the alchemy of their flights, to dedicate myself to a bird-filled future. Hoping someone had sighted or retrieved one of my falcons' leg bands, I repeatedly contacted The Peregrine Fund and the US Fish and Wildlife Service. But no one had any information for me. And perhaps it was better this way. I wanted to imagine a slate-backed, astonishingly successful Osiris bringing food to a remote cliffside aerie in the Rocky Mountains. And Gawain, that fearless knight, surveying his vast realm and chasing off wayward buteos. I wanted to picture Isis and Freya fiercely defending their territories and tenderly feeding their tiny nestlings. And I wanted to imagine Ra fulfilling his destiny as master of the skies, lord of his domain. After all, he had learned to fly with what I considered to be the finest of peregrines. In my mind, they all lived out full lives, streaking through the western firmament with lightning speed, as mountain landscapes unraveled beneath their streamlined bodies and cliffs echoed with the challenges they screamed to any intruders that dared to trespass through their patch of sky.

PART II

Hawaiian Crow

Haunted Forests

From prehistory to the present time, the mindless horsemen of the environmental apocalypse have been overkill, habitat destruction, introduction of animals such as rats and goats, and diseases carried by these exotic animals.

—E. O. **Wilson**, *The Diversity of Life*, 1992.

In the end, we will conserve only what we love, we will love only what we understand, and we will understand only what we are taught.

—**Baba Dioum**

CHAPTER EIGHT

Sheep Heads, Deer Guts, and Feedlots

I was suffocating. Sweat trickled down my legs, my back, my chest. I pulled the long-sleeved shirt that I'd wrapped around my head more tightly around my face, further obstructing my breathing. I checked the windows. They were rolled up. I had to prevent air from getting into the small pickup truck, which was parked in full sun on a sweltering August afternoon in southern Idaho. I was engaged in an exercise in futility, but desperation cloaked common sense; revulsion prompted irrationality. I had to block the smell—the unbearable, overpowering, sickly sweet, ferociously pungent smell. But it seeped into my truck through the air vents, permeated the vehicle's cloth seats, lodged itself in my clothes, imprinted itself on the inside of my nostrils. Sounds were equally pervasive, though less offensive. Gurgles, rumbles, whooshes, stamps, all accentuated with a steady accompaniment of lowing, a cacophony of mooing. I was surrounded by a veritable sea of bovines—and their attendant ocean of excrement—stretching to the horizon in every direction.

Raising my binoculars with one hand, I tried to focus on the bird I was monitoring, a captive-reared Common Raven that had been released to the wild and wore a prominent yellow tag—known as a patagial tag—affixed to its wing. For weeks now, it had made daily forays to the feedlot that currently threatened to overwhelm my senses: a 750-acre facility that could accommodate up to 150,000 cattle. Oblivious to my distress, the raven walked around an empty paddock, gleaning spilled feed and interacting—sometimes amicably, sometimes antagonistically—with a congregation of fellow corvids.

I tried to make a notation on my data sheet, and my head-wrapped shirt fell off my face. Unable to bear the stultifying heat and lack of oxygen any longer, I feverishly rolled down the window while exhaling a ragged breath as the already inconceivable stench intensified exponentially. What was it about corvids—that fascinating family of intelligent and entertaining birds that includes crows, ravens, jays, magpies, and nutcrackers? Whenever I worked with them, I seemed destined to be subjected to the unsavory. My mind drifted back to my earliest corvid experiences.

———

The spring semester before my peregrine summer, I had earned an opportunity to conduct a research project instead of taking the usual beginning biology lab. To my delight, rather than donning lab coats to work with a microbiologist, a fellow student and I would do a project with a famed organismal biologist who had a particular fascination with Common Ravens. Already recognized as an expert on insect temperature regulation, Bernd Heinrich had recently garnered additional fame with a popular book recounting his pioneering work with ravens. He had also captured my attention by regularly running between buildings on the University of Vermont campus where I was taking classes. Routinely seeing a renowned professor engage in the same rather undignified but indisputably time-saving way of getting from one place to another that I had always practiced made me feel a little less self-conscious about my own habitual dashes from point A to point B. But my similarities with the great man ended there. Heinrich was a world record–holding ultramarathoner, while I hadn't yet run a 10K. He was a respected biologist, whereas I was a brand-new convert to the biological sciences. Nevertheless, given Heinrich's reputation, his love for running, and his fascination with wildlife, I felt like I had won the lottery by getting a chance to work with him.

In our initial meeting, Heinrich tersely told my lab partner and me that we would be conducting experiments to test the intelligence of captive ravens. Our interactions with him after that were infrequent, though we made weekly forays to his wooded Vermont home to carry out our work. On our first visit, we were instructed to build a blind so that we could observe our experiments with the ravens, which Heinrich kept in a large aviary, without disturbing them. We knew virtually nothing about birds and even less about wildlife-viewing blinds. Worse, neither of us was particularly handy. So, it was hardly surprising that our first blind-building effort was met with what appeared to be barely masked contempt, but may just have been the inevitable weariness that came from decades of dealing with students whose enthusiasm surpassed their abilities. We had erected a screen with a small window so that we could tuck ourselves out of view while watching the birds, but we had erred in an important way. What was critical, Heinrich informed us, was to block the light *behind* us with a wall of sorts, so that we were sitting against a dark background instead of being silhouetted in a window. Once we'd constructed a proper blind, we were ready to start our experiments. Our clumsy tests sought to mimic the simple but elegant method that Heinrich had devised to evaluate raven intelligence by examining the birds' problem-solving skills and insight.

To determine whether Common Ravens and American Crows could visualize the solution to a problem and carry out the multiple steps necessary to achieving it (hallmarks of what we call intelligence), Heinrich had hung a piece of meat on a long string and attached it to an aviary perch. To obtain the meat, a bird had to pull up on the string with its beak, place a foot on the string to anchor it, reach down and pull up another loop of string, anchor it, and so on until the meat was in reach. Heinrich's crows were unable to obtain the meat, but within several hours of initiating the experiment with his ravens, one of the birds hopped over to the food morsel and flawlessly executed the sequence of steps necessary to obtaining it.

Not all of Heinrich's ravens managed to obtain the proffered food, nor did all of the wild-caught ravens he later tested. But the ravens that did solve the puzzle ignored a similarly sized stone that Heinrich attached to a string and hung beside the meat, and they ignored large food items that they couldn't possibly lift. They understood that their hunger would only be satisfied by manipulating the string that was attached to the meat. The birds' behavior was not learned. They had never encountered a problem like this. Nor was it instinctual. The ability of some individuals to solve the problem while others did not showed that consciousness and intelligence were not innate but varied among individuals.[1]

That my intelligence tests were far less elegant and fraught with confounding factors in no way diminished my delight at watching one of Heinrich's ravens pull on a string to extract the attached meat that I had placed in a long, opaque tube on the floor of the aviary. Less delightful, though, were some of the steps involved in setting up the experiments. On our second visit, I asked Heinrich if he had some meat for us to use as bait.

"There are some sheep heads in the freezer out back that you can use."

I blanched. Sheep heads? Surely he didn't mean sheep *heads*. I must have misheard him.

"Sheep heads?" I stammered.

"Yes. You can cut off some chunks of meat from the sheep heads."

I dragged my feet on the way to the freezer. I had only seen severed animal heads once before—in a slaughterhouse in Kenya—and that experience had cured me of eating red meat. I didn't hunt. I loved animals. I had never cut them up. But there seemed to be no way out of this unexpected dilemma. My lab partner and I pulled a heavy garbage bag out of the freezer, wincing when it thudded onto the ground and opened, unveiling its macabre contents. Sheep heads. We looked at each other in dismay. Where did one get meat off a sheep head? The skull? The cheeks? We set to work, slicing the sheep's wooly, bloodied skin, peeling it back, carving bits of cheek while trying to ignore the glazed, vacant eyes. I had never envisioned myself hacking away at a sheep cheek when I became interested in biology. *Sorry, sheep*, I said silently. *Thank you, sheep.* Then I forced myself to focus my thoughts on ravens, on their bright intelligent eyes, their glossy black plumage, their intriguing calls—rattles, knocks, croaks—and their captivating behavior: dominant birds raising hornlike feathers above their eyes, pairs gently preening each other. And I tried to accept that a degree of unpleasantness, of distress even, might be a price worth paying to discover more about the workings of a corvid's endlessly fascinating mind.

———

Two years later, as I steered a pickup truck through the vast volcanic plains of southern Idaho's sagebrush-steppe country, I tried to settle the nervousness I always felt on the first day of a new job. Surely my experiences thus far would give my colleagues the illusion that I was more competent than I felt. I had, after all, spent a summer helping young Peregrine Falcons reach independence. I had spent the following summer hiking hundreds of miles through remote and challenging terrain in Idaho's Targhee National Forest, searching for nests of the American Goshawk—that fiercely beautiful and elusive hawk, a gray-shaded denizen of deep woods—then watching the birds' nesting activities from blinds, and conducting habitat analyses of their territories.

My goshawk summer's more sublime moments had overshadowed the many challenges that I'd faced. Agitated ants had swarmed up my pants after I'd accidentally kneeled on an ant mound in the middle of the forest. I'd been stung a dozen times on the face and neck by yellow jackets after inadvertently disturbing their ground nest when searching for prey remains under a goshawk nest. I'd evaded an overly curious black bear, and hiked to exhaustion countless times. But I had also watched a female goshawk, deep in a quiet forest, shielding her small nestlings during a torrential rainstorm and shaking her head periodically to shed moisture from her waterlogged feathers. I'd observed a big-eyed Great Gray Owl feed three rodents in under an hour to his squat-bodied, fluff-covered owlets. I had walked through carpets of wildflowers under translucent canopies of trembling aspens. In the days before GPS units, I'd learned to rely on my compass to navigate for miles through conifer forests. And I had been inspired by my supervisor, Susan Patla, who thought nothing of walking thousands of forest miles alone while monitoring goshawks that were threatened by timber harvesting, who worked fourteen-hour days before heading out to monitor owls at nightfall, and who enthusiastically shared her knowledge of wildlife, plants, and field biology with me, a greenhorn from the East.

Now, I had come to southwestern Idaho to work on a project studying the impact of the military's training activities on Prairie Falcons and Golden Eagles in and around the famed Snake River Birds of Prey National Conservation Area. But first, I would spend a few weeks helping the project's principal investigator with a separate study that he was conducting on Common Ravens. It had been an unlikely coincidence that my supervisor-to-be for the raptor position, John Marzluff, was another highly respected corvid biologist who had been one of Bernd Heinrich's former postdoctoral students. The two scientists had discovered that young, vagrant ravens in Maine's boreal forest "yelled" to attract others of their ilk so that their subsequent aggregations could overpower dominant, territorial pairs of adult ravens at large animal carcasses.[2]

Ravens living in the West's wide-open sagebrush country also congregated at carcasses, but John suspected that the corvids discovered the area's smaller, more abundant carcasses independently by cueing on signals like the presence of feeding Black-billed Magpies and Golden Eagles. To test his hypothesis that western ravens aggregated at carcasses through local enhancement—the gradual accumulation of birds that spotted food in a particular area—rather than by responding to the alerting calls of other ravens or to cues shared by ravens assembled at communal roosts, John was placing animal carcasses out in the sagebrush. He and his colleagues then watched them to determine how ravens discovered them and who fed on them.[3] I was now following James, a fellow biologist, into the field to dump a road-killed deer, which I would observe.

The skiff of snow covering the ground highlighted the gray sagebrush that stretched to the horizon in every direction, lending a bright contrast to an otherwise uniform landscape. I turned up my truck's heater to ward off the February chill, then followed James as he turned onto a dirt track winding up to a slight rise. After parking, James opened the tailgate and grabbed one of the deer's back legs to drag the animal off the truck bed. I rushed to help, hoping my enthusiasm would mask my squeamishness and my inevitable sadness at seeing a dead animal. The deer dropped to the ground with a thunk and James started patting his pockets.

"Damn," he muttered. "I forgot my knife. You got one?"

"No," I answered lamely, making a show of searching my own pockets. "I don't." This had been my chance to look like a competent, prepared-for-any-eventuality biologist and I had blown it. I resolved to carry a Leatherman in the field from then on. I would not be unprepared the next time a knife or tool was needed.

"We gotta cut this carcass open so the ravens and magpies can get into it, otherwise they won't even try to feed on it. That would blow the whole experiment." James checked the truck's glove compartment but found nothing. "Well, there's a hoe in the back of the truck. I guess we'll have to make do with that."

I stood by helplessly, trying to maintain a supportive, ready-to-help stance as James began thwacking at the side of the bloated deer, swinging the hoe like an ax. Finally, the hoe bit into the deer's hide, making a small cut. James struck the area again. And again. And suddenly a geyser of putrid body fluids shot skyward as the hide burst open, releasing the pressurized contents. Before I could jump out of the way, the fetid mixture rained down on me, spattering my hair and coat.

"Thar she blows," James laughed. "Well, I guess that opened her up. You okay?"

"Oh sure. I'm fine." I choked out a laugh, trying not to grimace, trying not to recoil in disgust as the potent smell of decaying flesh settled heavily around me.

"Well, that should do it," James said cheerfully, tossing the hoe back into the truck. After a few quick instructions, he left me to watch the carcass.

The indescribable, unforgettable smell of rotting animal had nearly made me gag when the deer's guts had been exposed, and it threatened to overwhelm me as I sat in my truck, using the vehicle as a blind. Unable to bear the smell, I took off my offending coat and tossed it into the bed of the truck. But I could do little about the repugnant odor—and the vile fluids—that had settled in my hair. I cracked open the window to try to disperse the smell and was immediately assaulted by a frigid wind. I contemplated my choices. Endure the frightful smell or shiver in the truck on a bitter winter afternoon. The choice was clear. Opening the window all the way, I spent the next few hours—coatless—trying to breathe through chattering teeth as I scanned the sky for corvids until the sun sank below the horizon and darkness enveloped the sagebrush.

––––––––

And now I was sitting in the midst of a sprawling feedlot bordered by red rimrock, overwhelmed by a different kind of ghastly smell. But despite my discomfort, I felt incredibly lucky to be there. Monitoring color-marked, radio-tagged ravens was an unanticipated bonus that extended what had been a rich and rewarding summer. I had helped capture Prairie Falcons by enticing them into special nets known as dho-gazas that were set up around a captive Great Horned Owl—an enemy to falcons that reliably elicited an aggressive reaction because it sometimes targeted their young. I had also helped capture Golden Eagles by baiting them with road-killed jackrabbits into circular, folded ground nets—known as bow nets—that snapped closed over the bird when triggered remotely.

For months, I had spent hours in vehicles or small blinds, watching Prairie Falcons and Golden Eagles bring prey to their nestlings, interact with their mates, and trace their invisible territory boundaries through dramatic aerial flights. I had spent equally long hours alone in desiccated country using a radio-telemetry antenna—known as a Yagi—to detect and track the movements of transmitter-wearing falcons and eagles that searched for prey over a mosaic of public and private land and military training grounds. I had seen tank maneuvers and bombing practices. I had become familiar with signs proclaiming "Unexploded Ordnance." And I had witnessed unforgettable sights—such as nesting Prairie Falcons stooping, on two different occasions, at bobcats that were hunting marmots in the talus below the Snake River Canyon's cliffs, and baby badgers tumbling over each other in riotously exuberant play.

I had overcome countless personal challenges, too. While watching a Prairie Falcon nest, I'd spent an interminable eight hours in a small blind that had become home to a large black widow spider. I'd monitored her activities almost as closely as those of the falcons. I'd ushered a large (and not very congenial) rattlesnake away from a hole into which I needed to place a long pole to elevate my radio-tracking antenna so I could better receive raptor transmitter signals from a designated tracking area. And, despite my lingering fear of heights, I had learned to feel comfortable driving off a cliff. Or at least that's what it felt like I was doing each time I lurched over the lip of the Snake River Canyon, straining to see over the dashboard of my truck while navigating the frighteningly narrow track that clung to precipitous cliffs as it plunged down to the river corridor. Best of all, I had learned to embrace—and even cherish—solitude. As long as I was watching birds and the fascinating dramas of their daily lives, I rarely felt alone or lonely, no matter how remote or inaccessible my location.

It had all passed far too quickly, so I had leaped at the chance when John Marzluff asked if I would like to spend a month tracking ravens that some of his technicians had raised from eggs over the summer and recently released to the wild. In addition to supervising the raptor study, John was directing a vitally important effort to refine captive-rearing and reintroduction techniques that would inform efforts to conserve two critically endangered island crows—the Hawaiian Crow and the Mariana Crow. Successful efforts to bolster diminished populations of peregrines and other endangered birds had established both captive breeding and reintroduction as vital tools for recovering imperiled species, leading to a proliferation of such efforts.[4]

Captive rearing could boost the overall reproductive output of a species or supply individuals to replace a diminished population. The technique was designed to produce young in a sheltered environment protected from predation, diseases, inclement weather, food shortages, and other stressors that might reduce the birds' chance of success in the wild. But hand-reared chicks often grew more slowly than parent-reared ones and experienced more health problems, which ultimately affected their chances of survival once they had left their "nest" and gained their independence. So suitable captive-rearing techniques had to be developed and refined for each imperiled animal. Surrogate species were often used to develop these husbandry techniques for animals whose populations were too depleted to risk the death or poor development of young that were critical to the species' future survival.[5]

To give captive-reared Hawaiian and Mariana Crows every chance of success, John, his graduate students, and several technicians were testing different feeding regimens—manipulating the frequency of feedings and the amount of food given—for three abundant North American corvids whose body sizes and behaviors bore similarities to those of their unique island relatives: Black-billed Magpie, American Crow, and Common Raven.[6] It was a Herculean task. Several hundred nestlings that had been assigned to one of fourteen different feeding regimes had to be fed hourly. Once they were old enough to survive on their own, the captive-reared crows and ravens were marked with colored wing tags that folded over the front of each bird's wing and were secured by a small pin that pierced the patagium—a thin membrane forming the leading edge of the wing. The birds were also outfitted with radio transmitters before being released to the wild so that biologists could track them and evaluate their survival to inform reintroduction techniques for the island crows. Now, I was braving August's long days and unforgiving heat to monitor these ravens, observing their behavior and interactions with wild birds.

The raven with the yellow wing tag that I was watching suddenly took flight, joining a large group of ravens that had coalesced in the sky over the feedlot. Despite my silent pleas for yellow-wing to depart this odorous bovine hotbed, the flock settled again, and the ravens resumed their foraging. Another raven pecked at yellow-wing, which hopped nimbly out of harm's way before walking out of view. Twenty minutes later, I glimpsed it again as it took flight and headed toward an agricultural area where several hundred ravens were feeding. Other than inadvertently flashing its yellow patch, my captive-reared bird seemed to behave no differently than the wild ravens, which had fledged from nests in cliffs and trees; on windmills, utility poles, and abandoned buildings; under highway bridges; behind roadside billboards; or in any of the other locations that this opportunistic and versatile species had co-opted to raise its young. Wasting no time, I started my truck, only too eager to leave the dreaded feedlot and head into the sagebrush in search of other marked ravens.

Any hope that the sweet, strong smell of sage would lessen the feedlot odor that permeated my truck soon faded, and I eagerly parted company with my vehicle at day's end. After gathering my radio-tracking receiver and antenna, backpack, and empty water bottles in my arms, I headed for the field office. John Marzluff intercepted me as I approached the low-slung building, his dark eyes alight with the same gleam of intelligence, curiosity, and suppressed mischief that seemed to emanate from the eyes of his treasured corvids.

"Hey, how would you like to go to Hawaii to work with the Hawaiian Crows," he asked without preamble before I even had time to issue a cheerful evening greeting.

"What? Seriously? Really?" I stammered, as my brain struggled to absorb his question.

"The Peregrine Fund needs two short-term field assistants to care for and monitor the first batch of captive-reared crows that they're about to release."

Images of sheep heads, deer guts, and a colossal feedlot jumped into my mind, then morphed into a vision of white-sand beaches, sparkling surf, and tropical blooms.

"Really?" I said again, a grin splitting my face.

"Unless you'd *like* to keep tracking birds in the feedlot?" John teased with a returning smile, his nose crinkling pointedly at the unmistakable odor that emanated from my every pore.

"Hmm, tempting," I laughed, "but I'm all for new experiences. Wow. I'd love to work with the Hawaiian Crows."

"Well, you'll head to Hawaii in a couple of weeks then. We'll get you hooked up with The P. Fund tomorrow and they'll make your flight."

Hours later, I was still beaming. I couldn't believe my luck. I would be working with another endangered bird. I'd be traveling to an exotic place. I would be working with The Peregrine Fund again. Suddenly my somewhat unsavory, early corvid experiences seemed like an insufficient challenge, a too-meager rite of passage, an inadequate sacrifice for the disproportionately large reward I was about to reap by working in a vibrant tropical world inhabited by bright-eyed, ebony birds. After all, paradise was just days away.

CHAPTER NINE

Through the Mist

As the truck bumped and heaved up a mountainside that was shrouded in heavy mist and dripping with wet vegetation, I had my first inkling that paradise might not be quite what I had envisioned. It had started out promisingly. I had been greeted by heat, tropical flowers, and a surprising welcome committee composed of three young US Fish and Wildlife Service employees and Kurt, who I had worked with in Idaho and who would be my field partner. Outside the airport, I happily noted my first "Hawaiian" bird—a Common Myna. Introduced from southern Asia and closely associated with human dwellings, the brown-bodied, black-headed bird with a stout yellow bill and a featherless patch of bright-lemon skin behind its eye underscored how foreign a land Hawaii was compared to the continental United States.

The drive from the Hilo airport, on the east side of Hawaiʻi—or the Big Island as it is also known—to the town of Captain Cook, on the west coast, did not disappoint.[1] Verdant fields, black-sand beaches, and rough lava fields jutting into a dramatic ocean only whetted my appetite for the sun-and-surf tropics.

After a brief stop at our supervisor's house, where we would spend some of our days off, we began the arduous drive up to the field site. The sunny coastal viewscape quickly gave way to a steadily encroaching riot of exotic grasses, interspersed with scrub and scattered trees. This area had once been a diverse dry forest comprising species such as the fragrant sandalwood ('iliahi), which became the island's first cash crop, with millions of trees harvested for export to China in the early 1800s.[2] As we continued upward, trees began to predominate. The track that we followed morphed into a

101

steep and narrow tunnel of greenery amidst an eerie forest whose leaves, vines, and stems glistened with moisture. Tall, stately ʻōhiʻa and koa trees emerged from the omnipresent, swirling mist, like ghostly sentinels from a different age. Mosses and false staghorn (uluhe) ferns—one of the Hawaiian cloud forest's most prolific understory plants—blanketed the uneven ground, hiding long-ago lava flows. Hawaiian tree ferns (hāpuʻu) cast their broad, leafy umbrellas over dense groundcover. Though far removed from the sea-and-sand paradise I had envisioned, the forest through which we slowly navigated resonated with mystery, with enchantment.

The grass-lined, two-track dirt road up to the Hawaiian Crow reintroduction site ascended the side of Mauna Loa—the world's largest subaerial volcano—and climbed approximately 3,000 feet in 5 miles (900 m in 8 km) to the release area, which was nestled in a clearing 5,000 feet (1,500 m) above sea level. One of five shield volcanoes comprising the Big Island and among the world's most active volcanoes, Mauna Loa—standing 13,678 feet (4,169 m) above sea level—has been shaping local habitats for almost a million years and is expected to continue its eruptive activities for at least another half a million. As I peered through the drifting brume at the cathedral of greenery that surrounded us, my preconceived notions of volcanoes as bare, hostile environments devoid of life and reeking of toxic fumes were upended as irrevocably as those I'd had of the sunny Aloha State. I had no doubts that Hawaii was all that the tourist brochures claimed. But I was quickly discovering that it was infinitely more, too. And as we toiled upward through the drizzle, through the floating fog, through the primordial forest, I felt like the poet Robert Frost who, faced with two divergent roads in a yellow wood, chose the "one less traveled by."[3] And I hoped that my choice—like his—would make all the difference.

A dark, grouse-sized bird suddenly ran across our path, its bold red face a striking contrast to its black-with-hints-of-green-and-purple plumage. Its jaunty crest, shaggy gray breast feathers, black-and-white striped back and rump, and flamboyant bustlelike tail were further reminders that I was in a different world from any I had known.

"What was that?" I asked Mike, the head of the US Fish and Wildlife Service field team, as a more nondescript, red-faced bird with brown feathers tinged in cream rushed across the road in the wake of its more conspicuous companion.

"Kalij Pheasant," Mike answered. "A male followed by a female. It's an exotic species—like the myna you saw at the airport—that was introduced from Asia."

It would not be long before I understood that the presence of this pheasant was a relatively innocuous indicator that all was not well in this beguiling forest, nor on the island that harbored this timeless habitat. Introduced to a ranch on the Big Island in 1962, the Kalij Pheasant hailed from forests and thickets in the Himalayan foothills, ranging from Pakistan to western Thailand. The McCandless Ranch, which encompassed our release site, appeared to be a stronghold for the expanding pheasant population.[4] Less harmful than many of the species introduced—purposefully or accidentally—to the Hawaiian Islands, the Kalij Pheasant nevertheless distributes seeds from detrimental nonnative plants—such as the highly invasive banana poka vine—disrupting the forest's native complement of organisms and the ecosystem's natural function. The pheasants may also inadvertently bolster populations of nonnative mongooses and feral cats that prey on the bird's eggs and young.[5]

"The Fish and Wildlife Service camp is just down there," Mike continued as we passed a smaller track that broke away from the main road, "and you P. Fund people are based at the release site, a bit farther up the hill."

A minute later, we pulled off the road into a clearing that held an aviary so immense, so spectacular that I simply stared at it, wondering at the work that went into enclosing full-size trees in a fortress that was as impenetrable to outsiders as it had been accommodating to its prior occupants. We stepped out of the vehicle onto the wet grass.

And then, through the mist, came a strangely euphonic medley of throaty, echoing screams, intermixed with rumbled growls, and haunting, evocative wails. I felt my neck prickle. The eerie, commanding sounds seemed to drift out of the floating fog, seemed to embody the lost voices of the trees, the ferns, the vines, released at last from their cellulose prisons. I was hearing the most prominent and unforgettable voice of the Hawaiian cloud forest, that of the 'Alalā, as the Hawaiians dubbed their singular crow. The bird's native name (pronounced ah-lah-LAHH) means "cry like a child," but the crow's varied, penetrating, ventriloquistic vocalizations could only come from a child of an otherworldly forest, a creature of a more primal time and place.

I stood transfixed, irrevocably moved by the strange but captivating sounds, which washed over me like a turbulent wind that spoke of an uncertain future yet felt full of promise. And I thought: *Twelve*. Aside from the young reintroduced Hawaiian Crows that I was hearing and was about to meet for the first time, only twelve 'Alalā remained in the wild. Twelve birds to fill these forests with their astonishingly distinctive vocal repertoires. Twelve birds to embody the lost world that lingered in this enigmatic place. Twelve birds to carry this unique and intriguing species into the future. For a moment, I felt the full weight—and the privilege—of the task on which I was about to embark.

It is almost inconceivable to anyone familiar with our ubiquitous American Crow that any crow anywhere could be perilously close to extinction. The crows we know best—and indeed many of the world's most familiar crow species—are adaptable and opportunistic, and they readily associate with humanity and its heavily modified landscapes. Their prodigious numbers reflect these traits and the extensive time frame in which human-crow interactions have shaped crow evolution, culture, and populations. However, many island-living crows—particularly those that inhabit forests—are among the rarest birds in the world and are a surprising counterpoint to the crows that awaken us with their raucous cries in urban and rural neighborhoods.

The rarity of island crows is in part a function of where they live—on isolated islands surrounded by inhospitable watery habitats. In 1967, scientists Robert MacArthur and E. O. Wilson suggested that island biodiversity is determined by a balance between the rate at which new species colonize islands and the rate at which established species go extinct. These processes, in turn, are affected by an island's size and its distance from the mainland or similar habitats. Larger islands closer to mainland areas are likely to contain a greater diversity of habitats that harbor more species, since animals and plants can more easily immigrate to these islands. Organisms on large islands are also less likely to be driven to extinction by chance events. In contrast, smaller, more isolated islands are more likely to have fewer species whose numbers are not bolstered by immigration. These islands are also more likely to be affected by chance events that lead to extinctions.[6] Located in the central Pacific, more than 2,000 miles (3,200 km) from the closest continental landmass, the Hawaiian archipelago is the most isolated island system in the world, making its native fauna and flora particularly vulnerable to extinctions.

Because they are more vulnerable to habitat changes than their mainland counterparts are, island species have been particularly impacted by human colonization. People and their deadly arsenal of habitat-altering behaviors, tools, livestock, companion animals, and novel diseases led to the extinction of half the bird species that occupied tropical Pacific islands in a little over a thousand years.[7]

Rather than benefiting from anthropogenic changes to landscapes the way the opportunistic American Crow has, forest-dwelling island crows have been isolated and harmed by the conversion of native forests to agricultural lands and human settlements. In addition to being persecuted, island crows have also been devastated by the introduction of nonnative predators to which the birds haven't adapted suitable defenses, and exotic diseases to which the birds have little or no resistance. Two corvid species (the Deep-billed Crow and the Robust Crow) that once inhabited Hawaii were driven to extinction by Polynesian colonizers who first occupied the islands of O'ahu, Maui, and Moloka'i about 1,000 years ago.[8] Indeed, Polynesian settlers may have been responsible for the extinction of more than half of Hawaii's endemic avifauna.[9]

Today, populations of endemic Caribbean crows, such as the Hispaniolan Palm Crow that inhabits native forests but has adapted to treed settlements, are declining. But the White-necked Crow, which is found *only* in forests and pine woodlands on Haiti, Dominica, and Puerto Rico, is endangered because its forested habitat has been destroyed by timber harvesting, land conversion, and human development. Among the world's most imperiled corvids are the endangered Mariana Crow that inhabits limestone forests and agricultural edges on the Pacific islands of Guam and Rota, and the critically endangered Hawaiian Crow (*Corvus hawaiiensis*), whose only remaining habitat is the moist koa and 'ōhi'a forests of Hawaii's Big Island.[10]

More closely related to the Common Raven than it is to other crow species, the medium-sized, brown-black Hawaiian Crow historically inhabited dry and seasonally wet, mid-elevation montane forests in leeward areas of western and southeastern Hawai'i.[11] Although the crow may once have lived on the neighboring island of Maui prior to Polynesian colonization, naturalists have only ever documented it on the Big Island.[12] A tropical forest specialist and one of the most arboreal of crows, the 'Alalā feeds on a wide variety of fruits, nuts, insects, and occasionally, the eggs and nestlings of other birds.[13] Unlike most other corvids, the opportunistic Hawaiian Crow even probes flowers for nectar.[14]

Despite its limited distribution, the 'Alalā was fairly common in the late 1800s and still occurred throughout its known historical range, although the bird's decline was noted as early as 1896.[15] But throughout the 1800s and up to the 1970s, an ever-growing tide of settlers cleared, logged, burned, and grazed livestock in the crow's restricted habitat, taking a notable toll on the bird's population.[16] Coffee and fruit farmers, who further encroached on the crow's habitats during the 1890s to 1930s, readily shot the curious and confiding 'Alalā, which the native Hawaiians revered.[17] And the introduced animals and mosquito-borne diseases that accompanied the island's colonists further reduced crow numbers.

By the 1940s, the Hawaiian Crow had been extirpated from lower elevations. A decade later, the bird's remnant population was severely fragmented.
By the early 1970s, only about one hundred 'Alalā remained in three patches
of moist forest between about 3,200 and 6,500 feet (975 to 2,000 m) on
the western and southern flanks of Mauna Loa, and the western side of the
adjacent Hualālai volcano.[18] And still 'Alalā numbers continued to fall. A
forest-bird survey conducted by the US Fish and Wildlife Service in 1978
estimated there were seventy-six Hawaiian Crows. By the mid-1980s, fewer
than twenty-two still broadcast their singular calls. This last precipitous
population drop resulted, in part, from illegal shooting, particularly on
the slope of Hualālai, by opponents to the implementation of conservation measures for the beleaguered crow.[19] In the end, as John Marzluff later
noted, "Human challenges were too novel, too diverse, and too rapid for
this species to adapt."[20] And now the incomparable 'Alalā is best known for
its largely frugivorous diet, its strong ties to forested habitat, its astonishing
vocal repertoire, and its status as the most endangered corvid in the world.[21]

As I searched out the makers of the loud and varied calls that emanated from
the obscuring mist, I also examined the aviary that had served as both a hack
box and a release pen for the five crows that had recently been introduced to
the wild. Built by Peregrine Fund personnel, the aviary, which measured 100 ×
45 feet (30 × 14 m), had 15-foot (5 m) wire-mesh walls that were buried into
the ground to prevent predators from digging into the structure. Galvanized
steel siding covered the wire mesh from ground level to a height of over 3 feet
(1 m). Electrified wire ran along the tops of the steel siding and mesh walls.
The aviary, which reached a height of 60 feet (18 m), was topped by knitted
polypropylene netting that had been draped, using pulleys, over native trees
and was supported by disks attached to the tops of several stripped trees.

The south side of the aviary was comprised of a two-story wood facility with a mosquito-proof lower room that could temporarily house sick
or injured crows. The upper story consisted of a mosquito-proof hack
box flanked on each side by small observation booths with one-way-glass
portholes that allowed biologists to view crows that were in the hack box
or aviary. The hack box contained small doors that served as food ports
through which the nestling crows were initially fed. Two large, removable,
barred panels on the front and back of the box allowed the crows to be
released into the aviary first, then later into the wild. The hack box and

observation rooms were flanked by long decks where fruit-bearing branches had been attached to accustom the birds to native fruits after their release.

Earlier that year, in April and May of 1993, when wild ʻAlalā laid their eggs, a team of US Fish and Wildlife Service biologists had monitored three nesting pairs from blinds set up near each of the nest trees. To bolster the crow's rapidly dwindling numbers, the biologists would take eggs from the wild birds and hatch them in captivity. A few of the resulting young would join a captive-breeding population numbering eleven birds. The remainder would be released to the wild to increase the wild population. The biologists timed their egg removal to allow for some incubation by the parents, which increased the eggs' hatchability, and to give the adults enough time to lay another clutch that they would raise in the wild, thereby maximizing each pair's reproductive output.

Service biologists successfully removed eight pale blue-green, dark-speckled eggs from the moss-and-grass interiors of bulky stick nests built high in ʻōhiʻa trees, and transported them to a rented house—dubbed "the egg house"—that would serve as an incubation facility. There, staff from the Zoological Society of San Diego (now San Diego Zoo Wildlife Alliance)—a project partner with expertise in captive-rearing endangered species—placed the seven eggs they'd determined were fertile in specialized incubators. Two of the three ʻAlalā pairs subsequently renested, but, tragically, both nesting attempts failed when the chicks were about ten days old. Undeterred, one of these pairs—known as the Kalāhiki pair—renested a third time, but abruptly abandoned the nest during incubation. Alert biologists rescued the eggs, only one of which was fertile.[22]

Six of the seven fertile eggs first taken from the ʻAlalā hatched successfully. One nestling was sent to a captive-rearing facility on Maui, along with the successfully raised chick from the abandoned clutch. The remaining five chicks would be the first ʻAlalā to be reintroduced to the wild.

Of these, the three oldest chicks, which had come from the Kalāhiki pair, were all females. Each crow was given a Hawaiian name. Hiwahiwa (pronounced hee-vah hee-vah and meaning "precious or beloved"), Hoapili (hoa pili in Hawaiian, meaning "friend"), and Lōkahi (meaning "unity and agreement") were transported to the release site and placed in the hack box in the first week of June, when they were fifty, forty-eight, and forty-six days old, respectively. Slightly older than the age at which they would have left their nest had they remained in the wild, the crows were fully feathered but not yet feeding themselves. Service biologists and The Peregrine Fund's

Peter Harrity fed the females four times a day using a hand puppet designed to look like the head of a wild 'Alalā. In addition to feeding them an artificial diet, the biologists collected and fed the birds native fruits. Peter had attached fruiting branches to the box's walls before the females' arrival, and the youngsters unhesitatingly nibbled this proffered bounty from their first day in the box.[23]

In the middle of June, the two youngest chicks, both males, which had come from the Keālia pair, were placed into the hack box with the females. Mālama (meaning "to care for or honor") and Kēhau (meaning "mist, dew, or gentle breeze") were now fifty and forty-eight days old respectively.[24] The three females showed little interest in the younger males except to squabble with them and each other, while flying restlessly from perch to perch. It was time to release them into the aviary.

On June 23, while I was still in Idaho and scarcely aware of the 'Alalā's imperiled existence, Peregrine Fund personnel partitioned the hack box with a plexiglass divider. Mālama and Kēhau remained in one side of the box for some additional growing-up time, while the front panel on the other half of the box was removed to allow the females into the aviary. Releasing them there first rather than into the wild would give the crows a chance to develop their flight skills while staying safe from predators, and it would allow them to forage and interact in a safe environment where they could be monitored by biologists.[25] Since the aviary was located just over a mile (2 km) from the wild Kalāhiki pair's territory, biologists also wanted to protect the youngsters during their early encounters with the potentially antagonistic adults.

Hiwahiwa and Lōkahi were quick to explore their new surroundings, roosting in trees, and feeding at food trays on the hack box's front deck. More cautious and submissive, Hoapili remained in the hack box, venturing outside only for quick visits to nearby trees and the food trays. After almost a week, Peter replaced the front panel on the box, forcing Hoapili into the aviary, and giving Mālama and Kēhau free rein of the hack box. Later, the males were moved to the larger chamber on the ground floor to give them more room. The females, meanwhile, were closely monitored for symptoms of avian malaria and avian pox—two mosquito-borne diseases that were unintentionally introduced to the Hawaiian Islands and posed a serious risk to 'Alalā.

A month after their release into the aviary, the females showed no signs of illness or distress, so Mālama and Kēhau were released, as well. For another month, the crows improved their flight skills, explored the aviary's native trees and shrubs, learned how to maneuver among the treetops, and fed on both proffered and natural foods. Importantly, they were also visited periodically by the curious Kalāhiki adults. As biologist Bernd Heinrich has noted, exposure to what is important is an essential component of the learning process, particularly for organisms whose development depends, at least in part, on learned (rather than instinctual) behavior. Heinrich routinely exposed juvenile ravens that he had raised in captivity to wild ravens. Interestingly, his youngsters focused intently on raven calls and distant flying ravens, but ignored flying American Crows, showing an innate recognition of their own kind.[26] As the adult ʻAlalā foraged and broadcast their cacophonous calls nearby, they inadvertently served as mentors for the parentless youngsters, facilitating their upcoming transition to the wild.[27]

Despite their proximity to each other in the aviary, the young female and male ʻAlalā seemed to function as two distinct groups, rather than forming a hoped-for cohesive flock. Perhaps this was as much a function of their biology as it was of their separate upbringing. Wild ʻAlalā parents and their offspring, which depended on them for food for as long as a year, formed tight-knit family groups that fed and roosted together. During the nesting season, adults aggressively defended their territory from all other ʻAlalā, including their young from previous years.[28] Fortunately, as their release approached, the five captive-reared ʻAlalā began roosting together in a clump of aviary trees, though the males and females still parted during the day.

On August 21, the crows were enticed back into the hack box and recaptured to prepare them for their release to the wild. Blood was drawn from each bird and screened for avian malaria. Each crow was then outfitted with a tail-mounted transmitter so the birds could be tracked post-release. These small, cylindrical radio transmitters, weighing approximately one-third of an ounce (10 g) or about 2 percent of the birds' body weight, were sewn and glued to the base of each bird's central two tail feathers—known as deck feathers. Ideally, the transmitters would stay in place until each bird molted in new flight feathers the following summer.[29] The youngsters were then returned to the aviary and monitored to ensure that the transmitters were not hampering their movements. Fortunately, the birds were malaria-free and seemed unaffected by their transmitters. It was time for the first Hawaiian Crow reintroduction!

On August 23, biologists removed the back panels of the hack box and placed food trays on the aviary's exterior deck to entice the ʻAlalā out of their pen. The youngsters needed little encouragement. By late morning, all five juveniles had left the aviary, which opened into a large clearing, and had settled into the surrounding trees. But their initial boldness was followed, to varying degrees, by a retreat to the familiarity of their pen. Several crows roosted outside the aviary for the first two nights following their release, but on the third night, all five youngsters returned to the enclosure. Hoapili, in particular, seemed content to remain in the pen. Worried that she would fall behind in her development and become isolated from the other crows as they ventured farther afield, Peter capitalized on a moment when all five birds were outside the aviary, on August 28, and replaced the hack box panels, effectively barring the birds from their former home and forcing them to adapt to their new life as wild birds.

————

Now, on September 6, as fog swirled around me on my first day in Hawaii and mist settled in droplets on my hair, two dark forms emerged from the gloom, flapping toward the aviary and several food trays that had been placed high atop metal poles, where they were accessible only to flying frugivores. A modified scale, topped with a wooden perch, was attached to each tray so we could record a crow's weight as the bird landed and prepared to feed. Just as Lōkahi, the dominant female, and Hiwahiwa arrived at the food trays, a third bird flew into view, calling querulously in a high-pitched voice. Hoapili touched down by the other females, then fluffed up her head feathers and begged plaintively for a handout from her nestmates as they gulped down fruit.

I strained to read the distinct letter-and-number combinations on the crows' leg bands and familiarize myself with the birds. Each female wore a green alphanumeric band on her right leg and a silver US Fish and Wildlife Service band on her left leg. About the size of an American Crow, the 'Alalā were a duller, sootier black than their glossy mainland counterparts, with brownish primary flight feathers and a huskier bill. Their loose feathering made them look fluffier than the sleek American Crows I was familiar with. And each of the 'Alalā still had the endearing smoke-blue eyes and pink gape that marked them as juveniles. Minutes later, the females were joined by Mālama, the dominant male, and Kēhau his subordinate brother. Each wore a gold alphanumeric band on his right leg to help differentiate the males from the almost imperceptibly smaller females.

Long before I'd had my fill of watching this dynamic quintet, Peter Harrity was issuing instructions to Kurt, who had been in Hawaii monitoring the crows for a week, before heading down the mountain. After Peter left, Kurt led the way up the steps of the stained, single-wide white trailer that would serve as our home for the next six weeks. Although it had no running water or electricity, we were well equipped with lanterns for night reading, a camp stove for cooking, and folding cots on which to stretch our sleeping bags. In addition to providing us with a dry place to eat and sleep, the trailer served as a makeshift laboratory and as a facility for preparing the crows' daily, supplemental buffet. Shelves of miscellaneous equipment lined the small, cluttered room that would be mine for the duration of my stay. A short walk from the trailer, an insect-riddled pit toilet, tucked into a clump of trees, served as our restroom. Nearby, several large plastic barrels contained water that had to be transported up the mountain for drinking, washing dishes, and provisioning the crows. Though rustic, the accommodations seemed veritably luxurious compared to the alternative of living out of a tent in such a wet environment.

Later that night, I lay in my narrow cot, listening to a soft rain fall on the trailer's roof, grateful that I could succumb at last to the exhaustion that had dogged my steps from Idaho to Hawaii, around the Big Island, and part way up a volcano. Before I fell asleep, a comment that John Marzluff had made about corvids several months earlier drifted into my mind. Doubtless in response to my unbridled enthusiasm about the falcons and eagles I was monitoring, John had said, with a teasing glint in his eyes, "You know, the birds you raptor people focus on aren't nearly as interesting as corvids." Despite being a blatant bid to provoke interest in the engaging birds to which John had dedicated his career, his words highlighted an essential difference between raptors, which are hardwired with deep-seated instinctive behaviors, and corvids, whose legendary intelligence and complex social interactions reflect their learned behavior and the multidimensional, ever-changing environments in which they make a living. John's words gave me pause.

Although I had enjoyed innumerable entertaining hours observing birds of prey, I had also spent mind-numbing intervals watching stationary raptors survey the surrounding landscape. (I'd once monitored a Golden Eagle perched on a power pole for seven interminable hours.) My brief preview of the young 'Alalā's antics had given me a tantalizing glimpse into what John meant. The birds had been riveting as they clambered around on the aviary, tugged at the netting, nibbled fruits, rushed to investigate one another's findings, and gently preened each other's feathers, all the while narrating their activities with a strange medley of high-pitched yells, quizzical cries, and muted murmurings. I couldn't wait to get to know the crows and their curious habits better. And though my appreciation for raptors was unlikely to be diminished, if I was embarking on a path less traveled, I was quite sure that I was going to be accompanied by a troupe of endlessly beguiling companions.

CHAPTER TEN

Night Whispers

To my delight, darkness gave way to sunshine as I embarked at dawn on my first day of monitoring and caring for the recently released Hawaiian Crows. After the previous evening's damp gloom, I had wondered if the sun ever penetrated the cloud forest or if we would reside in a permanent fog. Garbed against the still-dripping vegetation in boots that weren't quite as waterproof as the manufacturer claimed, heavily patched rain pants, and a raincoat, I carried my binoculars strapped to my chest in a protective case, my radio-telemetry receiver looped over a shoulder, and a communication radio attached to my waist. In my hands, I clutched my field notebook, my Yagi antenna, and a cup of coffee. Tasked with monitoring the three female 'Alalā, I programmed one of their transmitter frequencies into my receiver.

Developed in the 1960s, VHF (very high frequency) radio telemetry has long been an essential tool for wildlife research and conservation. After a transmitting device with a unique radio frequency is attached to (or inserted in) an animal, a researcher uses a special receiver and an antenna to track the pulsed signals (VHF radio waves in the range of 30–300 MHz) emitted by the transmitter.[1] Knowing how important radio tracking was for wildlife biologists, I had relished learning how to track birds in Idaho, and I had been equally glad that the Snake River Canyon had provided such a challenging environment in which to perfect my new skill.

Holding my Yagi horizontally, I slowly moved it to the left and right. Listening intently, I noted the point where the beeps that emanated from my receiver were the loudest, since that would be the direction in which I would find my birds. When I moved the front of the Yagi away from the direction where the signal was the loudest, the sound faded. But radio waves bounce off trees and vegetation, as they do off cliffs and rock, which can mislead a radio tracker. So, for good measure, I checked for the signal behind me, as well. As expected, I heard a fairly strong signal from that direction, too, but the arc of sound was much narrower than the arc of sound in front of me, meaning that what I was hearing behind me was a "back signal" and that my crows were, indeed, up ahead. Judging by the strength of their signals, they were quite close, and I set off in pursuit.

The captive-reared 'Alalā were exploring their dense forest home ever more widely, continuously pushing the boundaries of their known universe. By tracking them and observing their activities, we could make sure they were behaving the way healthy, wild juvenile Hawaiian Crows should. As I clambered over moss-covered logs and pushed through wet ferns and grasses, heading toward my birds, two of the 'Alalā unexpectedly found me, winging over my head as they flew to the aviary for a morning meal. Lōkahi and Hiwahiwa landed on the aviary netting with a great flapping of wings, and then I heard Hoapili's high-pitched call as she arrived in the wake of the more dominant birds. Despite her slightly smaller size, Hoapili more than compensated for her inferior stature by having the loudest voice and using it the most often. She seemed to resonate with indignation at being left behind again. Not only were subordinate crows slower to mature vocally than more dominant birds, but they also retained their pink mouth linings—an indication of juvenile status that would turn black with age— longer than their more dominant counterparts.[2] Landing next to the other females, which were now tugging at the netting, Hoapili shuddered her folded wings, lowered her head, and begged plaintively for a handout from her industrious nestmates.

Ignoring Hoapili, Lōkahi walked around on top of the aviary then flew over to a feeding tray. Hiwahiwa followed, landing briefly on the adjacent scale perch before hopping onto the tray. Raising my binoculars, I managed to read the 495 grams (approximately 1 pound) that registered on the scale. To my relief, the busy 'Alalā paid no attention to my presence or my fumbling activities. After feeding on a mix of papaya, egg, dead mice (that thankfully were supplied to us), and dog kibble, which we had put out for them earlier that morning, Hiwa and Lōkahi flew over to the trees bordering the aviary's south side. Hoapili, meanwhile, flew to the food tray, landing on the scale just long enough for me to register her weight: 480 grams.

A faint but persistent beeping suddenly intruded on my absorption in the birds' activities. I looked around in confusion for a moment before realizing that the noise came from the timer on my watch, which I had programmed to go off at designated intervals so that I could conduct our requisite behavioral "scans." Rather than recording our birds' activities throughout the day, as I had done with peregrines in Wyoming, we were tasked only with monitoring our 'Alalā for several hours each morning and evening, documenting their behavior in ways that would allow researchers to quantify the amount of time that the birds engaged in specific activities.

Every fifteen minutes, for a two-hour period, we recorded the behavior of each crow that we had in view. Recording data at designated times reduced bias because it forced us to objectively document the activities in which our birds were engaged, rather than allowing us to inadvertently note those behaviors that most captured our attention. Biologists overseeing the 'Alalā's recovery would use this information to determine the percentage of time our introduced crows spent foraging, preening, playing, or engaged in other activities.[3] These data could then be compared to the behavior of wild 'Alalā to determine whether our youngsters were behaving as expected.

In addition to recording each crow's activities every fifteen minutes, we also estimated each bird's distances from the other 'Alalā and from the aviary. Our estimated distances between the 'Alalā would allow biologists to evaluate the juvenile crows' interactions with their flock members, as well as with any adult 'Alalā they encountered. And by noting our birds' distances from the aviary, researchers could quantify the young birds' eventual dispersal from the release site and better understand our 'Alalā's path to independence.

Jotting down the time and the word *scan* in my field notebook, I noted that Hoapili was still feeding at the food tray. A glance with my binoculars revealed that Lōkahi was clinging to a tree trunk, chipping away bark to reveal hidden spiders, pill bugs, or other insects. Hiwa, meanwhile, was poking her beak into a clump of leaves a few feet from Lōkahi. I was astonished at how instinctively and naturally the birds foraged despite having been in the wild such a short time, and without the benefit of parents whose behaviors they could mimic. After guesstimating the distances that separated the three females and the distances between each of the birds and the aviary, I returned, with some relief, to recording the birds' general activities. Between our every-fifteen-minute scans, we transcribed the 'Alalā's doings much as I had done with my peregrines.[4] This continuous (or *ad libitum*) data sampling during our observation periods allowed us to capture social interactions among 'Alalā as well as any interesting activities that might be missed by the scan samples.[5]

Shortly after I conducted my first scan, Lōkahi and Hiwa left their tree, calling loudly as they disappeared into the forest. Their vocalizations belatedly alerted Hoapili, who looked up abruptly from the food tray then launched herself in pursuit. Late again. But at least Hoapili could fly swiftly over the dense foliage and treacherous terrain through which I struggled. I had been warned to proceed carefully since the mosses and ferns that enveloped my path might cover the weak or pockmarked roof of an ancient lava tube, formed when molten lava flowed beneath the cooler, hardened surface of a lava flow. Falling into one of these cavernous tunnels was one of the lesser-known but no less forbidding ways a volcano could kill or maim. Those admonitions surfaced disquietingly in my mind as I followed the voice trails of the 'Alalā through the forest.

Periodically, the trees gave way to grassy clearings that eased my awkward pursuit of the crows and delighted me with their verdant, sunlit beauty. They seemed designed to provide respite and tranquility, the way a well-placed bench might prompt a welcome break on an arduous mountain trail. As a small child growing up in Switzerland, I had followed my American father on walks through forests that cloaked the slopes of the Alps, and discovered similar clearings. He had mesmerized his wide-eyed daughter with tales of the quiet spaces coming to life at night as fairies glided down moonbeams and small forest creatures gathered around rings of toadstools. And while I hardly expected the dust motes and pollen that danced in the Hawaiian sunbeams to come to life, the clearings held a touch of magic, and the 'Alalā that paused in them to walk single file along a moss-covered log or rest in an adjacent 'ōhi'a provided a special brand of enchantment.

Unbeknownst to me at the time, though, these moss-and-grass-carpeted glades were the unwelcome and inglorious work of introduced cattle and feral pigs that had once grazed, trampled, and uprooted the native vegetation. And while the subsequent barren areas had eventually developed into patches of pleasing greenery, the former ecological function of these areas, and the fruits and fibers that once fed and sheltered the forest's avian denizens, were irrevocably diminished.

Studies in the 1970s and 1980s found that 'Alalā occurred and nested in highest densities in closed-canopy koa–'ōhi'a forests with native understories—an already scarce habitat type that became ever rarer in the ensuing decades. 'Alalā densities were lower when understories had a mix of introduced and native plants, and lower still when understories were comprised solely of exotic species. Reduced 'Alalā densities likely reflected a loss of food plants in increasingly disturbed areas.[6]

Aside from providing less food, forest clearings also posed a hidden threat to the 'Alalā. The only predator to prey on full-grown crows—the Hawaiian Hawk, or 'Io—capitalized on the greater visibility of open areas when it hunted introduced rodents, game birds, and, occasionally, the native 'Alalā.[7] The 'Io's opportunism and the loss of the Hawaiian Crow's protective cover likely disrupted the historic balance between hunter and hunted. And while the hawk benefited, the crow skidded inexorably toward extinction.

Lulled by Hawai'i's illusory beauty and unaware of these concerns, I followed the female 'Alalā and recorded their activities for the rest of the observation period. The trio perched close together and gently preened one another's heads, hammered like woodpeckers at the giant hāpu'u ferns, and clambered around the treetops like primates, in search of insects and berries. Too soon, my morning observations ended, and I rejoined Kurt, who had been monitoring Kēhau and Mālama, for a hearty breakfast at the trailer.

Unlike the intensive field schedules I had become accustomed to over the previous few years, our daily routine was comparatively undemanding. Although we often observed the crows opportunistically throughout the day because of our proximity to them, our relatively short observation sessions were an attempt to balance keeping an eye on the birds with freeing them from disturbance. In between our formal observations, we cleaned the birds' feeding trays, prepared their food, filled up water containers, put out tubs so the birds could bathe, collected fruiting branches to attach to the aviary decking, entered our data, and had some personal time to rest, read, and, in my case, run.

As I prepared for my first run, after digesting a late breakfast, the sun gave way to clouds that materialized like a conclave of shadowy forms emerging from invisibility cloaks. The white vapor coalesced around me until I was ensconced in moving mist. A bit daunted by the vertical terrain, I set off slowly down the steep track that led away from camp, dreading the unforgiving uphill that would constitute my return journey. The opaque fog was slightly disorienting. I felt as if I'd taken up residence in a cloud. I'd imagined it would be softer—and drier, of course. But despite the penetrating dampness and the accompanying chill, being surrounded by the drifting brume was ineffably alluring. It felt surreal, evanescent, as though my surroundings—and the world—were in constant, gentle motion. Comforting rather than foreboding. Slightly bewildering, but subtly bewitching. With visibility reduced to the few feet of dirt roadway and bordering greenery that unfurled slowly in front of me, I ran as though in a dream, with tiny dewy droplets continually, almost imperceptibly sprinkling my eyelics. The chill, the fog, the misting rain lasted until nightfall—until after we'd watched our five murmuring crows settle down to sleep in adjacent trees—then bright stars appeared in the skies, preparing us for another sunny morning.

Among the world's most imperiled ecosystems, tropical montane and submontane cloud forests comprise less than 3 percent of tropical forests worldwide and about one-tenth of 1 percent of the Earth's land surface. These species-rich forests, which often are endemism hotspots, are disappearing at a faster rate than their better-known lowland rainforest counterparts. Cloud forests on isolated volcanic mountains—particularly on small oceanic islands—may be especially vulnerable to both overt human destruction and more subtle but pervasive anthropogenic pressures like climate change, which may promote shifts in atmospheric circulation patterns, along with concomitant changes in rainfall, cloud cover, humidity, and hurricane frequency.

Cloud forests occur where warm, wind-driven masses of air and water vapor—formed by water evaporating off the surface of the sea or land—cool and condense into clouds when they encounter mountains. Typically found in a relatively narrow altitudinal zone, cloud forests are characterized by low-level, persistent or seasonal cloud cover that reduces solar radiation. The clouds envelop the forests' distinctive vegetation, which often consists of dense stands of stunted, gnarled, broad-crowned trees and a profusion of epiphytes, including bromeliads, mosses, lichens, and ferns. Although cloud forests may receive significant rainfall, they are unique in receiving much of their water from the canopy's interception of cloud moisture—and are sometimes referred to as cloud strippers. Cloud moisture captured by leaves and branches drips to the ground, increasing the net precipitation the forests receive and adding to groundwater and streamflow levels. Often found at higher elevations and on steep slopes with waterlogged organic soils, cloud forests play an important role in erosion control and watershed protection.[8]

———

Although the infrequent sunshine and unfamiliar cloud-forest weather patterns were an ongoing adjustment for me, the vocal 'Alalā quickly lulled me into complacency. On my third morning, I left my bulky tracking gear behind since the birds seemed to commentate on their activities in such exuberantly loud voices that they were easy to find. I was following the males, while Kurt monitored the females. The two groups still spent portions of their days apart but were interacting more, particularly as they settled down to roost. After leaving the aviary area after their morning feed, Mālama and Kēhau set a blistering pace, moving every few minutes and flying farther and farther afield. Pushing through a web of vegetation, I stumbled after them.

Hearing the males nearby, I paused to catch my breath and spied the two compadres perched on a high branch. Kēhau held a large stick in his beak and jerked his head away from Mālama, who tried repeatedly to grab the irresistible toy. Hopping along the furrowed branch, Kēhau sought to put a little distance between his prize and his acquisitive sibling. Mālama hopped after him and then quickly opened his wings in pursuit when Kēhau left the tree to fly in a wide circle. Landing on a stout 'ōhi'a branch, Kēhau again brandished his stick, but, after a moment, Mālama got distracted by a cluster of the tree's red, fairy-duster blossoms. Perhaps tiring of a possession that was no longer in demand, Kēhau soon dropped his stick and nibbled leaves.

My own respite was short-lived. Soon the birds were on the move again. And again. Hastily, I followed until I was far enough away from the aviary to feel a twinge of fear. I had paid little attention to my surroundings in my heedless pursuit of Kēhau and Mālama. What if I couldn't find my way back? There were no obvious landmarks under the forest canopy, no boulders that might serve as beacons to guide me back along the floundering path I had forged. No distant mountains helped provide a direction. No trickling streams served as boundary markers. And now my crows had fallen uncharacteristically silent. I stood in a stillness broken only by the squeaking, chirping calls of a Hawai'i 'Elepaio—a diminutive, unbearably cute, brown-and-white Hawaiian flycatcher, with a rust-colored chest, checkered wings, and a stiffly cocked tail—that was one of the area's most endearing native songsters.

Rather than press on in search of the suddenly elusive 'Alalā, I decided to try to return to the aviary to retrieve my telemetry equipment. No matter how far the crows ranged, if I could pick up their transmitter signals, I knew they would eventually lead me back to camp when they returned to feed. The raucous sound of the wild 'Alalā calling nearby also reliably sent our charges fleeing back to the relative safety of their home base, since the adults often chased and harassed the youngsters. As I retraced my steps, I flushed a crimson 'Apapane from the low branch of an 'ōhi'a. The most abundant of Hawaii's unique honeycreepers, the nectar-sipping 'Apapane feeds from 'ōhi'a blossoms, whose vibrant red the bird mimics. Turning toward me, the bird displayed its sharp, slightly decurved bill, then raised its tail, flashed its dark wings and white undertail, and disappeared behind a tree fern. Lowering my binoculars, I resumed walking, and soon came across an old grassy road I hadn't seen before. It headed in the right direction and provided obstacle-free travel, so I followed it. To my surprise, it led back to camp.

As I approached the aviary, I heard quiet murmurs and broke into a smile when I spotted Kēhau and Mālama nibbling at the dark-purple berries of an 'ōlapa tree, an endemic plant whose fluttering, ovate leaves inspired both its Hawaiian name—meaning "dancer"—and hula dances.[9] Pleased as I was to be reunited with my birds, it was also gratifying to see the parentless crows eating native fruits like 'ōlapa, which reminded me of elderberries, and the lumpy white māmaki berries from another endemic tree whose slightly sandpapery leaves native Hawaiians used as a medicinal tea.[10]

Although we regularly attached fruiting branches of both trees to the hack-box decking to encourage the 'Alalā to feed on these and other native fruits, the birds also readily feasted on a wide variety of fruits that we hadn't first provided to them. In doing so, they inadvertently fulfilled a critical ecological role. Many fruit-eating birds disseminate plant seeds through their excreta, regurgitated pellets, or behavior. Since birds often transport seeds before eating or excreting them, such avian dispersal can minimize subsequent competition between the parent plant and its progeny. Birds can also ensure that seeds are placed in favorable locations by caching seeds that they later fail to retrieve. And birds can improve germination for certain plants when seeds that pass through the birds' digestive system are scarified. As the Hawaiian forests' largest songbird and frugivore, the 'Alalā was an essential seed disperser that had long sustained and shaped Hawai'i's forest-plant communities. Without the crow, many plants couldn't reproduce, leading to irrevocable changes in the structure, diversity, and species composition of the island's forests.[11]

Soon after I returned to the aviary, the three raucous females joined Kēhau and Mālama, and the group became increasingly vocal and animated. I was pleased to see them all together, walking along branches, cocking their heads when they spotted insects hidden under flakes of bark or within clusters of leaves, and rushing toward any crow that found an item of interest. As I watched Lōkahi chipping away at a horizontal branch, Hiwahiwa hopped over to her and almost bumped up against her. Raising a foot as if to protest her sibling's proximity, Lōkahi pushed it against Hiwa's chest. Unwilling to leave, Hiwa raised her own foot in defense. The two females grasped feet, then pushed and jostled each other. I was reminded of the legend of that beloved knave Robin Hood who, meeting his imposing future comrade, Little John, for the first time at a log bridge, sparred with him to see who would earn the right to cross the river first. Throwing their weight behind their clasped feet, the crows pushed and pulled. Then suddenly, Hiwa opened her foot and dropped away from the branch, gliding to another tree. Minutes later, the two females perched together amicably, and preened each other—a behavior known as allopreening.

Later that day, as we neared the end of our evening observations, the five ʻAlalā again congregated in two large trees to roost. Hoapili perched slightly apart from the others, as she often did, while the other crows perched in close contact with their respective sibling. As they relaxed, the ʻAlalā periodically elevated and separated their head feathers, giving them a fuzzy-headed look that I found irresistible. Even more endearing than their cotton-top heads and their habitual allopreening were their evening vocalizations.

Not surprisingly, given that the ʻAlalā's raucous cries accompanied their every activity, roost time for them was filled with their contact calls. Unlike the yells, whoops, and growls that ricocheted through the forest during the day, though, as daylight faded and night's chill crept through the undergrowth, the crows murmured quietly to each other, making soft croaks, squeaks, whispers, zipperlike sounds, and even muffled, kittenlike mewings. Their lengthy murmurations had a comforting quality, as though the birds were quietly reminiscing about the day or reassuring each other about the night to come.

Biologist Bernd Heinrich described similar roost contact calls made by his captive Common Ravens. "When the murmurs were very low, soft and long and almost whispered," he noted, "I learned the birds were at ease, as could also be seen from their relaxed postures. The calls seemed to be contact calls meaning, 'I hear you. Everything is fine.'"[12] The ʻAlalā's hushed sounds were as soothing to me as they seemed to be for the birds. That night and in the nights to come, I crouched amidst the wet vegetation transfixed by the heartwarming sounds. I envisioned, years from now, holding a conch shell to my ear. Many people do so with the romantic notion that they are listening to the ocean when they hear surrounding sounds resonate within the shell's cavity. But I knew that rather than hoping to hear the gentle sound of surf breaking on a beach, I would wish forevermore to be transported by the night whispers of the ʻAlalā.

CHAPTER ELEVEN

Malaria

Five days into my work with the ʻAlalā, I set out on a gray morning to follow Kēhau and Mālama. But I felt under the weather. To my surprise, it quickly became apparent that Mālama felt that way, too. As the other crows busily hopped along branches, poking their bills under bark and into ʻōhiʻa blossoms, Mālama rested, perching quietly and preening sporadically. Fortunately, the other ʻAlalā stayed nearby, moving from tree to tree as they foraged.

Upon completing yet another behavioral scan, I noted with growing concern that I had recorded Mālama sleeping several times—a highly unusual morning behavior. Kēhau appeared to be as puzzled as I was by his erstwhile companion's actions. Sidling up to Mālama, he made soft mewing calls, then gently preened his sibling's head and neck, and delicately nibbled his face. When his ministrations failed to arouse the subdued bird, Kēhau plucked an unopened koa flower bud—part of a berrylike cluster that would open into a creamy-white, powder-puff blossom—and pushed it carefully into Mālama's back feathers with his bill. No response. Kēhau cocked his head, watching Mālama, then stretched out his neck, plucked another globular koa bud, and again stashed it, with cautious deliberation, into Mālama's plumage. Still no response. Kēhau repeated the process again. And again. And still Mālama failed to react. Tilting his head, Kēhau looked at Mālama, murmuring quietly to him. Finally, Mālama fluffed out his body feathers and shook them—a comfort behavior known as rousing that often follows a bout of preening—and dislodged the buds. But he remained on his perch, staring at the ground.

As my bewilderment at Mālama's behavior escalated, thoughts of avian malaria reverberated in my mind like an ominous drum. We had been warned that this introduced disease posed one of the most serious threats to the 'Alalā, and so we needed to be diligent in noting any anomalous crow behavior. Avian malaria, a mosquito-transmitted disease caused by microscopic protozoan parasites in the genus *Plasmodium*, infects red blood cells. Although it occurs worldwide, it has caused particular harm in Hawaii, the Galápagos, and other archipelagoes where birds never evolved with the disease or long ago lost their resistance to it.

Originally lacking the common biting insects—mosquitoes, black flies, sand flies, and deer flies—that plague so many tropical places, Hawaii was a veritable paradise, once upon a time. Unfortunately, the southern house mosquito (*Culex quinquefasciatus*) was inadvertently introduced to the Hawaiian Islands in 1826, when sailors dumped ballast water—used to stabilize ships—from Mexico into a stream on the island of Maui. The mosquitoes that emerged from the larvae contaminating this water initially posed little threat to Hawaii's forest birds. But the blood-sucking insects became disease vectors that transmitted pathogens, such as the avian malaria parasite (*Plasmodium relictum*) and avian poxvirus (*Poxvirus avium*), when these deadly diseases were transported to Hawaii in the bodies of nonnative birds, which were introduced to the islands in the early twentieth century.[1] Although avian malaria can harm birds that evolved in close association with the disease by lowering their immune response or the number of offspring they produce, the illness is rarely fatal to them. However, after mosquitoes fed on infected birds that had been introduced to Hawaii, the invasive insects transmitted the disease-causing *Plasmodium* parasites to naive native birds. Malaria subsequently decimated native bird populations and has been implicated in the extinction, population decline, and altered distribution of many Hawaiian birds.[2]

The disappearance of Hawaii's endemic birds perplexed early naturalists. As early as 1902, ornithologist Henry Henshaw commented on the disappearance of birds such as 'Elepaio, 'Amakihi, 'Ōma'o, and 'I'iwi (pronounced ee-EE-vee) from seemingly intact forests during his six years in Hawaii. Henshaw also noted that dead birds were frequently found in the woods on the island of Hawai'i. Where once he had enjoyed the musical accompaniment of Hawaii's glorious songbirds, Henshaw now spent solitary hours without hearing the note of a single native bird.[3] Another notable ornithologist, George Munro—whose 1944 *Birds of Hawaii* is cherished for its descriptions of now-extinct birds—was convinced that introduced diseases had caused the disappearance of native birds on his adopted island of Lāna'i.[4]

Experiments in the late 1950s, which sought to illuminate the threat that avian malaria might pose to Hawaii's native birds, revealed how vulnerable the endemic birds could be. Laysan Finches that had never been exposed to the disease were transported to the island of Kaua'i and kept indoors for a month. Thirteen of these finches were then placed in an outdoor enclosure. After sixteen nights of potential exposure to biting mosquitoes, every one of the outdoor finches had died, whereas all the finches that were kept indoors, as a control, survived. Blood taken from the outdoor finches showed astonishingly high levels of *Plasmodium*.[5]

Follow-up studies in the early 1960s with three native Hawaiian honeycreeper species—'Amakihi, 'Anianiau, and 'Apapane—further revealed the extraordinary susceptibility of Hawaii's endemic birds to avian malaria. Ten days after previously uninfected honeycreepers were exposed to night-flying mosquitoes, every bird suffered from acute malaria. Weak and listless, they perched quietly with fluffed-up feathers. Periodically, the honeycreepers shivered and panted. Some seemed to be in severe discomfort. None survived.[6]

Introduced Warbling White-eyes were also exposed to the disease but showed far less susceptibility, with only one white-eye contracting malaria.[7] Interestingly, mosquitoes appeared to be more attracted to the native honeycreepers than to the nonnative white-eyes, which likely evolved alongside mosquitoes and the *Plasmodium* parasite. Five to ten times more mosquitoes were seen resting or feeding on the honeycreepers than were seen on the white-eyes. The unguarded way in which the native birds slept may have increased their vulnerability. Unlike the honeycreepers, the white-eyes typically tucked their bill and face into their back feathers, fluffed out their breast feathers, and crouched with their bellies nearly touching the perch, limiting the exposure of their legs, bill edges, and foreheads to mosquito attacks. Mosquitoes were regularly seen feeding on these exposed areas on the honeycreepers.[8]

Because mosquitoes thrive in warmer areas, lower-elevation Hawaiian birds were hit hardest by avian malaria, and the disease likely played a significant role in the disappearance of Hawaii's lowland-forest birds. Indeed, by the time additional malaria studies were conducted in the 1960s and 1970s, one-third of Hawaii's native forest-bird species were extinct and the ranges of the remaining species had contracted.[9] Since fewer mosquitoes frequented colder, higher-elevation forests, birds inhabiting these areas were less affected by avian malaria initially—unless they migrated to lower elevations, as the 'Alalā and some honeycreepers did to take advantage of seasonally fruiting and flowering plants.

Though mosquitoes were prevalent throughout the year at lower elevations, they were seasonally abundant where our five captive-reared crows foraged, played, and rested. Mosquito populations peaked in the warmer fall months of September through December between approximately 4,000 and 5,000 feet (1,200 to 1,500 m) in elevation, making this an especially risky period for our 'Alalā.[10]

Just as my morning observations came to an end, Kēhau spread his wings and headed toward the aviary. To my relief, Mālama raised his head then followed his energetic sibling. At breakfast later that morning, I voiced my concerns about Mālama's listlessness. Kurt had been at the aviary when the two males arrived. Both had fed at a food tray and were foraging in nearby trees when Kurt headed for the trailer, so he had seen nothing amiss. I tried to focus on his upbeat reassurances as we packed our backpacks and prepared to head down the mountain for a weekend off, but the disquieting feeling that all was not well with Mālama permeated my thoughts. When Peter arrived to take over crow-monitoring duties, I anxiously elaborated on my concerns, then tried to muster up some enthusiasm for my break.

———

As they so often do for the millions of bird-watchers around the globe (approximately forty-five million in the United States alone), birds provided me with a welcome distraction the following day, which alleviated my worries about Mālama.[11] For birders, one of the great joys of traveling is seeing new birds. An astonishing array of feather colors and patterns, an endless variety of forms, and a suite of unfamiliar songs, calls, and behaviors captivate, enthrall, and fuel an often-insatiable quest for the new, the never-before-seen. I was not immune to these temptations.

During my morning run, I saw Spotted Doves and Zebra Doves—introduced species from Southeast Asia—walking around in search of seeds. The Spotted Dove's black-and-white speckled mantle and the Zebra Dove's subtle-but-distinct striping made them readily identifiable, even without binoculars. After my run, a Yellow-fronted Canary—originally from Africa—captured my attention, along with a now-familiar Warbling White-eye (a green-gray bird from East Asia with a yellow throat and bold white spectacles), which was now widespread in Hawai'i and common in the 'Alalā's cloud-forest domain. Adjacent to a small beach near Kona later that morning,

I spotted a striking Java Sparrow, its massive pink bill a striking contrast to its black-and-white face, and two Saffron Finches—golden-yellow imports from South America.

Delightful as it was to see such colorful new birds, it was almost a relief to spot a Wandering Tattler preening on rocky lava that spilled into the sea. This little gray sandpiper, which still sported the barred undercarriage of its summer garb, had a legitimacy that the exotic birds I had seen lacked. It had arrived in Hawaii on its own wing power rather than being introduced by well-intended, but ecologically unenlightened human settlers who contributed to the destruction of Hawai'i's original lowland habitats and their native avifauna. As taken as I was with the Wandering Tattler's understated elegance, I was almost more enchanted by the bird's name, which spoke of faraway places, long journeys, and stories to be told. And indeed, the tattler had spent its summer breeding in Alaska or northwest Canada before winging across the Pacific to spend its winter on more clement shores.

My bird-filled day was capped when I stopped for coffee on my way back to The Peregrine Fund's rental house and spotted two male Yellow-billed Cardinals—native to South America. Each cardinal's glossy, dark-blue back and bold white front was capped with such a rich-red head that the bird seemed to be crowned in velvet. I had been especially eager to see this striking species. But as I made my way home, I couldn't help but wonder how many visitors to this illusory paradise realized that the bright and brilliant birds that surrounded them were an indication not of the biological wealth that Hawaii possessed, but of all that it had lost.

Concerns about Mālama returned with a vengeance the next morning, when Peter called and asked Kurt and I to return to the release site prematurely to help trap the still-despondent bird. Although each reintroduced 'Alalā had had an "off" day when it was less active than usual, Mālama's prolonged inactivity was seriously concerning. Peter had alerted the avian veterinarian for Hawaii's Endangered Species Propagation Facility, and he was due to arrive from Maui the next day.

We spent the afternoon devising a way to recapture our crows. After propping a roll of aviary netting above the hack box entrance and placing the birds' food trays inside the box, we rigged a pulley system to drop the net and trap Mālama inside when he entered to feed. Hiding in the observation areas flanking the hack box, we would watch through the one-way-glass peepholes and communicate via radio to simultaneously drop each side of the rolled netting. By the time we had everything ready, we had only enough daylight to check on our crows as they settled down to roost. My heart ached at the sight of Mālama's fluffed feathers and drooped head. And Kēhau's night murmurings to his subdued companion seemed all the more poignant.

If Mālama had contracted avian malaria, would we be able to trap him before the disease progressed enough to be fatal? As the *Plasmodium* parasite reproduces in a bird's oxygen-carrying red blood cells, infected cells rupture, disrupting the movement of oxygen and making the bird anemic. The ill-fated bird—with its watery blood and enlarged and discolored spleen and liver—loses its appetite and grows progressively weaker and more lethargic. Eventually, it succumbs to the disease or to starvation, predation, inclement weather, or other environmental stressors. The acute phase of the infection typically occurs six to twelve days after the parasites first appear in the bird's erythrocytes. The subsequent progression of the disease depends on the bird's ability to produce an immune response, the presence and level of particular hormones, and the parasite's ability to reproduce, given external factors such as temperature and rainfall.[12] I lay awake for hours on my cot that night, tormented by the thought that we were already too late.

At first light, we huddled in the observation boxes—Peter on one side, Kurt and me on the other. As the day brightened, the whistled song of a Northern Cardinal floated across the clearing. I had been astonished when I first glimpsed this jaunty-crested bird hopping around in a koa, Hawaii's largest native tree. There were few birds that I associated more with winter in Vermont than the brilliant-red cardinals that had graced my family's bird feeders and Christmas cards. But the species had carved out a successful existence in Mauna Loa's cloud-soaked realm, despite being more common in drier, more open lowland forests. First introduced to the Hawaiian Islands in 1929, when one escaped its cage in Honolulu and its mate was freed to accompany it, the species gained a firm footing on the Big Island after 163 cardinals were introduced to Hilo in a 1929 to 1930 Buy-a-Bird Campaign, during which school children raised money to import birds. This effort complemented more extensive introductions of colorful and melodious birds to the Hawaiian Islands, between the late 1920s and early 1960s, by affluent Honolulu society members who wanted to enliven gardens silenced by the extirpation of native landbirds.[13] Ultimately, more than fifty songbird species were brought to Hawaii from Southeast Asia, Africa, and North and South America. Almost two-thirds of those became established.[14]

Soon, we began hearing our crows' raucous calls as they awakened, stretched, and preened. In addition to the word 'Alalā meaning "the cry of a child or young animal," 'Alalā may also mean to arise with the sun (ala meaning "to rise up" and lā meaning "sun").[15] The 'Alalā greeted each new day with an exuberant medley of yells, whoops, shrieks, and growls. Minutes after we began hearing them, one of the females arrived at the aviary for an early morning meal. Mālama followed almost immediately. The two birds hesitated but then hopped into the box. As Peter whispered a quick countdown, I had a second to think how easy this had been before Peter and Kurt pulled their respective sides of the net. It crashed down, enclosing . . . nothing. The birds, infinitely quicker and warier than we had imagined they could be, had flown out of the sides before the net could be pulled flush against the box.

We reconvened and pondered our options. The birds had fled and were unlikely to return that morning. Soon after attaching rocks to the bottom of the net to make it drop faster, we heard a rumbling motor announcing the arrival of veterinarian Greg Massey.

In midafternoon, we again crouched in the confines of the observation boxes, Greg joining Peter on one side, and Kurt and me on the other. The birds usually returned to the aviary for an evening feed, and we didn't want to miss them if they arrived early. We waited. And waited. Muscles cramped. Boredom, tinged with undercurrents of anticipation, settled over us as interminable minutes coalesced into hours. Then, just before 1800 hours, Lōkahi and Hoapili landed at the hack box with a flourish, walked nonchalantly inside, and gulped down fruit. The other 'Alalā tantalized us with their calls but stayed away. Eventually, darkness pulled the curtain on our long wait and we stumbled to the trailer on stiff legs for a much-anticipated dinner. We would try again at first light.

Catching Mālama the next morning was almost anticlimactic after our arduous wait the day before. He arrived at the hack box early and entered it with little hesitation to feed. This time, the net dropped quickly and smoothly, and Peter soon had him in hand.

Greg carefully examined Mālama, then stretched out one of his wings as Peter held the wild-eyed bird. After dabbing alcohol on the large brachial vein that ran along Mālama's underwing, Greg drew a blood sample. He analyzed it in our trailer's makeshift lab, and looked at blood-smeared slides through a microscope. The results confirmed our worst fears: Mālama had contracted avian malaria. The malarial parasites were visible in his blood cells, and he showed physical effects of the disease, including a low red blood cell count.

Once Greg had finished his diagnostics, he injected the debilitated bird with chloroquine phosphate and primaquine phosphate—antimalarial agents that would destroy the rapidly reproducing parasites—giving Mālama's red blood cells a chance to recover and regenerate. Although Greg had now stopped the disease's progression, Mālama's recovery was far from certain. Reluctantly, we released our subdued patient into the screened holding facility below the hack box, so that we could monitor him and keep him hydrated in the coming days. Then we turned our attention to the other birds, since it was possible that they had contracted malaria, too.

Kurt and I spent the remainder of what began to feel like a farcical day being outsmarted by four young crows. On our first capture attempt, the birds escaped as we dropped the net. Fortunately, they soon returned. On our next attempt, Kurt started the countdown then suddenly stopped because he thought the birds were about to flee. Not hearing his count stall, I pulled my side of the pulley. The net fell crookedly, and the wily 'Alalā escaped with laughable ease. Later, one of the crows perched on the hack box, and then stealthily pushed one of the rocks weighting the net off its ledge. Without warning, the net tumbled down, making me jump. For a split second, I wondered whether the trickster crows were intentionally turning the tables on us.

Discouraged, we consulted with Peter, who had an ingenious idea that we quickly implemented. Removing the hack box's aviary-side panel, we draped netting over the opening and the adjacent deck, and placed food trays on this back side. When the birds arrived to feed, we would again drop the net in front of the box, but this time the birds would flush into the back netting and be trapped, or would have farther to fly out through the front of the box, buying us more time. Our new technique worked like a charm. Kēhau landed on the front deck, spied the food through the hack box door and, after glancing around, walked through the box and onto the back deck to feed. Before Kurt and I could react, Peter jumped out of the observation box holding our long-handled, bird-catching, fishing net, and dropped it over Kēhau.

To our relief, Kēhau was healthy and showed no indications of having contracted avian malaria. Nevertheless, Peter gently released him into the treatment facility so that Mālama would have company overnight. As I crouched in the dark observation box, waiting to trap the females, I listened with a smile to Mālama and Kēhau's soft contact calls.

While Kurt and I awaited the females, Peter went to monitor them in case they stayed away. Perhaps alarmed by the activity at the aviary earlier in the day, the trio had vanished. Pulling out his telemetry, Peter began his search. Hiwa seemed to be close by, so he headed in her direction searching the tree branches as her radio signal grew louder and louder. When the signal was booming, Peter realized he must be within feet of her, but he couldn't spot her. Perplexed, he scanned the undergrowth. And suddenly, sickeningly, he glimpsed the feathers.

Rushing forward, heart in his throat, eyes transfixed by a small pile of black feathers, Peter barely registered the images that flashed through his mind of an 'Io, a mongoose, a feral cat attacking our unsuspecting 'Alalā. What had felled our beautiful Hiwa? And then . . . waves of relief flooded him as he kneeled on the ground and found not a carcass, not a clump of feathers, but two deck feathers—the long, symmetrical, central tail feathers—and the dark cylindrical transmitter that was still attached to them. Hiwahiwa had merely shed her tail feathers and transmitter, and the absence of any other feathers suggested that she was alive and well.

As Kurt and I headed to the trailer at dusk, we met a still-shaken but hugely relieved Peter, who animatedly relayed the scare he'd had, followed by his intense satisfaction at seeing the three females murmuring to one another when they settled down to roost.

The following morning, we redoubled our efforts and trapped all three females. Given the importance of tracking our crows on a daily basis, we decided to hold Hiwa with Mālama until we received a new transmitter from project cooperator, John Marzluff. Now that she had no central tail feathers on which to attach a new transmitter, we would have to use a tarsal-mounted transmitter, which attached like a band around a bird's leg (tarsus).

Once we had Lōkahi, Hiwahiwa, and Hoapili in hand, Greg tested their blood for avian malaria. To our relief, all three females were healthy, though Lōkahi's red blood cell count suggested she had recovered from an ailment—possibly malaria. Based in part on Lōkahi's condition, Greg surmised that some healthy, well-provisioned 'Alalā might survive the disease, particularly if they were protected in an aviary or were monitored closely to forestall attacks by predators if they were in a weakened state.[16] Future reintroductions would confirm that individual 'Alalā responded differently to malarial infections, even when exposed to them under similar conditions. Some birds, such as Mālama, showed clinical signs of the infection, while others gave no indication that they were battling a potentially fatal disease.[17]

In avian malaria survivors, the acute disease phase is followed by a rapid decline in the infection's intensity to chronic levels, as the bird's immune system develops antibodies to counteract the parasite. Such chronic infections likely persist at low intensities for the rest of the bird's life, leaving the individual vulnerable to future infections, particularly if it becomes stressed or physically compromised.[18]

Before releasing Lōkahi and Kēhau, we glued rose-colored, heat-shrink tubing around their transmitter antennas to make the birds easier to identify in the field. Because of the thick vegetation, it was often difficult to secure a bird's identification from a leg band. Now Lōkahi and Kēhau's pink antennas would be readily distinguishable from Hoapili and Mālama's black ones (Hiwahiwa's tarsal mount had an antenna that was visible behind her leg), giving us additional identification clues during our behavioral scans, once the birds were back in the wild.

Surprisingly, given her subordinate status and submissive behavior with the other 'Alalā, Hoapili proved to be the fiercest of all the birds in the hand. No matter how I positioned her, she managed to clamp down on my hand with her sharp-edged bill. It was like being nipped by pliers that simultaneously delivered a paper cut. As I looked down at the red nicks that decorated my hands after releasing Lōkahi, Hoapili, and Kēhau, I could feel only pride at being marked with definitive evidence of our captive-reared crows' wildness.

———

Two evenings later, I set out to check on Lōkahi, Hoapili, and Kēhau, relieved that Mālama seemed slightly more energetic. I soon found Hoapili and Lōkahi perched close together, preening vigorously, downslope from the aviary. Suddenly, a third 'Alalā flew in and landed with a flourish near the two females, which immediately froze. Assuming the new arrival was Kēhau, I stared through my binoculars, searching for his transmitter antenna. But I quickly realized that the interloper wore only a silver US Fish and Wildlife Service band. It was the wild Kalāhiki female, which I had rarely seen, even though her territory encompassed the aviary.

Banded as a fledgling in 1977, the sixteen-year-old Kalāhiki bird looked sleek and powerful—an imposing matriarch in her cloud-forest domain. With deft, purposeful movements, she hopped quickly from branch to branch, moving closer and closer to her motionless offspring. Before she could reach the wary youngsters, the unbanded Kalāhiki male burst out of the surrounding greenery, catapulting toward Hoapili and Lōkahi with a speed and trajectory that communicated his aggressive intent. Launching into panicked flight, the young crows fled to the safety of the aviary, the Kalāhiki male in close pursuit.

Heart in my throat, I stumbled after them through the thick undergrowth. Upon reaching the aviary, I spotted the Kalāhiki female perched in a nearby koa, where she was soon joined by her puissant mate. Seconds later, the two birds flew off, disappearing back into the verdant growth from which they had so suddenly emerged. Still breathing hard, I quickly located Hoapili and Lōkahi, then spied Kēhau perched nearby. Their scattered locations the only indication of the fright they had experienced, the youngsters soon settled down to roost in separate trees. Adrenaline still coursing through me I crouched close by, until darkness obscured their feathered forms, then I headed to the trailer, awed. I knew that others had seen the Kalāhiki pair behave antagonistically toward our captive-reared juveniles, which the adults likely viewed as interlopers in their territory, but this was the first interaction I had witnessed. As capable as I had thought our young crows were, the wild adults had exhibited a markedly different level of flying, of being. Powerful, confident, skilled, they were the indisputable reigning monarchs of this forested realm.

On our own again the next day, Kurt and I took over hydration duties. Before his departure, Greg had determined that Mālama's packed cell volume—the percentage of red blood cells in his blood—was slightly high, indicating dehydration. He had shown us how to provide fluids to the debilitated bird before leaving the field site. Entering the treatment facility, we caught Mālama with our large fishing net and, after extricating him, stepped out of the enclosed room to minimize stress to Hiwahiwa.

I held Mālama gently, wings folded against his body, as Kurt opened Mālama's bill and inserted a thin tube into the back of his mouth and down into his throat. We had to make sure the hydration fluids went into his esophagus and down into his stomach, rather than into his trachea and on into his lungs. If we'd just delivered liquids to the back of his throat, the fluids might inadvertently go down the wrong tube and flood his lungs. After inserting the flexible hydration tube into Mālama's esophagus, Kurt attached a syringe and slowly injected several doses of lactated Ringer's solution to replenish our patient's fluids and electrolytes.

Mālama was subdued while being hydrated, and I wondered anxiously whether his lethargy was a symptom of his illness or a fear response to being handled. Several hours later, we repeated the procedure. This time, Kurt held the patient while I inserted the tube and another dose of fluids. And this time, we encountered a different bird. Struggling and feisty, Mālama had clearly regained some energy. Emitting heart-wrenching squawks, he voiced his displeasure with the proceedings. It was a relief to return him to his enclosure and to hear Hiwahiwa's soft greetings and comfort noises as the birds huddled together on their perch.

Still disturbed by the memory of Mālama's plaintive cries, I welcomed the distraction of a staggering sunset—a rare treat given the usual evening fog—when I went to look for the free-flying crows before they roosted. Glowing pink cumulus clouds floated on a golden layer of light, which rested on a band of aquamarine that blended into a deeper, oceanic blue. While I gloried in the celestial canvas, I was confronted, as I walked, with the less glorious reality of disturbed vegetation and muddy ground—signs that the feral pigs that now called this forest home had fed and wallowed in the area. In doing so, they had created perfect habitat for the mosquitoes that posed such a dire threat to Hawaii's forest birds.

Small pigs, which remained close to human habitations, were introduced to the Hawaiian Islands by Polynesian colonists. These domestic pigs were eventually supplanted by or hybridized with larger wild pigs (*Sus scrofa*) that were introduced to the islands by Captain Cook in 1778 and, subsequently, by other Europeans.[19] These animals ranged widely and reproduced prolifically. Today, in addition to consuming native plants, depredating the eggs and young of ground-nesting birds, and destroying vegetation, Hawaii's innumerable feral pigs create water-collecting depressions by rooting for tubers and introduced earthworms and wallowing in mud or wet areas. The pigs also uproot and chew holes in the starchy stems of tree ferns, creating troughlike cavities that fill with leaf litter and standing water. These areas are ideal habitat for gravid female mosquitoes, which oviposit their eggs in water that then harbors the emerging larvae.[20]

Goats were also introduced to Hawaii in 1778, followed in 1793 by sheep and cattle.[21] All of these ungulates destroyed native habitats, eradicated important wildlife food, and facilitated invasions by nonnative plants. But the biggest threat now posed to native birds comes from the water sources provided to these animals and the ongoing activities of feral pigs. The resulting increase in watery habitat for larval mosquitoes in forests where surface water was historically scarce has led to more mosquitoes, accelerating the prevalence and transmission of avian malaria in low- and mid-elevation forests (3,000 to 5,000 feet [about 900 to 1,500 m] above sea level).[22]

Climate change is exacerbating the threat malaria poses to native birds, since warming temperatures facilitate the upslope spread of mosquitoes and increase the duration and intensity of seasonal disease transmission.[23] Many of Hawaii's famed honeycreepers—renowned, like Darwin's Galápagos finches, for having diverged from a long-ago island colonizer into a panoply of extravagant forms—survive only at higher elevations because of habitat loss and the year-round presence of avian malaria at lower elevations. Although mosquitoes occur in cooler, high-elevation forests, the insects have long been rare, seasonal, and patchily distributed in these montane areas. Moreover, malaria transmission is less effective at higher elevations because cooler temperatures inhibit the development of the *Plasmodium* parasites that mosquitoes transmit to birds. The parasites' ability to reproduce (through a process known as sporogony) within their mosquito hosts slows below 59°F (15°C) and ceases altogether at 55°F (13°C). Avian malaria is therefore less prevalent above 5,000 feet (1,500 m), where these temperature and parasite-development thresholds occur. As a result, the biodiverse forests blanketing the upper slopes of Hawaii's volcanoes currently provide the last refugia for many native birds, including ten endangered species that are highly susceptible to this devastating disease.[24]

Climate models predict that a 3.6°F (2°C) increase in regional temperatures could increase mosquito prevalence and raise the upper elevational limit of malaria transmission from approximately 5,000 feet to well over 6,000 feet (about 1,500 to 1,800 m) within the next 100 years, encompassing virtually all the remaining forest-bird habitat in Hawaii. Mid-elevation birds such as Hawaiian Crows would be subject to malaria year-round instead of seasonally. High-elevation birds such as the endangered honeycreepers would be exposed to the disease seasonally rather than occasionally.[25] Whether highly susceptible native species can become resistant (meaning their immune systems clear the malarial infection before it ever reaches the acute phase) or tolerant (meaning that they survive the acute phase and reproduce successfully, but retain chronic levels of infection) in this evolutionary blink-of-an-eye will likely determine whether these unique and distinctive birds avoid extinction.[26]

As alarming as climate-change scenarios are for Hawaii's endemic birds, some researchers believe current models don't reflect an already disturbing reality. Malaria transmission rates and numbers of mosquito larvae have already increased significantly at all elevations since the 1990s. Malaria prevalence in native forest birds at ecologist Leonard Freed's high-elevation study site (6,460 feet; 1,969 m) on Mauna Kea—once considered a refugium from avian malaria—more than doubled between the early 1990s and 2002. The minimum temperature required for *Plasmodium* parasites to develop was predicted (in the early 2000s) to occur in this area in eighty years, but it had already been exceeded in 2001 during a major malarial outbreak (or epizootic) among forest birds.[27]

The tropical inversion layer that often forms on Hawaii's highest peaks may have caused the higher-than-predicted temperatures at Freed's research site. Visible as a thin layer of clouds, this band forms when warm, moist, rising air meets the cool, dry air of high elevations. Climate models don't always incorporate complexities such as this inversion layer, which creates warmer temperatures at higher elevations and has occurred more often in the past twenty-five years.[28]

Despite troubling climate trends and high mortality rates of native honeycreepers infected with avian malaria (approximately 65 percent of high-elevation 'Amakihi and 'Apapane, and 90 percent of 'I'iwi, die after a single infective mosquito bite), there are glimmers of hope for several of the more abundant and widespread species.[29] In particular, the yellow-green Hawai'i 'Amakihi appears to be recolonizing lowland areas from which it had largely been eliminated by avian malaria. Although these lowland 'Amakihi have high rates of avian malaria, most of the birds have low-grade, chronic infections, suggesting that they've developed some tolerance to the disease.[30]

Some higher-elevation 'Apapane and a few 'I'iwi have survived avian malaria, so perhaps these crimson Hawaiian endemics, both of which are more migratory in their habits than the 'Amakihi, will eventually develop some immunity to the disease and expand into areas from which they have been eliminated.[31] Unfortunately, though, these two species have less genetic variability, which facilitates adaptation and survival in the face of changing environmental conditions, so these birds are less able to adapt to virulent exotic diseases than are 'Amakihi. Nevertheless, higher elevation 'Apapane and 'I'iwi seem to have modified their *behavior* in response to avian malaria. Whereas the birds used to gradually move downslope in the fall to follow the flowering of nectar-producing trees, they now make daily circuits, flying to lower elevations to feed in the morning then returning each night to roost in the cooler, higher elevations where mosquitoes are scarce.[32] Climate change threatens to undermine this recently evolved disease-avoidance strategy. Moreover, despite the hope offered by the malaria tolerance of some of the more common honeycreepers, this adaptation comes at a cost. These chronically infected birds subsequently become reservoirs or carriers of the disease, thereby facilitating its transmission to imperiled birds that lack the genetic variability or population numbers to adapt to this dire threat.[33]

And avian malaria is not the only introduced disease to have exerted a viciously reductive force on Hawaii's native birds. Avian poxvirus—another introduced disease with a worldwide distribution, that has decimated Hawaii's endemic avifauna—further complicates the challenges posed by malaria. Unlike malaria, avian pox is a viral infection that comes in two forms: one that causes warty growths (dry lesions) on a bird's featherless areas and one that causes internal (wet) lesions in the mouth, trachea, and esophagus of infected birds.[34] The clinical term *lesions* hardly conveys the gruesome swellings that can ravage and disfigure a bird's head and feet.[35] The sometimes-fatal external lesions can impair a bird's vision, breathing, feeding, or perching, whereas the internal lesions can limit a bird's ability to breathe or swallow. Eye lesions can lead to blindness, and foot lesions sometimes cause the loss of toes.

Although mosquitoes are a major vector for avian pox, other biting insects such as mites and flies can also transmit the infection. The virus is transported from one hapless victim to another on the mouth parts (proboscis) of blood-sucking insects and is sufficiently virulent that a mosquito can infect susceptible birds for as long as two months after siphoning blood from an infected victim.[36] In addition, the virus can also be transmitted when a bird contacts infected objects or ingests food or water contaminated by dead or dying birds. Although the virus cannot penetrate unbroken skin, small abrasions enable its passage.[37]

The timing of avian pox's arrival in the Hawaiian Islands is unknown, but the disease was widespread by the early 1890s. Some believe it was responsible for the first wave of avian extinctions in Hawaii in the late nineteenth century, while avian malaria caused the second extinction wave in the early- and mid-1900s.[38] Nevertheless, it is difficult to separate the impacts of the two diseases since many birds with pox lesions also test positive for avian malaria. This frequent co-occurrence might come about because mosquitoes can transmit the two diseases simultaneously, because birds sickened by both diseases are more often found by biologists, or because older birds are more liable to have been exposed to both diseases. Alternatively, avian pox may suppress a bird's immune system, making a malaria recurrence more likely.[39] Regardless of why the two diseases often occur together, pox lesions are substantially less prevalent than avian malaria, and they appear to cause fewer fatalities (though some virus strains are more virulent than others).[40] A 1990s study found that about 40 percent of O'ahu 'Elepaio with active lesions died, but birds with milder infections consisting of one or two swollen toes frequently recovered. Their warty lumps eventually shrank or were sloughed off, and the birds healed.[41] Like avian malaria, though, avian pox is seen more often in native birds than in introduced species, and it inflicts more harm on Hawaii's vulnerable endemics than it does on the state's relative newcomers.[42]

Although avian poxvirus has been harming Hawaii's birds for well over a century, it wasn't officially recorded in 'Alalā until the 1980s, when two wild fledglings died of pox infections and five others had probable pox lesions. During this period, four of ten wild 'Alalā nestlings and fledglings died of problems related to exotic diseases.[43]

Fortunately, Hoapili did not succumb to avian pox when she contracted the disease several months after I left Hawaii. Peter Harrity managed to grab her after discovering her on the ground under a small coop near the release aviary, which contained several chickens (sentinel birds) whose blood was routinely tested for the presence of avian malaria. By the time veterinarian Greg Massey arrived to treat her, Hoapili had rubbed off the lesion that Peter had noticed near her beak. After being treated, Hoapili was freed and showed no further ill effects from the disease.[44]

————

Given the severity of the avian malaria and poxvirus threats, reducing mosquito numbers is critical to conserving Hawaii's forest birds. The task seems impossible without the broadcast spraying of mosquito-killing pesticides, which would endanger Hawaii's already-threatened invertebrate life. But other means of reducing mosquitoes, while not a panacea, offer hope. An obvious way to lower mosquito numbers is to eliminate the small pools in which the unwelcome insects breed. Computer modeling of avian malaria in Hawaii suggests that such a strategy could provide effective mosquito control, but more than 80 percent of the larval habitat in forested areas would need to be eliminated to effect a significant reduction in disease transmission and an increase in native bird populations.[45]

Removing cattle tanks and other objects that hold water is a critical first step, but this action is more easily achieved on public lands or natural areas than on private landholdings. Feral pig exclusion (with fencing) or removal (by shooting and trapping) is also essential, but this conservation measure is controversial since the exotic swine are ardently championed by game hunters. Larvicides that kill immature mosquitoes may be effective in certain areas and are unlikely to threaten native fauna. Despite the daunting task of trying to reduce mosquitoes by eliminating their breeding habitat, this strategy might work in areas such as the leeward side of Mauna Loa volcano—the last stronghold of the 'Alalā—where volcanic substrates are highly porous and don't naturally collect water, no well-defined stream drainages occur, and virtually all the mosquito larval habitat is associated with feral animals or ranching infrastructure.[46] But conservationists are not optimistic.[47]

Future measures to protect Hawaii's imperiled birds from introduced diseases might also include vaccines. Although vaccinating wild birds is impractical, vaccines might one day protect captive and reintroduced birds from avian pox and malaria. A more promising strategy involves the use of biocontrols, such as the *Wolbachia* bacteria, which can infect adult mosquitoes, reduce their lifespan, disrupt their reproduction, and interfere with the pathogens they carry.[48] Seeding forests with sterile male mosquitoes also could help reduce mosquito prevalence and disease transmission.[49] Perhaps the most promising solution, though, lies in a revolutionary but controversial technology that can transform mosquitoes by deleting, rearranging, or changing components of their DNA. Such gene editing could allow scientists to create mosquitoes that produce sterile offspring or are resistant to the malaria parasite. Alternatively, gene editing might confer malaria resistance to vulnerable birds such as the 'I'iwi.[50] Successfully used in 2015 to genetically modify mosquitoes so that they could no longer transmit the parasites that cause human malaria, this ground-breaking technology could transform people's lives and might one day, in combination with other conservation measures, change the calculus for Hawaii's beleaguered birds.[51]

All of these conservation measures to restore the health of Hawaii's forest ecosystems require significant funding and demand political will. And whether any of these options for reversing the extinction tide that has ensnared Hawaii's birds can sufficiently offset the threats posed by climate change will determine whether future generations experience the Hawaii in which I stood: a verdant cloud forest filled with native birds and ringing with the commanding, astonishingly varied, singularly captivating voice of the 'Alalā.

————

Thoughts of recovering ecosystems under siege were far from my mind as we finally prepared to return our malaria patient to the wild. Mālama's demeanor had improved dramatically. Between hydration bouts, he now hopped from perch to perch, pecked at the aviary netting, and tugged at a fern growing out of a log on the floor of his enclosure. Gone was the listlessness that had precipitated his capture and treatment. Hiwahiwa had been outfitted with her new tarsal-mounted transmitter but was still keeping Mālama company.

Although it had only been a week since we had captured Mālama to treat him for malaria, the crows' time in captivity had felt endless, and I was impatient to give the birds their freedom. On release day, I moved through our usual routines with barely suppressed excitement. The weather mirrored my exuberant mood, with the sun holding the usual clouds and mist at bay. Greg had asked us to hydrate Mālama twice more before freeing him, so we planned an afternoon release.

After tubing Mālama for the last time, I held him to make sure he had taken his fluids well. Meanwhile, Kurt caught Hiwahiwa so we could release the two captives simultaneously. Once we had the birds in hand, we stepped away from the aviary and then opened up our hands, lifting them slightly to give the birds a gentle boost into flight. Mālama flew clumsily to nearby 'ōhi'a trees and landed shakily amidst the foliage. Hiwa's flight was low but strong. After briefly tracing the grassy track that led upslope from the aviary, she landed on a downed tree trunk that crossed the path. Suddenly, Kēhau appeared beside Mālama with a dramatic flair of his wings. A chorus of burbled mewing filled the air. Lowering his head slightly, Kēhau partially opened his bill and repeatedly shifted his closed wings in partial imitation of the submissive food-begging posture used by nestling birds to solicit food. Mālama responded by putting his bill into Kēhau's and the two birds grasped bills in a mutual mouthing behavior known as allobilling, which is often seen in ravens and, less commonly, in American Crows.[52]

Before Kēhau and Mālama's allobilling could deteriorate into jabbing, as it sometimes did, the males' sweet reunion was overshadowed by raucous cries that emanated from the forest near where I had last seen Hiwahiwa. In total contrast to the quiet murmurings of the males, the female 'Alalā seemed to be voicing their pleasure with boisterous ebullience. Their yells and whoops echoed through the afternoon stillness, and they made me smile as much as the male's more subtle greetings.

No allowances were made for the newly freed birds by their companions that evening. All five 'Alalā were almost constantly on the move, busily foraging for berries and insects, and flying from perch to perch. When the adult crows' throaty cries drowned out our youngsters' comparatively vulnerable-sounding calls, the juveniles high-tailed it to the aviary. Mālama and Hiwahiwa kept up with their companions, giving us no hint of their recent incarceration. As nightfall silenced the crows' murmurations, I headed back to the trailer in a quiet that was broken only by my own footfalls, my head filled with the day's unforgettable reunions, my heart full of hope for Mālama's future.

CHAPTER TWELVE

Forest Intruders

An abrupt shout pierced the enveloping gloom of a midmorning mist that insinuated itself around dripping leaves, crept over branches, and spirited away the tenuous warmth bestowed by early sunshine. Startled out of my distracted trudge up the hill to the 'Alalā aviary, I paused and looked toward the US Fish and Wildlife Service camp, where the crew leader, Mike, was gesturing to me as he leaned over what looked like a black garbage bag. Reluctantly, I headed over to him, smiling in greeting but dreading what he was likely to show me I knew the Service personnel were responsible for trapping and "removing" nonnative mongooses and rats from the forest comprising the 'Alalā reintroduction site. But while the practical biologist in me understood the importance of ridding the crows' forest home of its alien invaders, I struggled with the reality of killing individual animals, even though the unsavory task was essential to conserving Hawaii's endemic birds.

Although Hawaii—with its black-sand beaches, lush forests, tropical flowers, and glowing lava flows—felt like a foreign land to me, its inclusion in the United States made it feel closer to my world geographically than it actually was. Lying over 2,000 miles (3,200 km) from the nearest continent, Hawaii's splendid isolation led to the evolution of thousands of unique life forms. Home to more than 10,000 native species, 90 percent of which are found nowhere else on Earth, Hawaii's remoteness prevented any terrestrial mammals—with the exception of the volant Hawaiian hoary bat—from naturally colonizing the islands.[1] As a result, over time, the birds that colonized Hawaii not only evolved their often-fantastic colors and forms, but also lost their natural defenses to mammalian predators. Baby 'Alalā, for example, left the nest at about six weeks of age, and they often perched on or near the ground, begging loudly for food from their parents, for as long as two weeks before they were able to fly.[2] This delayed-development strategy was eminently successful in a world with no terrestrial predators, but ever since sharp-toothed mammals first came to Hawaii alongside human colonizers, these animals have posed a serious threat to vulnerable young crows.

Flightless Hawaiian Crows are not alone in their vulnerability to introduced organisms. Worldwide, such invasive species—organisms that have been moved from their native ranges to new areas where their introduction causes environmental or economic harm—have played a leading role in causing environmental change and a loss of biodiversity. In the United States, invasive plants, animals, and microbes are believed to be responsible for the decline of over 40 percent of the species now listed as endangered or threatened.[3] And nowhere are endangered species more prevalent than in Hawaii, which has been ignominiously dubbed the "extinction capital of the world" for having more of its endemic flora and fauna go extinct in historical times than any comparable region. Although Hawaii represents only 0.02 percent of the United States land area, the state has sustained nearly three-quarters of the nation's documented plant and animal extinctions.[4] Since humans first arrived in Hawaii, 71 of 113 endemic bird species have gone extinct (48 following Polynesian colonization and 23 following the arrival of Europeans). Of the remaining species, three-quarters are listed as endangered.[5] Given the outsized role invasive species have played in the extinction and endangerment of native flora and fauna, Hawaii's legislature declared in 2015 that invasive species are "the single greatest threat to [the state's] economy and natural environment and to the health and lifestyle of [its] people."[6]

The vulnerability of Hawaii's endemic wildlife to introduced organisms is characteristic of island ecosystems the world over. Evolving in isolation, island species are often poorly equipped to cope with the influx of predators and competitors that have accompanied humanity's erosion of geographic barriers through travel and commerce. Indeed, 93 percent of the 176 documented bird extinctions worldwide since the 1600s occurred on islands.[7] The introduction of predatory animals is second only to habitat loss in causing island bird extinctions and population declines.[8] In addition to preying on island birds, nonnative predators compete with them for scarce resources, destroy native habitat, and subsidize other predators. For example, numbers of feral cats and native Hawaiian Hawks likely have increased because of introduced rats.[9]

Worldwide, rats have ravaged island ecosystems, causing more bird extinctions than any other organism.[10] The first predatory hitchhiker to the Hawaiian Islands was the relatively small, brown-coated, white-bellied Polynesian rat (*Rattus exulans*) that accompanied early human settlers on journeys around the Pacific. Now the third most widely distributed rat in the world, the Polynesian rat, which has been in Hawaii for perhaps 1,200 years, lives in a wide variety of habitats, from the lowlands to timberline. Having few natural predators or competitors when it arrived in Hawaii, this pernicious omnivore multiplied exponentially, and its consumption of native vegetation likely played an instrumental role in destroying low-elevation Hawaiian forests and their avifauna.[11] Preying on ground-nesting seabirds and forest birds, as well as on seeds and vegetation, the Polynesian rat—a skilled climber—is now rarely found in trees, possibly because of competition with a more recent arrival to the Hawaiian Islands: the black rat.[12]

Also aptly known as the ship rat (as well as the house rat or roof rat), the black rat (*Rattus rattus*) originated in Asia but came to Hawaii along with European explorers. First recorded in Hawaii in 1899, the prolific invader may have arrived as early as 1870. Now the most common rat in Hawaii, this predatory scourge has devastated native birds because of its superlative climbing skills and arboreal lifestyle.[13] Forest-dwelling black rats often nest in tree cavities and in the tops of hāpu'u tree ferns, where they threaten roosting and nesting birds. The rats also pose a severe threat to ground-nesting seabirds. The invasion and proliferation of black rats on oceanic islands has led to widespread declines and extinctions of island birds and other animals.[14] Indeed, this opportunistic rat, considered the world's most widespread invasive animal, has been implicated in the decline or extinction of approximately sixty native vertebrate species worldwide.[15]

In Hawaii, black rats contributed to the extinction of the flightless Laysan Rail in the first half of the 1900s, and (along with exotic diseases) to the probable extinction of the Kaua'i 'Ō'ō—a dark-bodied, yellow-thighed honeycreeper endemic to Kaua'i, whose sweet, haunting whistles were last heard in 1987.[16] Black rats may also have played a role in the probable extinction in the late 1980s of the 'Ō'ū, a hook-billed finch with a chunky olive body and sharply contrasting yellow head, whose resonant call was once heard on every major Hawaiian island.[17]

Two other rodents introduced to Hawaii in the late 1700s to early 1800s—the house mouse (*Mus musculus*) and the Norway rat (*Rattus norvegicus*)—also successfully colonized the islands, but they pose a lesser threat to Hawaii's forest birds than does the black rat. Nevertheless, in addition to preying on native invertebrates, plants, and nestling birds and eggs, these rodents may harm endemic birds by competing for limited food resources and bolstering populations of predators such as the small Indian mongoose.

Considered one of the world's most destructive invaders, the small Indian mongoose was intentionally introduced to Hawaii in 1883 to control the teeming rats that had infested sugarcane fields. A native of southern Asia and the Middle East, the mongoose is an opportunistic carnivore whose ferocious predatory habits (it can kill venomous snakes) led to its introduction to many of the world's sugarcane-growing areas, including sixty-four islands and northern South America.[18] Looking like a cross between a fluffy-tailed squirrel and a slinky-bodied weasel, the mongoose's lithe, low-slung form is covered in a gold-flecked coat that earned the animal its former species name (*Herpestes auropunctatus*—meaning "the golden-spotted one who creeps").[19] More closely related to cats than to dogs, the mongoose's russet eyes and pinkish nose give the animal a feral, malevolent look that is wholly lacking in the unrelated but superficially similar weasels of North America and Eurasia. Or perhaps the mongoose just appeared more sinister to me, since I quickly became aware of its lurking threat to Hawaii's birds.

Following their introduction to Hawaii, mongooses undoubtedly ate sugarcane-invading rats when they could catch them. But this ground-dwelling predator is active during the day, and its ratty prey is usually nocturnal and often arboreal, so the ill-conceived introduction scheme didn't exact a meaningful toll on the agricultural pests. Conversely, growing numbers of mongooses depleted native bird populations. Spreading into wetlands, upland habitats, and even forests, the adaptable mongoose voraciously consumed the eggs and nestlings of ground-nesting birds. These invaders will even kill adult birds, given the opportunity. Mongooses have preyed on at least eight endangered bird species in Hawaii, including the Hawaiian Goose (or Nēnē) and forest birds such as the 'Alalā, all of which lack defenses to mammalian predators.[20]

At least one still-flightless 'Alalā fledgling was killed by a mongoose in the early 1980s, while a second was killed by either a mongoose or a feral cat. Such predation events were probably not uncommon after the introduction of these predators, but documenting them is rare. As Peter Harrity watched one of his 'Alalā charges hopping along the ground looking for insects in the mid-1990s, a mongoose suddenly leaped out from some obscuring foliage and latched on to the bird's head. Peter lunged toward the predator, and the struggling 'Alalā wrested itself free of the mongoose's toothy grasp and flew to a safe perch.[21]

I rarely saw mongooses in the 'Alalā's forest, but I knew these furtive creatures were an ever-present threat. That had been abundantly clear when, on several of my visits to the Service's camp, I'd seen the wily creatures in the agency's humane, wire-mesh traps. While Kurt and I monitored the newly released 'Alalās, Service personnel monitored the wild adults and worked to improve the crows' habitat by removing non-native predators. Mike had shown me one of the victims of these trapping efforts on one of my first days in Hawaii. I had been saddened to see the mongoose's snaky movements, its barely contained energy, its bluster and fear as we looked at the trap into which the ill-fated animal had been baited. My instinct was to help anything in distress. But helping these hapless invaders whose forebearers had been introduced to lands where they did not belong and on which they perpetrated endless harm meant condemning 'Alalā and other birds to certain extinction.

Even though it would be years before I fully understood the terrible toll exotic species inflicted on native animals the world over, I still felt a reluctant gratitude for the work the Service personnel did to protect the crows in ways that I couldn't. Nevertheless, I thought, as I slowly walked over to where Mike was peeling away a garbage bag that had temporarily covered one of his wire-mesh traps, I needed to tell him how difficult it was for me to see these doomed creatures.

"Look what we caught this morning," Mike said, interrupting my runaway thoughts. I'd hoped to avoid looking into the trap, but his words drew my gaze to it. I drew in a sharp breath as my eyes met the unblinking eyes of a tabby cat. A wild, feral tabby cat.

I felt sucker punched—as though the wind had been knocked out of me. I *loved* cats. I had always had them as pets and considered myself a bona fide "cat person." My own beloved cat, Roo, a long-haired gray tabby with a leonine nose and green eyes that I had found abandoned as a kitten in a parking lot in Philadelphia and smuggled into my college dorm room, awaited me at home.

I swallowed uncomfortably, looked wide-eyed at Mike, and said stupidly, "It's a cat."

"Yeah," he responded, grinning at my perspicacity. "It's quite a catch. Cats can be difficult to trap and they're deadly on the birds."

"Are you . . ." I hesitated, ". . . going to kill it?"

"We have to," Mike said gently, seeming for the first time to recognize my dismay. "These feral cats are completely wild. They're not adoptable. We euthanize them. It's humane," he said, trying to reassure me.

"Hmmm," I murmured noncommittally, struggling to accept what I instinctively wanted to fight against. And then I gushed, "I love cats. I miss my cat. This is really hard to see."

"I know," Mike agreed. "It's really hard to do. But Hawaii's birds aren't going to survive if we don't."

"I know," I agreed numbly. "I know," I said again, as though trying to convince myself.

Soon after, I walked back to camp, footsteps heavy, my head full of visions of feline eyes that seemed to stare into my soul with reproachful intensity. I hated the unfairness of an individual animal paying the ultimate price for what we humans were responsible for. And yet, I didn't ever want to stand in a Hawaiian cloud forest and listen in vain for the clamorous calls of an 'Alalā. I didn't want the flame-red 'I'iwi to be something that existed only in photographs. I didn't want to never again be stopped in my tracks by the unbearable cuteness of a tiny 'Elepaio.

I pined for Roo more than usual that night as I lay in my cot listening to the rain. But competing visions of the 'Alalā playing follow-the-leader along a moss-covered log, chasing stick-carrying siblings through the treetops, and murmuring softly to each other when roosting counteracted my heartache. A world devoid of these charismatic corvids, the brilliant honeycreepers, and other Hawaiian birds would be impoverished beyond measure, would diminish us all.

Over the ensuing years, my love for cats never faltered, but the more I learned about the devastating impacts of free-ranging cats on native wildlife, the more convinced I became that ridding islands of these introduced predators was imperative, as was controlling the wantonly destructive behavior of our own cherished pets. Outdoor cats in the United States kill between 1.3 and 4 *billion* birds per year (with unowned or feral cats responsible for almost 70 percent of these deaths).[22] The extraordinary predatory prowess of our favorite felines makes cat predation the greatest human-caused mortality threat to our native birdlife. More birds are killed by cats than die from vehicle strikes, pesticides and poisons, and collisions with windows, skyscrapers, communication towers, electrical lines, and wind turbines combined.[23]

The damage inflicted by free-ranging cats has been particularly dramatic on islands. Introduced to about 10,000 islands (nearly 5 percent of the

Earth's total number), cats have negatively impacted at least 123 threatened bird species or subspecies, causing or contributing to the extinction of 22 of them.[24] As prolific as they are predatory, introduced cats have overwhelmed insular ecosystems, taking a heavy toll on naive native wildlife. In the Kerguelen Islands (also known as the Desolation Islands) in the southern Indian Ocean, the introduction of a cat and her three kittens led to a population of 3,500 cats that killed 1.2 million birds per year.[25] An undetermined number of cats that were introduced to Herekopare Island in New Zealand extirpated six landbird species and large colonies of seabirds over several decades, leaving only a few thousand of the nearly half a million birds that originally inhabited the island.[26] The toll inflicted on island birds by introduced felines has been eclipsed only by alien rats.[27]

Cats likely were introduced to Hawaii by the earliest European visitors and were considered common in some of its forests in the late 1800s.[28] Today, free-ranging feral and domesticated cats are widespread and abundant in Hawaii, and their ongoing impacts are severe. In 2015, cats killed at least 237 endangered birds, including Hawaiian Stilts and Hawaiian Common Gallinules, at a wildlife refuge. Cats have also killed endangered forest birds such as the Palila—a finch-billed honeycreeper dressed in dandelion yellow, white, and gray, with hints of olive green—and regularly kill endangered seabirds such as the Newell's Shearwater and Hawaiian Petrel.[29] Alarmingly, cats have a more negative impact on insular birds (particularly on endemics) when alternative prey is available. Abundant introduced rats and mice allow feral cats to breed and proliferate, exacerbating the cats' impacts on vulnerable island birds.[30]

Aside from the obvious threat that feline paws, claws, and teeth pose to birds, cats also pose a more insidious danger to Hawaii's native wildlife. Along with carrying and transmitting diseases such as plague and rabies, outdoor (and, particularly, feral) cats can be carriers of a minute parasite whose astonishing ability to transform its intermediate host's behavior seems more akin to science fiction than to biology.

Toxoplasma gondii is a single-celled parasite that reproduces in feline intestines. During its development, it is excreted (in the form of oocysts—single, egglike cells) in cat feces. When the parasite is inadvertently consumed by an intermediate host—a rat, mouse, bird, or even a person—it multiplies rapidly, infecting and destroying healthy cells and tissues. Eventually, the parasite settles in the form of cysts in muscle and nerve tissues, especially in the limbic portion of the brain, which controls mood, instinct, defensive behavior, and sexual attraction. There, *Toxoplasma gondii* disrupts the brain's neural circuitry in still–poorly understood ways, altering the behavior of the host. Once infected, rats that instinctively fear and avoid cat urine begin responding to it as though it were an aphrodisiac. Losing their innate fear of cats, the dysfunctional rats become easy prey, and the ingenious *Toxoplasma* parasite is ingested by a cat, beginning its reproductive cycle all over again.[31]

Worldwide, *Toxoplasma gondii* infects animals ranging from birds to whales. Toxoplasmosis—the disease caused by this parasite—is one of the most common human parasitic infections, occurring in up to a third of the global population, including more than forty million people in the United States.[32] Runoff after rains can flush the extraordinarily persistent and pervasive *Toxoplasma* oocysts into freshwater and marine environments, contaminating reservoirs used for drinking water and irrigation water used for agricultural crops. People become infected with the parasite by eating contaminated vegetables or undercooked meat from domesticated animals that ingested tainted food or water. Children may inadvertently ingest oocysts when playing in sandboxes and other areas where cats defecate. Gardeners may encounter the omnipresent oocysts when weeding areas visited by outdoor felines. And pregnant women who clean their kitty's litter box may inadvertently transmit the infection to their fetus, which can lead to fetal abnormalities and miscarriages.

People infected with toxoplasmosis may experience fever, headaches, fatigue, and, less commonly, death if they have a compromised immune system. Healthy immune systems typically neutralize toxoplasmosis, and the disease enters what was once considered to be a latent phase, with cysts surviving in muscle and neural tissues but having little measurable impact on the human host. This optimistic conclusion is increasingly being challenged by research that is as compelling as it is disturbing. *Toxoplasma* cysts that settle in the eyes can inflame retinas, leading to glaucoma and, eventually, blindness. The parasite can also cause changes in human behavior that are similar to those seen in rodents: reducing anxiety and fearfulness (prompting risk-taking), and promoting an attraction to cat urine. Worse, a substantial body of evidence now suggests that many of those diagnosed with latent toxoplasmosis show symptoms associated with a wide range of mental illnesses, including depression, obsessive-compulsive disorder, bipolar disorder, and schizophrenia.[33] Studies in Europe found that women infected with toxoplasmosis were twice as likely to commit suicide as those who weren't.[34]

As is often the case with prevalent parasites, wildlife has also been harmed by *Toxoplasma gondii*. In Hawaii, where feral cats are particularly abundant, and well-intentioned but misguided caretakers support innumerable outdoor-cat colonies, the parasite has been flushed into nearshore marine waters, killing at least fourteen critically endangered Hawaiian monk seals in the last two decades.[35] On land, toxoplasmosis is now the most prevalent infectious disease found in the endangered Nēnē (Hawaiian Goose), an elegant bird with furrowed neck feathers and reduced foot webbing that allows it to walk on rough lava flows. In addition to succumbing to the parasitic infection, Nēnē—and people—with toxoplasmosis are more prone to various types of trauma, and so may be killed indirectly.[36] And, as the feral tabby that balefully eyed me from a trap unwittingly hinted, toxoplasmosis has invaded even Hawaii's forests, killing or contributing to the deaths of at least three and possibly four of the twenty-seven critically endangered 'Alalā that were reintroduced to the wild in the 1990s, and sickening a fifth that was recaptured and successfully treated.[37]

Unaware of the toxoplasmosis threat and the damage outdoor cats inflicted on wildlife worldwide, I continued to let Roo roam outside, as I had always done, when we were finally reunited. I was saddened by the dead rodents he brought me, but I convinced myself, as so many pet owners do, that he wasn't damaging rodent populations. I didn't fully appreciate that my cat was just one of ninety million pet cats and sixty to one hundred million unowned cats in the United States. Nor had scientists yet determined that outdoor cats in the United States kill between six and twenty-two *billion* small mammals each year—food that is no longer available to hawks, owls, shrikes, herons, foxes, coyotes, and other wildlife.[38] Exhibiting the bias of a budding ornithologist, I comforted myself with the fact that I had never known Roo to kill a bird.

And then, in 1996, at the age of ten, Roo brought me the remains of a Pine Siskin—a tiny, feathered puff of cream and streaky brown, with hints of gold in the wings and tail—that had been visiting my finch feeder. From that day forth, Roo became an indoor cat. Doubtless this was an unwelcome change for a cat that had roamed the streets of Philadelphia and Washington, DC, and hunted fields in Vermont, Idaho, and Montana. But Roo adapted—and thrived. And I no longer worried about him being killed by cars, coyotes, or Great Horned Owls. Roo lived happily indoors until he was seventeen years old. Regrettably, he didn't benefit from the enclosed outdoor pens (catios) that I provided to the felines that followed in his pawprints. Like Roo, my future cats became bird-watchers instead of bird killers, but they also enjoyed lying outside in grass and sunshine without harming the world around them. As a cat lover devoted to birds, I was taking what seemed like a small step toward combating one of the most serious threats that we humans inflict on wildlife. But keeping cats indoors, which benefits both cats (increasing their longevity) and wildlife, was one of the most effective—and least taxing—individual actions I could take to help birds and small mammals in the face of habitat loss, climate change, and other more intransigent pressures.

For now, though, in Hawaii, my overarching concern for the world's beleaguered wildlife was tempered by my more immediate need to protect five critically endangered crows whose current enthusiasm for taking a bath catapulted me from somber reflection to grinning delight. From my vantage point under the trees by the aviary, I could hear wings flapping and see water spraying off the hack box deck, where we had placed a shallow water tub. The birds used their heads like ladles, dunking them into the tub, then lifting them so that water poured onto their backs. Lowering their bodies, the crows cupped their wings by their sides, then shook them, flicking water skyward as they soaked their feathers. Each bird seemed drawn to the activities of its siblings, and the birds splashed around together with the unbridled exuberance of children playing in a fountain on a hot summer day.

While watching their enthusiastic ablutions, I was struck yet again by how very much corvids seemed to enjoy themselves. Like the ravens that had tumbled acrobatically through the skies over peregrine cliffs in Wyoming, the 'Alalā seemed to imbue even the most mundane tasks with a measure of playfulness.

After bathing, the three females convened on the aviary where they were soon joined by Mālama and Kēhau, who engaged in a mock battle. As Kēhau lifted a foot to ward off his sibling's advances, Mālama fell backward, reclining in the aviary netting like it was a comfortable hammock. Still prone, Mālama raised a foot in defense, locking feet with Kēhau. The two pushed back and forth like lumberjacks powering a two-man saw. Gaining an advantage despite his more vulnerable position, Mālama stood up and held Kēhau in place by putting a foot on his brother's back. Dominance asserted, Mālama flew to one of the feed trays and gulped down dog kibble. Kēhau landed on the hack box deck and picked up a dead mouse. Before he could swallow it, Mālama took off, heading toward a forest opening we had dubbed the amphitheater. Quick to follow his sibling, Kēhau took flight, the dead mouse dangling from his beak. I turned my attention to the females.

Lōkahi and Hiwahiwa had flown to a koa and were thrusting their bills under the tree's bark, searching for insects. Hoapili joined them, quivering her wings and begging for a handout. Hiwa responded by shaking her own wings and food-begging from Hoapili, who appeared to pause in surprise, her smoke-blue eyes and pink gape reminding me how immature these youngsters still were despite their proficient foraging. Hopping down a branch, Hiwahiwa perched within millimeters of Hoapili and Lōkahi, and the three preened, seeming to relish each other's close presence.

Their endearing companionship was short-lived, though, since Lōkahi soon reached over to Hiwahiwa with her foot and began pushing at her. Grasping Lōkahi's foot in her own, Hiwa engaged in a gentle push-me-pull-you foot fight before following Lōkahi into flight. Hoapili waited her usual half a minute and then pursued her siblings with a plaintive cry. As I followed my charges through the forest as unobtrusively as possible, my awkwardness was underscored by the Kalāhiki adults winging over the forest canopy with the synchronous grace of ice dancers.

When I found the young females, Lōkahi and Hoapili were mouthing clusters of 'ōlapa berries, while Hiwahiwa foraged near the ground on māmaki fruit. After several minutes, the three flew again, disappearing into the forest. I looked down at my watch and smiled wryly. Twenty seconds until my next behavioral scan. Typical. My birds regularly departed right before I had to record their activities. Sighing, I noted the time and wrote: "All females out of view" in my field notebook.

Again, I pushed through the wet vegetation and, within minutes, heard two crows foraging low to the ground. I identified Lōkahi just before she took off for the aviary. For once, Hoapili had preceded her sisters, since I could hear her distinctive squawk coming from near the enclosure. I expected Hiwahiwa to follow her siblings, but I suddenly realized that *she* was watching *me*. Cocking her head, she gazed at me fixedly, as though trying to divine the intentions of the clumsy creature, crouched amidst the ferns, that followed her but did her no harm. Hopping along a moss-covered log, Hiwa moved toward me, pausing every few hops to cock her head and stare at me. I remained frozen, glancing away to appear unthreatening but keeping her in my peripheral vision. Closer she hopped until she stood about 6 feet from me, almost level with my face. And then, curiosity apparently satisfied, Hiwa flew up into an 'ōlapa and resumed feeding.

She was joined minutes later by the males. Kēhau flew on to the aviary, while Mālama joined Hiwa, feeding on berries before appearing to notice my huddled form. Seeming as captivated by me as Hiwa had been, Mālama flew toward me and landed on a branch above my head. But where Hiwa had observed me with gentle curiosity, Mālama seemed to view me as a hostile intruder. Staring fixedly at me with a partially opened bill, he hunched forward flaring his throat feathers, and scolded me with short, sharp, barklike vocalizations. My watch alarm beeped softly and I jotted down the birds' behavior: Hiwahiwa preening and Mālama perched alertly, scolding me. Exhibiting classic displacement behavior that stemmed either from his conflicting curiosity and alarm about my presence or his desire to assert himself despite fearing me, Mālama swiped his bill on the branch before leveling several hard blows at it with his bill, scattering bark chips. Seeming unconcerned, Hiwa snapped at a passing bee. Then, without any discernable signal, the two birds spread their wings and headed to the aviary.

As I followed after them, I thought about their different reactions to me and wondered whether Mālama recognized me as the tormentor who had wrapped hands around him and put tubes down his throat. Even though we had held Hiwahiwa captive with Mālama, our interactions with her had been far more benign than those we had inflicted on our malaria patient, and the different ways the two birds responded to me seemed suggestive of the different experiences each had had with me.

Years later, John Marzluff confirmed that American Crows could recognize people's faces, particularly those belonging to humans who had "harmed" them. After trapping and banding scores of crows during a long-term study at the University of Washington, Marzluff and his assistants marveled that whenever they walked across campus, crows singled them out amidst thousands of students and colleagues, alarm calling and warning other crows that the trappers were about. Nearby crows fled, "while uttering a call that [sounded] like vocal disgust." Meanwhile, the crows ignored the people who hadn't captured and handled them.[39]

Marzluff tested these observations experimentally by wearing a caveman mask when trapping and handling crows. Later, whenever he and his assistants wore the mask as they strolled across campus, they were targeted by agitated crows. Even those they hadn't trapped learned from their crow relatives and neighbors that the caveman was dangerous. When Marzluff wore a Vice President Dick Cheney mask, which hadn't been used when trapping crows, his disguise aroused abundant human commentary, but the local crows ignored him. Further experimentation confirmed that American Crows do indeed distinguish "dangerous" and neutral human faces.[40]

150

If Mālama's behavior toward me was any indication, Hawaiian Crows appeared to be equally discerning. After poking, prodding, and restraining him, I had lost the immunity that I still enjoyed with Hiwahiwa and the other 'Alalā. Though he sometimes still ignored me, Mālama regularly subjected me to vociferous scoldings as I made my way through the forest, while I monitored his behavior and tried to keep him safe.

Dawn was the suggestion of light filtering through trees when I heard the Kalāhiki male's first morning call. *Owww!* The cry reverberated through the forest, giving me a caffeinelike jolt. John Marzluff dubbed this particular 'Alalā vocalization the James Brown call, recalling as it did the Godfather of Soul's throaty cries. I lay in my cot, listening avidly, as the male broadcast a medley of cacophonous wake-up calls—cloud-forest church bells—and wondered at how the woods must once have rung with sound as neighboring 'Alalā announced their territorial boundaries with whoops, yells, croaks, and howls. Solitary as the Kalāhiki pair now was, the male nonetheless seemed to follow his resonant James Brown call with the singer's famed "I feel good!" proclamation, putting an exclamation point on the new day with his powerful voice.

In an echo of times past, our juvenile 'Alalā soon began broadcasting their mewing calls, ebullient screams, and upslurred growls, which seemed more akin to questions than proclamations—as though they too felt compelled to greet the new day but were asking the territory holder's permission to join in the festivities. I couldn't conceive of the pall that would enshroud the Big Island's cloud forest if the 'Alalā singular calls were ever silenced by the minatory specter of extinction. Jumping to my feet, I followed the 'Alalā's example and embraced the day, determined to do my part to prevent that from happening.

Four of the juveniles arrived at the aviary to feed before 0600 hours. Before Hoapili could join them, the youngsters flew off again, setting my day in motion as I followed in their wake. I soon located the females resting and preening in the tree they had roosted in the night before. Lōkahi and Hoapili grasped each other's bills, vigorously allobilling before Lōkahi returned to the aviary with Hoapili close behind her.

Standing in a beam of sunshine, I relished the rare warmth and the peaceful forest while watching Hiwa preen for a few minutes. As I headed after the other females, I heard a clear whistle and saw an I'iwi bury its scarlet head and sickle-shaped orange-red bill into a pom-pom-like 'ōhi'a blossom.

Pausing in its search for nectar to proclaim its territory, the resplendent bird broadcast a peculiar medley of whistles, trills, and squeaks that sounded like a rusty hinge.

I arrived back at the aviary at the same time as Kēhau and Mālama, who flew in and perched on its highest point. Within seconds, Mālama took off and circled over the release area, followed by Lōkahi who flapped hard to join him in his overflight. The two circled overhead, briefly grasping feet as they flew. Launching into flight, Kēhau joined Hiwahiwa, who had just flown into view, then chased Lōkahi before landing back on the aviary. And now it was Kēhau and Hiwa's turn to squabble. Clinging to the netting, the two lunged at each other with open bills. After raising a foot in protest, Hiwa fled. Kēhau pursued her briefly then veered into a nearby koa. Hiwa circled then joined Hoapili and Lōkahi at the top of the aviary and the trio played king of the mountain—one of their favorite games. Pecking at each other and pushing with their feet, the females vied for the choice spot on the highest disk supporting the aviary netting. No sooner had one attained the mountaintop before another nibbled at her legs, causing her to relinquish the hard-won position. And then it was time for the youngsters to eat, bathe, and preen again, before flying farther afield to explore the cloud forest's plenitude.

I wondered about the ʻAlalā's daily forays as I watched the females settle down to roost that night, and wished I could be privy to their activities as they increasingly ranged beyond our capacity to follow them. Kēhau and Mālama had roosted far enough downslope that their soft contact calls were barely audible. An adult ʻAlalā called briefly from a distant roost upslope. Tucked behind leaves high in an ʻōhiʻa, the females perched in close proximity to one another, vocalizing softly—a safe, feathered unit sheltered by obscuring vegetation and a darkness that intensified imperceptibly as their murmurations quieted.

A scurry of movement suddenly caught my eye as a black rat rushed down the trunk of the ʻōhiʻa in which the females were roosting and disappeared amidst the ferns cloaking the forest floor. Its relatively small, seemingly innocuous form seemed disproportionate to the weighty reminder it provided that all was not well with the ʻAlalā's world, despite the preternatural peace of the gloaming. Nor is all well wherever introduced organisms erode the biodiversity they have invaded, threatening an ecological homogenization that can be measured by the lost voices and lost forms that haunt so many imperiled habitats worldwide. The economic costs of invasive organisms are staggering—a 2011 study estimated that the economic damage and control costs associated with alien invasive species in the United States amounted to approximately $219 billion per year.[41] A more recent study, released in 2023, estimated the yearly global cost of invasive species to be $423 billion.[42] Such price tags clearly argue for preventing invasions and controlling invaders early on. But these estimates don't include the incalculable costs of lost biodiversity, lost ecosystem services, and species extinctions. For how can one assign an economic cost to the yellow-and-black brilliance of the now-extinct Hawaii Mamo, which once flickered like sunlight in the dark forest? Or to the never-to-be-seen-again, silver-capped Poʻo-uli that once held endemic Hawaiian snails aloft in its beak? Or to the voice of an ʻAlalā whispering a reassuring accompaniment to nightfall or raucously announcing the dawn of a new day?

CHAPTER THIRTEEN

Extinct

Giving casual thought to the crisp weather and colorful foliage that usually spelled October for me, I pushed my way through cloud-forest vegetation in dogged pursuit of our itinerant crows. Our 'Alalā had traveled farther afield this morning than they had ever gone before, and keeping up with them had begun to feel hopeless. Despite the distance I had walked, their faint radio signals told me the birds were still far off. The undergrowth thinned as I approached a clearing where several grassy paths converged. Catching sight of two Hawaiian Crows in an adjacent tree, I stopped in surprise and glanced at my tracking receiver in puzzlement before realizing that I had stumbled upon the Kalāhiki adults.

Although I regularly heard the pair's resonant calls and witnessed their interactions with our young crows, I had never seen the adults immersed in their daily activities. The male clung to a tree trunk several feet off the ground and poked under the bark, looking for insects. The female conducted a similar search on a low branch. Even though the birds were a mere 15 feet (5 m) from me and registered my presence with an almost imperceptible pause in their foraging, they ignored me and continued tapping and probing the tree bark.

I kept absolutely still, admiring the crows' sleek plumage and dark eyes. Without any suggestion of fear or alarm, the birds glanced between me and the task at hand. As always, the pair looked powerful, confident, and in command of its environment. But with a touch of sadness, I recognized that this was yet another illusion woven by this magical island. The world that these wild, beautiful 'Alalā had known for so long and adapted to so brilliantly had been transformed, and these creatures now faced an army of insidious threats against which they were powerless.

Scarcely breathing, I watched them until first the female and then the male moved unhurriedly away, feeding quietly amidst a bounty of fruits, flowers, and insects. Reluctantly, I moved on, too, continuing the quest for my wide-ranging charges. I had started up one of the grassy byways that slashed through the forest, following a faint signal from Kēhau's transmitter, when I suddenly heard strong wingbeats behind me. Turning abruptly, I came face to face with the Kalāhiki male, who was flying straight toward me. Without pause, he flew past then landed on a branch just beyond where I stood, almost level with my eyes. Cocking his head, he looked at me inquisitively and called quietly. He didn't appear to be scolding me as Mālama did; the tenor of his vocalization was softer—less a reprimand than an indeterminate declaration. In the wonder of that moment, time seemed to pause as he studied me. And then, with no wasted motion, he was flying back the way he had come and disappearing into the tangled green canvas of his known universe. Awed and inexplicably humbled, I resumed my search, knowing I had been irrevocably transformed by the acknowledgment of a bright-eyed bird whose life, for one too-brief moment, had intersected mine.

———

Still chilled from the pervasive dampness of a fog-cloaked morning, I forked down breakfast while perusing job listings in the Ornithological Societies of North America newsletter. Reluctant as I was to accept that my time in Hawaii was drawing to a close, I hoped that the anticipation of new adventures would help alleviate my heartache whenever I contemplated leaving the 'Alalā. I skimmed descriptions of several avian research and conservation projects before my attention was arrested by a request for help studying the Yellow-naped Amazon—a parrot—in Guatemala. The work sounded warm. Exactly the opposite of the penetrating chill I'd endured for weeks. I had never been especially captivated by parrots, but the thought of immersing myself in the habits and haunts of intelligent birds that most people only associated with pet stores seemed like a worthy sequel to my emotional investment in the incomparable 'Alalā. I vowed to apply for the position upon returning to my home in Idaho.

Later that day, I cast aside thoughts about the future, relishing the moment and delighting in the 'Alalā's more mundane activities as much as in their entertaining antics. As dusk whispered away light and lengthened shadows, Hoapili and Lōkahi perched together at the tip of a dead koa branch. Mālama, who was perched below Lōkahi, reached up and pecked repeatedly at her legs, forcing her to relocate. Hiwa engaged in her own form of harassment by nibbling on Kēhau's transmitter antenna as the two perched together in a neighboring tree. After a bit more shuffling, the quintet settled down to roost in the same tree. The mist swirled, conspiring with nightfall and a light rain to obscure my view, and the 'Alalā's dark forms faded into the koa's sheltering arms.

————

Although we were only providing our crows with dog kibble now, to encourage them to feed exclusively on the forest's bounty, the youngsters returned to the aviary as usual the next morning, amidst a flurry of exploratory flights. Following the females downslope soon afterward, I found the trio arrayed along a koa's sloped trunk. Quivering their wings, the three 'Alalā food-begged from one another while holding chips of bark in their bills. Amused and perplexed by this puzzling performance, I couldn't begin to guess what it meant or how it had come about.

Shortly thereafter, the 'Alalā returned to pecking and prodding the loose tree bark. Congregating near what might have been a productive patch of grubs, Lōkahi and Hiwahiwa engaged in a brief foot fight, from which Lōkahi emerged victorious after pushing Hiwa off the branch. Since bold Lōkahi was the more dominant bird, I wasn't surprised by this outcome. But the juveniles' behavior often confounded me since I had also witnessed submissive Hoapili push Lōkahi off a branch during a foot fight. Perhaps their dominance dynamics were still evolving. Or maybe a crow's strength varied with its motivation to defend a resource. The birds' interactions invariably provided food for thought, as well as entertainment. Abandoning their insect search, Lōkahi and Hiwahiwa joined the males in feeding on māmaki fruit. Hiwa's focus was short-lived. Flying over to a hāpu'u tree fern, she pecked at the stem and extracted a grub, assuring me that the crows had no shortage of natural substitutes to the food we supplied them.

Later, all five crows foraged close to the ground, delighting me with their fuzzy-headed proximity. Crouching amidst fern fronds, I watched Mālama and Hoapili play tug-of-war with what looked like a small stick. Unable to wrestle the object from the smaller female, Mālama grabbed Hoapili's foot in his and pushed at her. Rather than ceding her "toy," Hoapili wrested her foot free and stepped away from him, holding her prize aloft in her bill. Now, however, she found herself between the two males, who simultaneously pushed her with their feet. Their bullying precipitated a squabbling free-for-all. Hiwahiwa and Lōkahi quickly defended their sibling by attacking the two males, who turned on each other.

Commenting on the playfulness and intelligence of Common Ravens, animal behaviorist Millicent Ficken asked: "Do ravens exhibit complex play because they are intelligent or are they intelligent because they play? The answer is probably both. Because they are intelligent, they are capable of diverse and complex play activities, and it is also probable that through play they learn relationships with the environment that contribute to their plasticity of behavior . . ."[1] Biologically, play has been defined as a behavior that has no immediately discernible function but whose purpose ultimately proves to be useful.[2] Play is particularly evident in animals that rely on extensive learned—versus instinctual—behavior. It likely helps an animal develop food acquisition and manipulation skills, and also influences its relationship with conspecifics and other elements of its environment.

Corvids exhibit the most complex play behavior of any birds, and they routinely engage in both solitary and social play. Ravens, for example, may engage in playful locomotory activities, such as sliding down snowy slopes on their backs, or they may manipulate objects, even throwing and catching sticks and other items.[3] The 'Alalā frequently chased each other and played king of the mountain, keep-away, and tug-of-war. Their play likely had adaptive value, improving their foraging techniques, allowing them to evaluate each other's skills and strengths, and helping them develop their relationships with each other, with other 'Alalā, and with their cloud-forest home.

Our 'Alalā's playfulness may well have been linked to the legendary intelligence of corvids. Relative to their body size, corvids have larger brains, on average, than any other group of birds (only two macaw species have larger brains relative to body size). Endowed with exceptional memory and smarts, the superior corvid brain allows Clark's Nutcrackers and Pinyon Jays to retrieve thousands of cached seeds each year with an accuracy rate exceeding 90 percent, Carrion Crows in Japan to deposit walnuts in front of stopped cars whose tires crack open the nuts when the vehicles move, Common Ravens to use logic to solve problems, and New Caledonian Crows to manufacture tools to retrieve food.[4]

The use of tools, which was once considered the exclusive purview of big-brained mammals such as humans and chimpanzees, has long been considered a sign of animal intelligence. The New Caledonian Crow—a resident of two South Pacific Islands, east of Queensland, Australia—is thought to be the master toolmaker of the bird world. In the wild, this resourceful jet-black corvid shapes a variety of spears and hooks from leaves and twigs, then uses these tools to fish insects out of crevices. Researchers have tested the bird's intelligence and toolmaking skills in an elaborate series of experiments. Aside from bending wire into a hook to retrieve a food-filled bucket placed out of reach at the bottom of a tube, New Caledonian Crows exhibited insight by solving complex food-acquisition problems. Confronted with an assortment of seemingly random tools, wild crows that were captured and held temporarily in captivity retrieved a short stick dangling from a string, used the short stick to extract a long stick from a box, and then used the long stick to push food out of the box (many of them accomplished this on their first try). The crows completed a series of steps that followed a mental plan they had envisioned to solve a problem.[5] Rather than overtly trying out different solutions, the birds evaluated their choices then made intelligent decisions about how to achieve their goal.

One New Caledonian Crow, nicknamed 007, became an internet sensation in 2014, when it solved a novel eight-step puzzle to acquire a piece of meat. Within minutes, the wily crow retrieved a short stick dangling from a string, used the stick to retrieve three stones (each of which had been placed in a different box), dropped the stones individually down a chute to weight a lever that released a long stick, and then used the long stick to reach a piece of meat placed in the back of a fourth box. Clearly, corvids provide resounding evidence that the derogatory term *bird-brain* to denote stupidity is wildly inappropriate.[6]

After working with New Caledonian Crows for over a decade, biologist Christian Rutz wondered whether any other corvids—many of which are little studied because they live in remote tropical areas—were skilled tool users. He suspected that the New Caledonian Crow's straight bill and large, highly mobile eyes (which give the bird its impressive depth perception) might be adaptations for handling tools, so he looked for these traits among the world's more than forty species of crows and ravens. The 'Alalā stood out to Rutz because its relatively straight bill and forward-facing eyes seemed characteristic of a tool user. Contacting the program manager for one of Hawaii's captive-breeding facilities, Rutz suggested that 'Alalā might use tools given the opportunity. To his surprise, the manager responded, "Oh yeah, they do all sorts of funny things with sticks."[7]

Rutz and his colleagues subsequently collaborated with captive-breeding personnel to study the ʻAlalā's tool-using skills. Testing most of the captive Hawaiian Crows, the researchers found that more than three-quarters of the birds spontaneously held sticks in their beaks and used them, with astonishing dexterity, to poke or pull food out of holes in wooden logs. Even juveniles used tools to probe for food, without ever having been exposed to tool-using adults. Since Hawaiian and New Caledonian Crows are only distantly related, the researchers speculated that their ability to use tools arose convergently—that is, each species developed its tool-use capability separately—in response to similarities in their environments. Both crows evolved on islands with few native predators, giving the birds the time and freedom to use tools without having to be constantly alert to potential attackers. The two island crows also faced little competition for embedded prey from birds such as woodpeckers, which might have eliminated hidden, difficult-to-reach insects that the large-billed crows had to target more creatively.[8]

Upon hearing about the Hawaiian Crow's tool use when Rutz and his colleagues published their results in 2016, I was delighted to see ʻAlalā demonstrating their skills on internet videos. As I watched the dexterous crows use twigs and, on occasion, flakes of bark to extract food from holes in a log, I was flooded with memories of Hoapili, Lōkahi, and Hiwahiwa food-begging from each other while holding chips of bark in their beaks, and of Mālama and Hoapili playing tug-of-war with a small stick. Perhaps these behaviors were precursors to manipulating and using tools to extricate difficult-to-reach insect larvae. Biologists in Hawaii never observed ʻAlalā using tools in the wild, though. US Fish and Wildlife Service biologist Donna Ball speculated that food was so readily available to the last remaining wild ʻAlalā that they had no need for tools. Clearly, though, the birds were intelligent and resourceful enough to devise ways to extricate food if they needed to.

A few days before my departure from Hawaii, our five ʻAlalā achieved another important milestone when they all headed downhill and out of radio contact first thing in the morning rather than coming to the aviary to feed. Trudging downslope, Kurt and I eventually received faint signals from their radio transmitters, but the birds were almost constantly on the move, traveling farther than ever before, and exploring new areas that were virtually impenetrable to us. After several hours, the birds headed back upslope. Thrashing through the wet vegetation, Kurt and I made for a large dirt road (dubbed the North-South Road), which had recently been bulldozed through the forest, in an attempt to intercept our adventurous crows. They eluded us, but we finally caught up to them near the aviary, where they preened and lounged like veritable homebodies. Later that morning, they were joined by the adult ʻAlalā, who surprised us by perching in an ʻōhiʻa near our trailer. Despite their occasional proximity, though, the wild adults never touched our youngsters' supplemental food.

The next morning, our ʻAlalā again disappeared into the "black hole" they had vanished into the day before. Once they returned to the aviary, Hoapili, Hiwahiwa, Lōkahi, and Mālama perched together at a food tray. As often happened when they were close together or interacting, the crows had erected their head feathers, which gave the birds their fuzzy-headed look.[9] Soon, the youngsters dispersed, and they spent the remainder of the day resting, preening, and foraging in hāpuʻu ferns, koas, ilex (Hawaiian holly), and other native vegetation. Though they still returned periodically to their home base, our ʻAlalā were becoming independent and were less interested in supplemental food. As they ranged more widely, it would be more serendipity than routine to spot them amidst obscuring leaves and drifting mist, to watch them preen and play, or listen to their night whispers at roost. And for me, heartbreakingly, such moments soon would be the stuff of memories and daydreams.

My last morning with the ʻAlalā began at 0544 hours, when the juveniles began calling from their roost trees. For a moment, I closed my eyes and relished the exuberant sound, knowing no reveille would ever bring me so much pleasure. To my delight, the females soon arrived at the aviary, followed within minutes by Kēhau and Mālama. None of the birds lingered. A short while later, I caught up with Lōkahi and Hoapili as they foraged on a favorite ridge southeast of the release area. Hiwahiwa arrived moments later and Lōkahi flew over to join her. The two greeted each other by quivering their wings and food-begging from one another. Nearly independent though they were, they still retained some juvenile behaviors.

Landing on a koa branch, Hiwa plucked off a long seedpod that resembled the seed casings of other flowering trees in the legume or pea family, to which the koa belonged. Pinning the dark, mature seedpod against the branch, Hiwa worked at it with her beak, trying to open it to extract insects or seeds. Lōkahi and Hoapili, who were perched on one side of her, cocked their heads curiously and tried repeatedly to grab the enticing object. Apparently frustrated by her inability to get close to it, Lōkahi hopped over Hoapili and Hiwahiwa and landed on Hiwa's free side. Whatever Hiwa extricated from the seedpod was obscured from my view by the three crows' beaks, which hovered within millimeters of each other as each female prepared to grab and abscond with the prize. Launching into flight in a sudden flurry of jumbled wings and bodies, the winner fled with the spoils, sisters in close pursuit.

Leading me through the sun-dappled forest as they moved downslope in short bursts of flight, the trio convened in another koa where Hiwahiwa, exhibiting admirable expertise, brandished another seedpod. Unable to resist its magnetic attraction, Lōkahi and Hoapili tumbled over themselves to get closer to Hiwa. This time, though, Lōkahi tilted the playing field by pushing Hoapili with her foot, knocking the unfortunate bird off the branch. In the ensuing chaos, the birds took flight again—a discordant, feathered commotion careening through the trees.

Scrambling after them, I eventually met up with the females back at the aviary. But before I could record their activities, they reversed course. Sadly, I listened as their radio signals joined those of Kēhau and Mālama then faded into silence as the birds headed away from me for the last time. Following a final fruitless search for the vagabond youngsters, I turned back toward the trailer to finish packing for home. Brushing tears from my eyes, I passed through a forest that still held mysteries, fervently hoping for a last glimpse of the ʻAlalā while appreciating their newly acquired independence.

When I finally drove away from the release site, I clutched a cassette tape that the Service employees had made for me. Interspersed with classic Hawaiian songs, the team had recorded the riotous, raucous, unforgettable voices of Hoapili, Lōkahi, Hiwahiwa, Kēhau, Mālama, and the wild adults. Adding to the memory trove, my coworkers also included recordings of our radio chatter as we discussed ʻAlalā locations and behaviors. Trying to imprint snapshots of the forest in my mind as I lurched down the rough track, I contemplated goodbyes. Difficult as it was to leave coworkers and friends, memory-filled forest glades, and scarlet birds and flowers, nothing compared to the ineffable sadness of saying a silent goodbye to the captivating creatures I had come to know so intimately through countless hours of observation. I could write to the friends I had made and might even see them someday, but I was unlikely to see the comic, charismatic ʻAlalā ever again.

That night, still reluctant to embrace future plans, my thoughts drifted instead to some of the unforgettable birds that had written parts of my history: five young peregrines chasing Wyoming's wind; fledgling goshawks disappearing like ghosts into somber woods; young Prairie Falcons taking first flights from towering cliff walls. And above all, the 'Alalā: little Hoapili, always the last to leave a particular spot, and calling repeatedly upon being left behind once again; intrepid Lōkahi, ever ready to explore new areas and lead the little band of females on their feeding forays; independent Hiwahiwa, approaching me with gentle curiosity and brandishing treasures as her sisters fell over themselves to share in the bounty; husky Kēhau, yielding to his smaller sibling and greeting him with gentle *mews* after any separation; and bold Mālama, invincible survivor of avian malaria, scolding my dangerous presence, and, on occasion, even standing up to the powerful Kalāhiki adults.

A rush of indelible images that would stay with me in perpetuity flooded my mind. 'Alalā flying away, one after the other, seconds before I recorded a behavioral scan. Comical battles, with one bird pushing another with its foot while the other tumbled off the branch as it lost its balance and the fight. Crows, foraging like primates high in the treetops, clinging to thin branches that bent perilously beneath the birds' weight. Sibling 'Alalā perched in a row—tight-knit units sharing warmth and companionship. Bold, bright eyes, cocked heads, curiosity incarnate. And soft whispers at day's end, when the forest seemed to hold its breath, swirling mist confounded leaf and shadow, and five dark shapes settled into the enveloping darkness and forever into my heart.

———————

Several months after leaving Hawaii, I flew to Guatemala to work with the Yellow-naped Amazon—a large parrot the color of early spring leaves, with a splash of yellow across the back of its neck and bold dashes of red and indigo in its wings. I relished the region's warmth and its spectacular bird life. I learned to use a machete to cut and strip branches, and build tepee-shaped blinds from which to observe the parrots' nesting activities. I learned to run whenever cavity-dwelling killer bees attacked, and suffered only a single sting to the face. I accompanied the valiant climbing crew who launched ropes high over the branches of giant trees using a bow and arrow, and then climbed to perilous heights—garbed in bulky beekeeping suits—to retrieve baby parrots so we could band them and instrument them with small radio-tracking collars. With an unapologetic lack of scientific objectivity, I delighted in the way adult parrots flew from one destination to another in pairs, and I ached at how long they searched for their chicks when their nests were plundered by poachers.

The Yellow-naped Amazons were engaging and the dedicated Guatemalans with whom I worked were warm, generous, and funny. To my amazement and to their credit, they made me, a Swiss-born American, feel at home in an environment plagued by poverty, poachers, guerillas, and darker, more nefarious forces. I learned to walk without a headlamp at night so I wouldn't reveal my presence to dangerous men. I discovered that the police who stopped me at roadblocks and threatened to impound my vehicle could be won over with the gift of a few sodas. I administered first aid to a crewmember who was beaten with the butt of a soldier's rifle. I heard the gunshots that killed a driver on a nearby highway. I narrowly avoided encounters with guerillas and the Guatemalan secret police. And I chased off a poacher who was trying to supplement his meager income by stealing baby parrots from a tree cavity I was watching. Through it all, I became a passionate advocate for the conservation of tropical birds. I laughed, birded, and made lifelong friends.

I returned to Idaho for another season of working with Golden Eagles and Prairie Falcons, taking on greater responsibilities and gaining incremental confidence in my growing abilities as a field biologist, before heading to Argentina. There, I assisted with research on the Yellow-billed Teal, a diminutive duck whose understated, speckled elegance was only enhanced by its intriguing courtship displays and its use of large, arboreal stick nests—built by Monk Parakeets—to raise its young. I learned to catch ducks using mist nets and to mark the birds with small, plastic nasal discs.[10] I spent countless hours observing ducks from a wooden tower near wetlands that teemed with exotic waterfowl, while ostrichlike rheas and ungainly Southern Screamers—goose-chicken hybrids on stilts—ambled past. I watched day-old Yellow-billed Teal ducklings—downy puffs of chocolate and vanilla—leap to the ground from their tall tree nests and tracked their radio-transmittered mothers to determine each brood's fate.

After returning to Guatemala for another field season, I headed to Arizona's Mogollon Rim, on the southern edge of the Colorado Plateau, to spend a summer monitoring songbird and woodpecker nests before beginning graduate school at the University of Montana, in the fall of 1995. And all along, my thoughts returned often to the 'Alalā that had so enthralled me with their playful antics, gregarious interactions, and astonishing voices. Unable to bear the thought that any of my quintet had perished, I avoided trying to learn about their fates. Working in remote locations and immersed, eventually, in my own graduate research on American Dippers—North America's only aquatic songbird—it was easy to miss what little news about Hawaii's endangered birds trickled out of the mainland media. And it was easier still to delude myself that all was well with the 'Alalā that had captivated me in a way that no other birds had done.

But as I began my graduate studies in organismal biology and ecology, the Kalāhiki pair that had so impressed and delighted me disappeared. The male vanished in early 1995, about six months after being struck, but not killed, by an 'Io. The female—who holds the known longevity record of eighteen years for wild 'Alalā—disappeared about seven months later. Without the vigilance and companionship of her mate, she may have wandered more widely and been more vulnerable to attacks by gun-toting people or predatory 'Io. Or perhaps she just reached the end of a long, productive life. With the disappearance of the Kalāhiki pair, only two wild 'Alalā pairs remained: the Ho'okena pair that claimed the territory adjacent to the Kalāhiki birds, and the Keālia pair (Kēhau and Mālama's parents) that nested farther west. The Kalāhiki pair's last wild fledgling (from 1992), known as The Bachelor, also wandered the forest widely, never being associated with a territory, until he disappeared sometime after 1997.[11]

In 1998, my fervent hopes that the 'Alalā recovery program was proceeding apace—with more captive-raised juveniles being released to supplement the wild crow population—were dashed when I heard that seven reintroduced 'Alalā had died in a four-month period, prompting biologists to recapture all the remaining reintroduced birds and return them to captivity.

While searching the news to learn more details about the fates of the birds I had helped reintroduce, I came across Kēhau's name. He had survived. In the wild for five years, he had chosen a mate. They had built a rudimentary nest, but had failed to raise young. And now, stalwart Kēhau had been returned to captivity. As elated as I was to hear that Kēhau was a survivor, I was dismayed to picture him confined to an aviary, no longer free to hang from a koa's outermost branches or to chase his mate over cloud-shrouded treetops. And since Kēhau had been returned to captivity with only five other 'Alalā out of the twenty-seven that had been released to the wild between 1993 and 1998, I could no longer be very hopeful about the fates of Lōkahi, Hiwahiwa, Hoapili, and Mālama. Several of them had certainly died. Not wanting to confront the inevitable bad news, I made only ineffectual efforts to learn what had befallen them.

With no more reintroduced Hawaiian Crows to bolster the declining wild population and no successful reproduction by wild birds, the ʻAlalā hurtled toward extinction. By the year 2000, only three ʻAlalā remained in the wild, the Keālia pair and one member of the Hoʻokena pair. It may be possible—perhaps even easy—to dismiss the extinction of an unfamiliar species, to hear that one more of a handful of remaining birds has disappeared without viscerally feeling the loss. But for those working with the species as its outlook becomes increasingly bleak, dispassionate objectivity is nigh impossible, particularly when dealing with a creature whose social nature, intelligence, and charisma make each individual so unique and so knowable.

US Fish and Wildlife Service biologist Glenn Klingler, who had moved from Washington State to work with the ʻAlalā, poignantly expressed the heartache of those who fought for the Hawaiian Crow's survival. Speaking to author Mark Jerome Walters, who chronicled the ʻAlalā's decline, Klingler said,

> It's been hard to watch what's happening . . . It's been hard emotionally. I remember most of all the Hoʻokena bird, how after it lost its mate it cried out for weeks. I'd wake up every morning and hear the mourning—a terribly high-pitched sound, like an inconsolable moaning, from the top of an ʻōhiʻa. First it would be here, and then a moment of silence, and then it would be there. The sound just went straight through my heart.[12]

The lone Hoʻokena bird seemed to be seeking companionship as it took to disturbing the Keālia pair. In the summer of 2000, though, it too disappeared. Only two ʻAlalā now remained in the wild.

In the summer of 2002, Service biologist Jeff Burgett visited the Keālia pair's territory, after another unsuccessful nesting season, and found them in their usual haunts. The birds were scruffy since they were in the midst of molting their feathers, as they did each year after nesting. In addition, the female looked listless. In a manner reminiscent of Kēhau trying to prod his sibling into motion when Mālama was sickened with avian malaria, the

Keālia male prodded the female and fed her a couple of fruits.[13] Despite countless searches, no one ever saw the pair again.

————

I don't know when I first heard the devastating news that the Hawaiian Crow had gone extinct in the wild, when I first learned that the singular 'Alalā was no longer broadcasting its clarion calls among stately koas and mist-shrouded tree ferns. I often coped with life's hurts by forcing them into the far recesses of my mind until I felt less raw and better equipped to deal with them. It was a long time before I felt able to handle the heartbreak of the 'Alalā's demise. I had been so captivated by my irrepressible blue-eyed, pink-gaped charges. I had been so thrilled to observe the seemingly invincible Kalāhiki pair. I had felt so privileged to mingle with the 'Alalā in their native forest and to play a small role in furthering their unique legacy. I had been so full of hope for the 'Alalā when I left Hawaii and so certain that its recovery would succeed. I was bereft upon hearing that extinction had silenced the bird's inimitable voice. Others lamented the likely repercussions of the 'Alalā's absence.

Biologist John Marzluff, who continued to advise the Hawaiian Crow recovery project for several years after my departure, remembered exactly what he was doing the day he learned of the 'Alalā's extinction in the wild. After spending a morning studying opportunistic American Crows that were colonizing a subdivision that had recently replaced a patch of Pacific Northwest forest, Marzluff received a fax from Hawaii informing him that recent searches for the Keālia 'Alalā had yielded no sign of the birds. Nor had the pair been seen in the previous two months. For Marzluff, the loss of the Hawaiian Crow was a glaring example of the impossible challenges humans have imposed on small, specialized populations of island corvids, and of how profoundly humans were "shaping the evolutionary tree of corvid life through careless and indifferent pruning."[14]

From an ecological perspective, the disappearance of the 'Alalā from Hawai'i's forests meant diminished biodiversity, an end to the bird's interactions with other organisms in its community, and the loss of the ecological services the crow provided, such as dispersing plant seeds and controlling harmful insects. Birds are often the only native seed-dispersing animal on oceanic islands. When an important seed disperser disappears, the plants that relied on it to disseminate their progeny face an uncertain future. As a result of these "lost mutualisms"—or the loss of cooperative interactions between organisms—ecological communities may experience cascading (or secondary) extinctions. Alternatively, if the plants survive, they may be left with what scientists call "ghosts of past mutualisms." Certain large-fruited trees that persist in South America lost their major seed dispersers with the extinction of giant ground sloths, elephantlike gomphotheres, and other Pleistocene megafauna. Why these plants developed the types of fruits and seeds that they did makes sense only when one considers the now-absent seed dispersers with which they evolved eons ago.[15]

In addition to causing lost mutualisms, the 'Alalā's absence could exacerbate the impact of nonnative organisms that have colonized the crow's former habitat. Even seemingly innocuous invaders, such as songbirds like the resplendent Red-billed Leiothrix and the cryptic Warbling White-eye, might precipitate profound and unanticipated changes to the 'Alalā's forest. Because of their small bills and bodies, these birds consume only the smaller fruits and seeds traditionally consumed by the Hawaiian Crow. As a result, these birds could facilitate the dispersal of small-seeded plants while inadvertently forestalling future reproduction of larger-seeded plants that were dispersed by the stout-billed 'Alalā. In doing so, small exotic songbirds could irrevocably alter the crow's former haunts, fostering very different forests than those that colonized Mauna Loa's irruptive slopes through the corvid-accompanied ages.[16]

Another less tangible but still significant threat to the future of Hawaii's cloud forest in the 'Alalā's absence is embodied in a phenomenon known as "shifting baseline syndrome," first coined by fisheries scientist Daniel Pauly in 1995.[17] Pauly suggested that successive generations of people view whatever they first encounter in nature as the norm, thereby redefining the "baseline" components of an ecosystem. What Hawai'i's endemic forests look like to a child today is quite different from the forests viewed by a child in the early 1900s, and each bases her perception of what is gained or lost during her lifetime on her very different baseline perspective. Those who first view the Hawaiian cloud forest with its nonnative birds, its diminished cast of native honeycreepers, and its absent 'Alalā are likely to view that depauperate ecosystem as the norm. They may see a beautiful, green forest with some colorful birds, just as I initially viewed the forest's damaged understory patches as lovely, restful glades before learning that they represented lost native habitat. Without knowing what once was, we are more likely to settle for what now is, unless those who experienced a richer, more natural environment can convince newcomers and new generations alike that a Hawaiian cloud forest that no longer contains a panoply of native birds, including the vibrant and voluble 'Alalā, really is no forest at all. Unless we are aware of what has been lost, we cannot begin to understand or convey what we should work to save.

Many native Hawaiians are unsettled by the altered state of their forests today since they and their ancestors revered 'Alalā as family gods or spiritual guardians ('aumākua) that provided protection and guidance, and that inspired a style of chanting once used by warriors in battle.[18] For them, the 'Alalā's disappearance from its native forests was a cultural loss as well as an ecological one, an erosion of traditional beliefs and identities, a sign of a troubled universe.

For all those whose lives were enriched by the world's rarest corvid, who delighted in its antics, relished its curiosity and playfulness, and were captivated by its diverse calls and alluring behavior, the 'Alalā's extinction in the wild was, above all, an emotional loss. It was the loss of a vital, resonant, timeless force that animated and shaped Hawai'i's endemic forests and left a disquieting void in its wake.

With the disappearance of the Kalāhiki, Ho'okena, and Keālia 'Alalā, years of silence ensued. Evanescent clouds still permeated Mauna Loa's forests, and a fickle sun still illuminated ephemeral water droplets that coalesced on leafy surfaces. It is tempting to anthropomorphize that the forest was waiting. But for those whose lives had been touched by the 'Alalā, it felt like just that. For, unlike the 150 or so bird species catapulted into extinction by humanity's careless and often indifferent hand in the last five centuries, and unlike the nearly 500 vertebrates rendered extinct since 1900, the extinct-in-the-wild 'Alalā lived on—in captivity. Nurtured like a smoldering ember by a dedicated cadre of scientists, conservationists, and wildlife managers, a growing flock of captive 'Alalā awaited its chance to rise like a Phoenix, recolonize its former haunts, and breathe life into Hawai'i's forests once again.

CHAPTER FOURTEEN

Captive

Captive breeding—the conservation measure that ultimately offered the 'Alalā its best hope for survival—began almost by accident on the Big Island in 1970. In June of that year, US Fish and Wildlife Service biologist Winston Banko, who was serving as the nation's first federal endangered species research biologist and was tasked with investigating the decline of Hawaii's birds, found his first fledgling 'Alalā. The crow had sores on its head and by its bill that were symptomatic of avian poxvirus. Banko found a similarly afflicted 'Alalā fledgling in an adjacent territory. Hoping to nurse the sick birds back to health, he easily captured the two 'Alalā and ensconced them in his residence at Hawai'i Volcanoes National Park.[1]

Banko's efforts to care for the crows were unsurprising given his youthful passion for the wild creatures that inhabited his childhood haunts near Washington's Cascade Range. By the age of twelve, Banko was signing notes to his family with "The Naturalist" in lieu of his signature. Receiving intellectual enrichment from wildlife artist and author Ernest Thompson Seton and later from legendary biologist Aldo Leopold, Banko made a name for himself as a game biologist, a national wildlife refuge manager, and the author of the first definitive work on the Trumpeter Swan before moving to Hawaii in 1965.[2] For the next half a century, he devoted himself to the study and conservation of Hawaii's birds.

Concerned by the low number of 'Alalā he was finding when he discovered the sickened fledglings, Banko hoped to return the young crows to the forest once they had recovered. His mainland supervisors, though, requested that he ship the rare birds to the Service's Patuxent Research Refuge (formerly known as the Patuxent Research Center) in Maryland. Heartbreakingly, one died soon after arriving in the continental United States, while the other survived for only three years. Although neither crow fulfilled hopes for preserving their species, they did bolster the idea of breeding 'Alalā in captivity. But their demise foreshadowed the challenges to come.

The nascent captive-breeding program received its next feathered infusion in 1973, when Winston Banko's son Paul captured three more fledglings that appeared to be debilitated by disease and parasites. Two of the youngsters actually may have been unable to fly because they had only recently fledged. (At the time, biologists were unaware that 'Alalā leave the nest before they can fly.) Paul had been captivated by the 'Alalā since accompanying his father into the field as a teenager, during their early days in Hawaii. Both Bankos found the charismatic crow's curiosity, comic behavior, striking vocalizations, notable boldness, and surprising stealth irresistible.[3] In much the same way as 'Alalā formed tight-knit family groups, the Bankos made Hawaii's imperiled birds a family affair, with Paul dedicating a lifetime of study and conservation efforts to the 'Alalā, the Nēnē, the Palila, and other endangered forest birds.

After caring for the three captive crows for several years under the aegis of the US Fish and Wildlife Service, with the hope of reintroducing their offspring to the 'Alalā's historic range in Hawai'i Volcanoes National Park, the Bankos reluctantly ceded their charges, which were on the cusp of initiating breeding activities, to the State of Hawaii in 1976. In a time-worn battle between the states and the federal government, Hawaii was eager to assume responsibility for 'Alalā propagation, and ensconced the crows in a facility that had housed the Nēnē captive-breeding program. Located high on Mauna Kea's cold, windswept slopes, above the 'Alalā's warmer forest habitat and next to an active US Army bombing range, the facility was ill-suited for propagating the intelligent, behaviorally complex 'Alalā. Indeed, few bird species would have thrived given that nearby artillery fire and explosions, nighttime flares, helicopter overflights, and earth-shaking detonations regularly rocked the breeding facility.[4]

Eight more wild fledglings were added to the captive flock between 1977 and 1983, and three fledglings were produced in captivity in 1981, but the 'Alalā languished (and several died) in a program that struggled with limited oversight, resources, and staffing.[5] To address these challenges, Hawaii's

Division of Forestry and Wildlife eventually transferred the captive flock to a renovated breeding facility—a former minimum-security prison—on the island of Maui in 1986. Located on state land, the Maui Bird Conservation Center later received additional upgrades and expansions, and the facility now houses dozens of aviaries that support captive-breeding efforts for Hawaiian forest birds, including the ʻAlalā.

With improvements in their location and care, captive ʻAlalā produced more eggs but were still plagued by low hatching rates. The breeding program finally enjoyed a much-needed boost when eggs taken from wild ʻAlalā in 1993 and 1994 were successfully incubated and hatched in captivity with the help of new program partners that included the San Diego Zoo and The Peregrine Fund. Lōkahi, Hiwahiwa, Hoapili, Mālama, and Kēhau were products of this success, becoming the first cohort of ʻAlalā released to the wild, in 1993. In 1994, seven more captive-reared juveniles were released from the same site where I had watched over my charges. (Four of the 1994 cohort were the hand-raised offspring of captive parents; three were hand-raised from eggs taken from the wild Keālia pair, and so were Mālama and Kēhau's siblings.) The 1994 cohort was held longer in the aviary so that if they contracted malaria, they would do so *before* being released to the wild and could be treated in an environment that offered abundant food and safety from predators. One of these birds developed pox lesions and tested positive for malaria prior to his release but gave no indication of being ill.[6] While Peregrine Fund biologist Peter Harrity and others monitored the 1994 ʻAlalā post-release, as we had done with the 1993 cohort, plans were underway to grow the captive flock to stave off what many viewed as the ʻAlalā's inevitable extinction.

In 1996, a second captive-breeding facility—the Keauhou Bird Conservation Center—was constructed just outside Hawaiʻi Volcanoes National Park, on the Big Island, to ensure that captive ʻAlalā would survive if a disease or natural disaster destroyed one of the captive populations.[7] Located on 155 acres of regenerating ʻōhiʻa forest, the facility included incubation and hand-rearing rooms, and spacious enclosures that eventually would house two-thirds of the ʻAlalā population.[8] Open to the elements, the outdoor aviaries had roofed sections under which the birds could shelter in inclement weather. Patches of live vegetation were planted amidst the lava stones that covered the ground in each aviary, and animal keepers provided the birds with branches of native fruit trees to supplement a carefully designed captive diet.

Shortly after construction of the Keauhou Bird Conservation Center, the captive flock was divided between the two breeding facilities, and Hawaii's Division of Forestry and Wildlife turned over the propagation program's leadership first to The Peregrine Fund and later to the San Diego Zoo.[9] Benefitting from improved husbandry techniques and expert advisement, both facilities adopted standardized holding and breeding protocols. Nevertheless, raising ʻAlalā in captivity proved to be challenging. Breeding pairs required separate aviaries, many pairs were incompatible, and some males interfered with their mate's egg-laying and incubation attempts.[10] Because the captive flock was descended from only nine individuals or founders, with almost half of the flock descended from a single pair, ʻAlalā pairings were managed to maximize the genetic contribution of the other founders and reduce additional inbreeding. Despite chronic low-hatching rates (only 46 ʻAlalā fledged from 375 eggs laid in captivity between 2000 and 2008) that may have been related to inbreeding and low genetic diversity—common dilemmas plaguing small, remnant populations—propagation improved at first fitfully, and then steadily over the ensuing years.[11] After foundering for so long, the captive ʻAlalā population finally began to grow.

For a time, the prognosis for wild ʻAlalā seemed as hopeful as it did for the captive flock, with eight of the twelve birds reintroduced in 1993 and 1994 still alive in 1996. But as with many endangered species programs, the Hawaiian Crow's path to recovery traveled a roller coaster of successes and failures, buffeted by the headwinds of agency, land ownership, and personality conflicts; inadequate funding; and unforeseen pitfalls.

Years later, in the fall of 2015, I read Mark Jerome Walters's tendentious account of the missteps and milestones in trying to save this charismatic corvid, hoping to discover—despite my trepidation—the fates of the ʻAlalā that had made such an indelible impression on me. Walters's book was illustrated with silhouettes of individual ʻAlalā arrayed on family trees at different stages of the recovery program.[12] To my delight Kēhau, Lōkahi, and Hiwahiwa were depicted on one of the earliest trees. To my dismay, tenacious Mālama and vociferous Hoapili were not. Although losses such as theirs were a typical outcome for most young animals and are expected in reintroduction programs, images of Mālama scolding me after recovering from avian malaria and Hoapili broadcasting her outsized cries while chasing her sisters immediately came to my mind. And I belatedly mourned those birds the way one mourns the loss of a cherished pet.

Mālama had disappeared at the end of February 1994, about six months after his release to the wild. When we reminisced, in 2017, about the first 'Alalā release, US Fish and Wildlife Service biologist Donna Ball, who spent virtually her whole career working to effect 'Alalā recovery, told me that dominant juveniles like Mālama seemed vulnerable to 'Io attacks because of their tendency to loudly proclaim their status from exposed, elevated perches. Wild adults that engaged in such behavior likely had a greater awareness of the danger posed by 'Io than did naive youngsters. Adult pairs may also have been less vulnerable to 'Io attacks by working together to ward off hawks, by engaging in distraction displays and evading attacks, and by being more vigilant and protective of one another. Indeed, when they were two or three years old, Kēhau successfully distracted an 'Io that had pinned his chosen mate, Lōkahi(!), in a tree, allowing her to escape without injury.[13]

Peter Harrity also expressed concerns regarding Mālama's bold behavior in the face of danger. After hearing the Kalāhiki adults and young 'Alalā alarm calling at an 'Io near camp, Peter had observed the crows briefly swarming the flying raptor before returning to their perches. Only Mālama continued his bold but foolhardy pursuit of the 'Io, disappearing downslope in the predator's wake. Alternatively, despite his dominant demeanor, Mālama might have been vulnerable to predation, hunger, or other threats if his health was compromised by having contracted avian malaria. Whatever ultimately killed him, Mālama's fate, like Hoapili's, would forevermore be listed as "unknown."[14] Given her submissive status and tendency to lag behind her sisters, I was less surprised to hear of Hoapili's disappearance in August 1994. Having worried about her prospects because of her smaller stature and slower development, I could only be glad that she had spent a year in the wild with her siblings.

Days after learning about Hoapili and Mālama, I sat in my garden in Wyoming, surrounded by late-summer blooms and the husky chip notes of migrating Wilson's Warblers, turning the pages of Walters's book. With little preamble, the author recalled receiving an email from biologist Donna Ball in August 1997. "The remains of HiwaHiwa were recovered near the south boundary of Kai Malino," Ball wrote. "Her transmitter was located 6 feet off the ground in an uluhe [false staghorn fern]. . . . Shortly after the carcass was discovered a juvenile 'Io flew into the area with its parents."[15]

I was blindsided. Although time had tiptoed along, leaving only the faintest suggestion of its breadth, over two decades had passed since I traipsed after the first cohort of captive-reared 'Alalā, laughing at their antics, delighting in their unique personalities, and marveling at their vocalizations. Yet tears sprang to my eyes and I abruptly laid the book aside.

I spent the rest of the day trying to distract myself with an escapist book, walks with my border collie, anything that took my mind off cascading thoughts of a curious, confiding, bright-eyed bird that had captured my heart and then thrived in the wild for four years before being struck down by a Hawaiian Hawk. Although Hiwa's cause of death couldn't be verified, the presence of 'Io and the condition and location of her remains were suggestive.

In addition to being a profound emotional loss to those who had known her—Peter Harrity told me with quiet sincerity when we spoke in 2017 that Hiwa's death broke his heart—her untimely demise was a serious blow to the recovery program.[16] Having paired with a male known as The Bachelor—believed to be the only juvenile in the remnant population of twelve wild 'Alalā when I arrived in Hawaii—Hiwa had been nest-building with her mate and was one of the recovery program's best hopes for sustaining the wild population.[17] Her death foreshadowed a growing threat to the reintroduced 'Alalā.

———

No Hawaiian Crows were released to the wild in 1995 because the captive flock produced no young and the wild birds produced only one infertile egg that year.[18] But nine juveniles—hatched in 1996—were released in 1997, joining the eight captive-reared crows that still survived in the wild. Rather than being freed from the same release pen—dubbed the Waiea aviary—from which the 1993 and 1994 'Alalā cohorts had taken flight, the 1997 birds were released in two separate groups (one in January, the other in September) from a newly constructed satellite aviary, located about a mile away. Unfortunately, the habitat in this area proved to be of poorer quality, with more pastures and forest openings than were present near the Waiea aviary. Food was abundant near the satellite aviary, but feral cats were more prevalent. The wide North-South Road, on which I had often run, slashed through the forest nearby. Although the dirt road saw little human or vehicular traffic, biologists were nonetheless concerned that the newly released 'Alalā walked along the road, picking insects out of cow patties that littered the area.[19] My 1993 cohort had also spent a great deal of time foraging on the ground, but they did so in thicker forest with more protective ground cover.

The Kalāhiki and Keālia adults, Hiwahiwa, Kēhau, and Lōkahi soon discovered the satellite youngsters and treated them as aggressively as the Kalāhiki pair had once treated my quintet. But the older birds visited the satellite aviary infrequently, and the resulting absence of vigilant adults made the juveniles more vulnerable. Ultimately, the suboptimal habitat around the satellite aviary provided inadequate cover to protect the naive 'Alalā.[20] Opportunistic 'Io—bird-killing specialists that occur only on the Big Island—became a serious threat, as they capitalized on the unnatural openness of the habitat and the vulnerability of the parentless crows. Reintroduced 'Alalā also succumbed to toxoplasmosis for the first time, revealing that feral cats posed a devastating double-threat to the birds.

Seven of the reintroduced Hawaiian Crows, including Hiwahiwa, died in 1997. Hopes for the 'Alalā's future suffered an even crueler blow the following year with the deaths of eight more birds.[21] Stunned by this unsustainable level of mortality, biologists recaptured the remaining crows—including Kēhau and Lōkahi—and returned them to captivity. Their genes were too important, their lives too valuable, their future in the wild too precarious to place them at continued risk, given the high rate of attrition experienced by the reintroduced 'Alalā.[22]

During the reintroductions in the 1990s, 'Io—themselves a threatened species—killed seven of the twenty-seven 'Alalā that were released to the wild. Another 'Alalā was killed by a feral cat or small mammal. At least three (and possibly four) youngsters succumbed to toxoplasmosis, and two died of bacterial or fungal infections. Eight of the reintroduced juveniles died of unknown causes or simply disappeared, like Mālama and Hoapili. And, finally, the six remaining crows were returned to captivity.[23]

Considering the high number of fatalities, it was all the more remarkable that Lōkahi and Kēhau survived in the wild for five years. 'Alalā from later release cohorts perished more quickly and in greater numbers. In addition to expressing concerns about poor habitat quality, biologist Paul Banko speculated that sequentially reintroducing groups of 'Alalā into the same general area created an unnatural and disruptive social environment that compromised the crows' well-being.[24] Even historically, when wild 'Alalā were abundant, they were not seen in the large flocks that often characterize American Crows and Common Ravens. 'Alalā were highly territorial and family oriented. Newly released crows may have disrupted attempts by older reintroduced crows to establish territories and breed. And, as intruders into a growing flock of established crows, the newly released juveniles may have been pushed into ever-less-suitable habitat by dominant older birds. Such pressures likely compounded the ever-present threats of native and exotic predators and introduced diseases.

Donna Ball recaptured Kēhau in September 1998, six days after a coworker captured Lōkahi. The two five-year-olds had constructed nests, but Lōkahi had not laid eggs in the wild, even though 'Alalā are considered sexually mature at two to three years of age. Despite the heartbreak of returning these now-wild 'Alalā to captivity, Donna and others consoled themselves with the hope that these survivors might one day serve as mentors for future 'Alalā when reintroductions were reinitiated. First, though, additional habitat needed to be secured and threats to 'Alalā had to be better understood and mitigated.

After spending a few days back in their old release pen, Lōkahi and Kēhau were transferred to the Keauhou Bird Conservation Center. There, they were housed together in a large aviary. The flight cages housing breeding pairs of 'Alalā are widely spaced and visually obscured from one another by trees and shrubs so that pairs can hear each other's calls but will not be stressed by seeing crows that they might view as territorial intruders.[25] Captive-breeding personnel hoped that once Lōkahi and Kēhau adjusted to captivity, they would resume their breeding activities. But although the pair constructed stick-nests, Lōkahi still failed to lay any eggs. Finally, a surgical examination revealed that she had oviduct abnormalities that prevented her from producing and laying eggs.[26]

Because she couldn't produce young, Lōkahi was separated from Kēhau and transferred to a large education aviary where she could be viewed by Hawaii's growing cadre of 'Alalā fans.[27] Doubtless, she provided hope and inspiration to many regarding the 'Alalā's future, but when Peter Harrity visited her and saw the intrepid bird he had once pursued through the forest confined by netted walls, he felt only heartbreak. Future 'Alalā reintroductions were repeatedly delayed, and Lōkahi never again flew through the cloud forest, teaching new generations how to secure hidden insects or hang upside down from drooping branches while feeding on succulent fruit. She died on February 4, 2013, at nearly twenty years of age.

Kēhau, meanwhile, received a new mate. Rather than allowing captive 'Alalā to choose their own mates, biologists create pairings that minimize inbreeding and maximize the species' genetic diversity. If members of a juvenile flock show a preference for one another, such pairings may be maintained if they meet genetic criteria. When genetically desirable pairs are not particularly compatible, animal keepers may restrict the male's role only to "conjugal visits" or temporarily separate a pair if a male is overly domineering.[28] The crows are monitored from a control room containing screens that live-stream video from cameras housed in each aviary. By watching the birds' behavior, researchers can ensure pairs are compatible, monitor nest-building activities, determine when the females are preparing to lay eggs (females frequently nestle into their nest cup prior to laying), and intervene, if necessary, to ensure that eggs hatch successfully or are removed for artificial incubation at the right time.[29]

Sadly, Kēhau's new pairing did not bear fruit and he never produced any offspring. Instead, he succumbed to a debilitating illness after only a few years in captivity. Despite months of effort—including intensive drug therapy—by veterinarians and staff, who labored to restore him to good health, Kēhau was finally euthanized on July 18, 2002, at the age of nine.

A necropsy revealed that he'd had a large brain tumor. Even more troubling, given his four-year tenure in captivity, a histopathological examination revealed that Kēhau was infected with toxoplasmosis.[30] Upon learning about Kēhau's fate, I once again saw, in my mind's eye, the baleful, feral eyes of a trapped tabby cat. Although I unequivocally supported the removal of introduced predators, I'd been dismayed to witness it and hated that such measures were necessary. Now, though, I simply felt overwhelmed with heartache for all that we humans had wrought with our cavalier and misguided introductions of exotic pests. Until we managed to restore the habitats we had so badly damaged, there would be no winners in what once had been an avian Eden.

––––––––

Under the aegis of the Hawai'i Endangered Bird Conservation Program, overseen by the San Diego Zoo Wildlife Alliance in partnership with Hawaii's Department of Land and Natural Resources–Division of Forestry and Wildlife and the US Fish and Wildlife Service, the captive 'Alalā population grew steadily. Numbering forty-one crows in 2003, it climbed to sixty by 2009.[31] Small milestones prompted exuberance in those who worked in thankless anonymity to keep the population growing, help it retain its fading wildness, and prepare it for a better future. In April 2014, keepers celebrated the first unassisted hatching and rearing of Hawaiian Crow nestlings by a captive female in over twenty years. The breeding program had long focused on maximizing the 'Alalā's reproductive output by artificially incubating eggs and hand-raising chicks. But with 114 'Alalā now in captivity, biologists could let some birds raise their own young.

Scientists, conservationists, and a public that had a growing awareness of the 'Alalā's uniqueness and plight, celebrated the expanding captive population. But most recognized that it was a means to an end, an attempt to preserve a charismatic species until its habitat could be rendered hospitable again. Biologists, in particular, recognized that the 'Alalā population would continue to evolve in captivity just as it had in the wild, and traits that allowed a bird to thrive in an aviary were unlikely to help its offspring once they were introduced to life in the forest.

Time was needed to grow a genetically diverse captive population that was large enough to withstand the fatalities that inevitably occur during reintroductions. But the longer the birds remained in captivity, the more likely they were to lose their wildness—their ability to avoid predators, seek out appropriate foods, and interact with each other and with their ecological community the way their species had done successfully for eons. Captive 'Alalā were also likely to lose "cultural" traditions—such as knowing when and where to take advantage of seasonally abundant foods—that were important to a species dependent on learning rather than instinct for its survival in the wild. On the other hand, returning 'Alalā to the wild prematurely meant reintroducing birds to habitats in which the species had failed to thrive. Reintroductions were unlikely to succeed in forests where rats destroyed eggs and nests, mongooses and cats attacked unwary fledglings, mosquitoes transmitted avian pox and malaria, cat feces harbored toxoplasmosis, or cows and pigs destroyed vegetation, leaving crows with no refuge from predatory hawks.

So while caretakers worked to grow the captive population, biologists and others tackled the unglamorous but vital tasks of securing land, managing and improving habitat, and laying the groundwork for the 'Alalā's much-anticipated return to the wild.

CHAPTER FIFTEEN

Second Chances

Sometime in 2014, biologist Donna Ball navigated the rough track that wound up to the site where captive-reared ʻAlalā had been released to the wild two decades earlier. Approaching the Waiea aviary, she passed trees where her feathered charges had once chipped bark with their powerful beaks, plucked berries from swaying branches, and preened one another's fuzzy heads. Its hack-box prow still looming out of mist-wrapped trees, the aviary looked like a ghost ship adrift in a sea of greenery, its wooden sides worn and weathered by innumerable rains, its netting torn and tattered by the unrelenting press of growing branches, its course stalled by a maelstrom of external impediments. But what struck Donna most was not what remained of the structure or the memories it held, but the silence that surrounded it—an eerie, deafening silence that spoke volumes about loss, absence, and unwanted change.

In the 1970s, Winston Banko, who dedicated much of his life to conserving Hawaii's birds, warned presciently that midnight was nigh for the ʻAlalā, and only a comprehensive recovery program that included the establishment of sanctuaries, captive breeding, and increased conservation and research would forestall the bird's path to extinction.[1] Dragging the crow back from the precipice—an effort now still in the earliest stages despite the passage of decades—has taken all that and more. The ʻAlalā's future is uncertain. The bird's forest remains largely silent.

Legally protected by the State of Hawaii since 1931 and among the first species to be listed as endangered under the federal Endangered Species Preservation Act of 1966 (a precursor to the Endangered Species Act of

1973), the Hawaiian Crow has endured years of conservation efforts that didn't always achieve their objectives.[2] As biologist Paul Banko stated:

> Perhaps because the field of players in Hawaiʻi is relatively small, individual efforts, especially when motivated by personal beliefs or agency affiliations, have sometimes strongly affected outcomes—in both positive and negative ways. . . . At the same time, management action is often delayed or thwarted as government agencies struggle to provide funding and leadership for a cause that, by local standards, may seem hopelessly expensive, technically difficult, controversial, and distracting from other urgently needed conservation.[3]

The US Fish and Wildlife Service's first recovery plan for the ʻAlalā was completed in 1982, though it wasn't until the agency commissioned a body of experts to review the bird's status and recommend conservation measures a decade later that recovery efforts began in earnest and the tide turned on the species' falling numbers.[4] Subsequent successes in the long-stalled captive-breeding program and the first reintroductions of captive-reared birds to the wild presaged future successes that, bewilderingly, failed to materialize. Some viewed the 1990s reintroductions as a failure. Even the Service, in its revised 2009 ʻAlalā recovery plan, conceded: "Twenty-one of the 27 released birds died from disease, were depredated, or disappeared. The remaining six were returned to captivity in 1998 and 1999. The prediction that released birds would integrate into the wild population was not borne out—there was no reproduction, and only limited reproductive behavior was observed in the released birds."[5]

An optimist, however, might view these early efforts as a beginning—and a reasonably successful one at that. One-fifth of the reintroduced birds survived before being brought back into captivity. Hiwahiwa spent four years in the wild and appeared to be engaging in courtship activities with a wild ʻAlalā when she met her untimely end. Lōkahi and Kēhau survived for five years in the wild and were initiating breeding activities that might have been successful had not Lōkahi had a problem with her oviduct. Aggression toward introduced juveniles by territorial adults, the social disruption caused by releasing successive cohorts in the same general area, and perhaps even inherent incest avoidance mechanisms triggered by rearing birds together may have forestalled the reproductive activities of other reintroduced ʻAlalā that reached breeding age. Had wild adults been breeding successfully during these reintroductions, perhaps their offspring would have paired with captive-reared birds.

The 1990s reintroductions also provided critical information about the factors leading to the species' decline, the characteristics of quality 'Alalā habitat, and the challenges of releasing intelligent, behaviorally complex birds. Threats such as toxoplasmosis and 'Io predation were unknown before these efforts.[6] Ultimately, what was learned informed habitat restoration in the ensuing decades that would benefit future 'Alalā.

The positive outcomes of the 1990s reintroductions may be better appreciated when viewed in the context of other avian recovery efforts. Restoring peregrines in North America was bolstered by the release of nearly 7,000 captive-reared falcons over a period of decades. Hundreds of reintroduced peregrines were killed by Great Horned Owls, collisions, and other causes before the species was successfully recovered. Hack sites whose peregrines became the targets of predatory owls were usually abandoned in favor of ones that posed a lesser threat to vulnerable, parentless birds. In Hawaii, in the 1990s, such adjustments were impossible since few suitable areas were available in which to release 'Alalā. Moreover, operational constraints were significant even in the one area where wild 'Alalā remained: the private McCandless Ranch, which leased the state land on which the Waiea aviary was built after lawsuits were settled over access for biologists.[7]

Further, reintroducing peregrines, which have relatively simple social systems and innate hunting and predator-avoidance behaviors, proved to be simpler than reintroducing a socially complex species whose survival depends on lengthy parental care and learned behaviors. Peregrines also faced one overriding threat—pesticides—that could be mitigated through legislation, whereas 'Alalā face a complicated assortment of human-caused problems that undermine the quality of their habitat, leaving the bird vulnerable to introduced diseases and predation pressures for which it has not evolved suitable defenses. Once contaminants were reduced or eliminated from the peregrine's environment, the bird recolonized vacant habitat that was still suitable. But none of the 'Alalā's remaining habitat was untouched, and the crows were reintroduced to the same compromised habitat that had failed to sustain the wild birds.[8]

Captive breeding can be a vital stopgap measure to forestall extinction, retain genetic diversity, and bolster population numbers. But the 1990s 'Alalā reintroductions provided irrefutable evidence that the recovery of imperiled species can succeed only if the underlying reasons for the species' endangerment are adequately addressed and mitigated. Nevertheless, treating the often-intractable environmental perturbations responsible for a species' decline can be time-consuming, logistically challenging, and expensive.[9] Indeed, acquiring and restoring suitable habitat—the most vital measure to recovering the 'Alalā—has taken decades.

The first sanctuary for Hawaiian Crows and other endangered birds was designated on state-owned land in 1984. Located near the dry northwestern edge of the 'Alalā's historical range and isolated by surrounding ranching activities, the sanctuary's habitat was compromised by feral ungulates, exotic vegetation, and a fire that burned part of the area. Even so, the sanctuary provided important habitat for the crows until their disappearance in the 1990s. A second prospective 'Alalā sanctuary—the nearby Kona Unit of the Hakalau Forest National Wildlife Refuge, a seasonally wet forest acquired by the US Fish and Wildlife Service in 1997—seemed more promising. Unfortunately, protracted legal disputes over access to the Kona Unit and other former-landowner concerns impeded refuge management for years. As a result, plans to fence then eradicate feral ungulates from this area—located in the Hawaiian Crow's core range and home to the last wild 'Alalā—were delayed.[10] Unlimited access to the refuge was finally won in 2005, and habitat restoration work began in 2008, over a decade after the area was acquired and more than half a decade after the 'Alalā went extinct in the wild.

In the meantime, suitable 'Alalā habitat continued to dwindle as conservation efforts were undermined by limited funding, competing land-use priorities, and the inherent difficulties in restoring the Big Island's forests. Mid-elevation forests in the Kona area were fragmented by housing lots and agriculture. Old-growth koa was logged on private lands, though reforestation became more commercially attractive. Cattle grazing declined in native forests, but sheep and feral pigs—avidly sought by hunters—remained common. Mosquitoes, rats, cats, and mongooses were prevalent, even though intensive control efforts reduced their numbers in targeted areas.[11]

Cognizant of the drawbacks to maintaining wildlife in small islands of suitable habitat surrounded by disturbed landscapes, biologists looked beyond the fragmented forests that had sustained the last 'Alalā to larger, more contiguous habitat patches within the bird's historical range. They identified areas that encompassed wide elevational ranges and diverse habitats, which would allow reintroduced 'Alalā to seek out seasonally available foods. The best of these options, the Kūlani-Keauhou area, which ultimately was selected as the next reintroduction site, was subsequently managed to promote the native forest. The site, which is now fenced and has been free of ungulates for several decades, has a dense understory and an abundance of 'Alalā food plants.[12]

Improving degraded lands in Hawaii has increasingly involved fencing. Because the state has no native terrestrial mammals, reptiles, or amphibians, fences don't obstruct native wildlife. Instead, fencing keeps alien animals out of protected areas and creates workable spaces within which invaders, from feral pigs to mice, can be eliminated. Feral ungulates have now been eradicated from many fenced areas by hunting, while exotic predators have been reduced by trapping and poisoning.

Sophisticated traps are now being used to eliminate introduced rodents at local scales. In particular, Goodnature traps, which make humane kills and reset themselves up to twenty-four times before being rebaited, have transformed rodent control in remote forest-bird habitat and seabird colonies, where biologists often have limited access.[13] Rat poisons such as diphacinone—an anticoagulant that doesn't persist for long in the environment and appears to pose a relatively low risk to nontarget species like feral pigs and introduced birds—have also been effective at eliminating rodents and mongooses from natural areas.[14] In addition to being used at bait stations, such rodenticides have also been aerially broadcast from helicopters—a less labor intensive, more effective, and more practical way to eliminate nonnative predators from remote areas.[15]

Eliminating rodents decreases nest predation, which benefits Hawaii's endemic birds by allowing more nestlings to survive. Since rodents will eat even adult birds as they are incubating eggs or brooding young, controlling rodents also increases adult survival.[16] When more adult and young birds survive, more genetic diversity is preserved in the population, which can foster enhanced disease resistance. Indeed, controlling predatory rodents at mid-elevations, where avian malaria is prevalent, may allow endemic bird populations to persist, helping to facilitate the evolution of malaria resistance in some currently endangered Hawaiian birds.[17]

Perhaps the most promising lesson learned from the 1990s reintroductions was that avian malaria poses less of a risk to 'Alalā than it does to Hawaii's other endemic birds. Despite the severity of Mālama's illness and the treatment of several other 'Alalā that were weakened by apparent malarial infections prior to their release, avian malaria and pox are not known to have killed any reintroduced Hawaiian Crows. Even though other native forest birds have been less resilient, protecting and restoring future 'Alalā habitat may help sustain these other species until the application of new technologies can disrupt the mosquito's unrelenting obliteration of Hawaii's native avifauna.

———

Although the battle to combat the worldwide scourge of introduced organisms is still in its infancy, conservation actions *are* having a positive effect. For every five species that have become rarer globally because of the impact of exotics, two species now show positive population trends because of programs to control or eradicate these invaders.[18] The magnitude of the threat still outweighs the conservation actions that are being marshaled against it, but better funding, new technologies, and more targeted and widespread control efforts may yet bolster native species that have been harmed by the introduction of exotic organisms.

Conservation efforts have significantly reduced the threat of extinction in North America since the enactment of the Endangered Species Act in 1973, improving population trajectories for 20 percent of threatened mammals and birds.[19] Single-species conservation efforts are sometimes criticized for being too resource intensive and focused on charismatic species to the exclusion of organisms that are less likely to capture the public's imagination. Yet recovering the 'Alalā—whose presence is an indicator of forest health—would not only secure quality habitat for other forest-dependent species but would also restore ecological processes by reestablishing a critical seed disperser that long shaped the structure and diversity of Hawaiian forests.[20] Protecting and restoring these forests in recent decades already has improved prospects for native birds, with numbers of some species stabilizing or increasing despite seemingly insurmountable odds.[21]

Habitat improvement efforts are now sustained through a growing recognition of the threat posed by exotic organisms and from stricter policies to keep new invaders out of Hawaii. In 1999, President Clinton established the National Invasive Species Council. Cochaired by the United States Secretaries of Agriculture, Commerce, and Interior, the Council mobilizes and coordinates the government's efforts to defend against harmful alien species by collaborating with state, local, tribal, and territorial governments, along with stakeholders and the private sector.[22] In December

2016, President Obama, who grew up in Hawaii, reaffirmed the federal government's commitment to eradicate and control invasive species, and called for environmental- and human-health concerns, climate change considerations, and technological innovations to be incorporated into federal efforts to address the costly impacts of introduced organisms.[23]

To help coordinate statewide efforts to prevent and address invasive species infestations, the State of Hawaii formed the Hawaii Invasive Species Council in 2003, which engages in proactive efforts to identify and control alien pests.[24] Nevertheless, managing invasive species and conserving the state's unique biota could be undermined if insufficient funding or inadequate regulations on international commerce allow future introductions of pathogens (such as West Nile virus), disease vectors, or other harmful organisms into the Aloha State.

Funding scarcities have stymied the conservation of Hawaii's natural legacy, particularly its imperiled fauna and flora. Nationwide, federal funding plays a critical role in reversing declining bird populations. As a 2016 study found, threatened or endangered birds that received more conservation funding were more likely to have growing population trends.[25] At-risk island birds in the United States receive significantly *less* funding than do mainland birds and are more likely to show declining population trends. Resources directed to imperiled Hawaiian birds have long fallen far short of their need.[26]

Even though Hawaii had more than a quarter of all the birds listed as endangered in the United States in 2014, the state received less than 7 percent of the federal recovery funds that were slated for birds.[27] Such funding inequities were underscored between 1996 and 2004 when expenditures for the Whooping Crane and the California Condor were twenty-four and six times greater, respectively, than expenditures for the Hawaiian Crow, even though the 'Alalā's population status was more dire and the threats facing the crow were more difficult to mitigate.[28] These discrepancies may be partially blamed on perceived charisma. The general public, unfamiliar with the 'Alalā and its desperate plight, has been quicker to rally around cranes and condors than around crows. Between 1996 and 2004, approximately $1.7 million was dedicated to conserving the 'Alalā.[29] Meanwhile, the best-funded imperiled bird in the United States, the Red-cockaded Woodpecker, received more than $112 million—over three times what was available to *all* thirty-one endangered and threatened Hawaiian species. Despite their dire need, not one Hawaiian species was among the twenty top-funded birds, which received 85 percent of all the available funds for listed birds.[30]

Because of the severe pressures facing Hawaii's endangered avifauna and the insufficient funding for dealing with those stressors, the recovery rate for imperiled Hawaiian birds is significantly lower than it is for mainland birds. Whereas more than three-quarters of mainland endangered bird populations are stable, increasing, or have recovered enough to be delisted thanks to the Endangered Species Act, only about half of Hawaiian bird populations show such positive trends.[31] And for those who bear witness to the declines, for those who feel impoverished by the erosion of the natural world, for those who fight in ways large and small to hold fast to unique and beloved life forms, such declines—such losses—are immeasurable and eviscerating.

As Hawaiian biologists Sheila Conant and David Leonard stated poignantly in 2008, "People in Hawai'i personally witnessed [the 'Alalā's] decline year after year for nearly five decades—a period spanning entire professional careers or more. During these years, countless petitions and pleas from government biologists, academic scientists and conservation advocates for support to save the species went virtually unanswered, as do similar requests for other endangered Hawaiian bird species today."[32] For Conant and others who fought to conserve the 'Alalā, funding inadequacies and inequities were nothing short of tragic. At long last, the plight of Hawaii's imperiled forest birds finally garnered sufficient attention that the US Fish and Wildlife Service expanded its recovery efforts, spending just over 18 percent of its avian recovery funds on Hawaii's birds in 2016.[33] In recent years, federal, state, and private entities have rallied to forestall Hawaii's ongoing extinction crisis, with the Department of Interior announcing in June 2023 a historic commitment of nearly $16 million to restore Hawaiian birds.[34]

————

In 2016, while searching for a job after returning to Montana's mountain country, I came across a listing for "Research Assistants—'Alalā/Hawaiian Crow Reintroduction." As vibrant memories of my Hawaii sojourn flooded my mind, I read that:

Beginning in September 2016, approximately 12 birds will [be] reintroduced annually to suitably restored habitat. The

Research Assistant will work as part of a team responsible
for the post-release monitoring and husbandry of 'Alalā.
Primary duties include: caring for birds in pre-release
aviaries, radio tracking released individuals, preparing and
provisioning supplementary food, assisting in recapture and
transmitter attachment, monitoring behavior and condition
of birds both during captivity and following release, detailed
record keeping, GIS data management; and other duties as
assigned. Staying overnight at remote sites will be required.

Aside from managing Geographic Information System (GIS) data, the ad
perfectly described my 1993 'Alalā duties. Thoughts of Lōkahi, Hiwahiwa,
Hoapili, Kēhau, and Mālama, who still felt like cherished friends, spurred
an intense longing to abandon my responsibilities, return to the Hawaiian cloud forest, and lose myself in the riotous and uplifting company of
the world's rarest corvid. Regrettably, succumbing to that temptation was
impossible, but I could still delight in knowing that 'Alalā would once again
broadcast their resounding calls from the tops of ancient koas.

After years of planning and preparation, the Hawaiian Crow recovery
team was finally ready to return 'Alalā to the wild. Before resuming reintroductions, biologists had wanted the captive population to number at least
75 'Alalā to prevent further losses of the population's genetic diversity. Over
the years, the captive population's productivity had finally increased. Now,
15 or more fledglings could be produced each year, providing a sufficient
number of birds to release to the wild without diminishing the captive
population.[35] More than 115 'Alalā now resided in captivity. Habitat had
been protected and restored. Release facilities had been constructed. Protocols had been debated and agreed upon.[36]

Juvenile 'Alalā would be introduced to the wild—fourteen years after the
last wild adults disappeared—in the Pu'u Maka'ala Natural Area Reserve.
Located on state land on the southeast slope of Mauna Loa near the
Keauhou Forest and Hawai'i Volcanoes National Park, the reserve had been
managed for years to restore forest health. Fenced and free of feral ungulates
and introduced predators (despite the costly destruction of 2 miles of fence
by vandals in 2015), the proposed release area—on the edge of the 'Alalā's
historical range—received more rainfall than did the last site the Hawaiian
Crows had inhabited, but was deemed high-quality habitat, with a healthy
understory and abundant fruiting plants.[37]

In October 2016, five juvenile male Hawaiian Crows were transferred to a large aviary in the reserve to acclimate the birds to the forest.[38] (Since corvids may be reluctant to pair with birds they regard as siblings, females would be released later.) In November, the 'Alalā Project—overseen by Hawaii's Division of Land and Natural Resources—held a community event celebrating the iconic bird's imminent return to the wild. Over 600 people attended.[39] In the intervening years since the 'Alalā's extinction in the wild, project partners had worked tirelessly to inform Hawaiians about the plight of their native crow. Long revered in traditional Hawaiian culture, the 'Alalā had now acquired a more universal appreciation, with many Hawaiians viewing their celebrated endemic the way mainlanders view our national symbol, the Bald Eagle.[40] Hawaiians were particularly proud that the 'Alalā occurred nowhere else in the world and had garnered recognition as a highly intelligent tool user. Local fourth graders engaged in a naming competition, bestowing Hawaiian names on the soon-to-be-released birds. An artist worked with children to paint a community mural of the 'Alalā and its forested habitat on a historic building in Hilo, Hawaii.[41] And Hawaiians, young and old, avidly followed news of the birds' impending release.

A week before the scheduled reintroduction, the juvenile 'Alalā were transferred to a smaller release aviary. Finally, on December 14, 2016, biologists opened the aviary doors, allowing the groundbreaking quintet to fly free. The 'Alalā emerged slowly into the adjacent forest, cocking their heads curiously and scanning their surroundings with bright, intelligent eyes. Soon, they were exploring, gaining strength, and moving farther afield in search of insects and berries. After nearly fifteen years of relative quiet, the forest rang once again with the 'Alalā's raucous, exuberant calls. And observers around the world celebrated the 'Alalā's second chance at rejoining Hawaii's forest-bird community.[42]

Then shockingly, shatteringly, tragedy struck. Within two weeks, three of the youngsters were dead and biologists were scrambling to recapture the two survivors, return them to captivity, and field a thousand questions about what had gone wrong. Necropsies revealed that two of the crows had been killed by 'Io. The third appeared to be in poor physical condition, suggesting that he hadn't been meeting his nutritional needs.

The loss was an unimaginable blow to those who had labored long and tirelessly to bring the 'Alalā recovery program to this point. Particularly stricken were the caretakers who had raised the youngsters in captivity and relinquished them with a mixture of trepidation and hope to recolonize the wild. Heartbroken school children asked baffled parents why the bird *they* had named died. Condolences poured in from around the world. Plans to release a second cohort of females were shelved. Biologists, conservationists, and spokespeople tamped down their own bitter disappointment and assured 'Alalā fans that successfully reintroducing captive-bred animals was a lengthy process that often experienced setbacks. Indeed, recovering the endemic Nēnē had taken over five decades. Then the 'Alalā recovery team began the difficult discussions about what had gone wrong and what could be done differently the next time around—in 2017.

The most significant change would be to release the 'Alalā at higher elevations within the reserve. Although individual 'Io might occur above about 5,200 feet (1,600 m), they rarely defended territories at such high elevations and would be less likely to discover reintroduced 'Alalā in this part of the forest.[43] Since Hawaiian Hawks weren't known to be a significant danger to wild adult 'Alalā, the severity of the threat they now posed to reintroduced crows was puzzling. Many believed that the 'Io depredations of reintroduced juveniles in the 1990s were linked to poor-quality habitat that lacked protective cover.[44] But other factors may also have been at play. Low 'Alalā densities as the population declined may have left individual crows more susceptible to attack. A lack of wariness in captive-reared birds coupled with the absence of protective parents to spot and ward off predators may also have increased their vulnerability.

Upon discovering the danger 'Io posed to reintroduced 'Alalā in the 1990s, biologists had captured and relocated a number of the endangered hawks. But adjacent territory holders or young, nonbreeders invariably filled vacant 'Io territories. Recognizing that 'Ios were themselves an important component of Hawai'i's forests, biologists had subsequently managed habitat rather than hawks to tackle the challenges 'Io posed to 'Alalā recovery.[45]

Now, though, the failure of the 2016 release suggested that habitat restoration wasn't a panacea for restoring the balance between 'Io and 'Alalā. As a result, when biologists next returned Hawaiian Crows to the wild, in the fall of 2017, they released eleven juveniles (in two groups that included the survivors from the 2016 reintroduction) in the hopes that a larger cohort of birds would be more vigilant and better able to fend off predators. Prior to their release, the youngsters received intensive predator "aversion" training to better equip them to cope with hawks. Juveniles were subjected to recorded 'Io hunting calls, followed by a simulated flyover by a taxidermy 'Io and broadcasts of 'Alalā warning calls. Next the juveniles witnessed a simulated predation event when they saw a live, flapping 'Io in an adjacent cage and heard 'Alalā distress calls as a hidden handler put a stuffed American Crow under the 'Io's talons.[46]

Incredibly, all eleven birds from the 2017 reintroduction survived their first year in the wild, providing much-needed hope that the challenges of reintroducing 'Alalā had finally been resolved. Biologists reintroduced ten more 'Alalā in the fall of 2018. Despite the subsequent loss of several reintroduced birds, which disappeared or were killed by 'Io, 'Alalā fans celebrated the formation of three wild pairs in the spring of 2019. Two pairs built nests and one female appeared to incubate eggs but produced no young. At long last, the 'Alalā seemed to be on a successful path to recovery.

Seven more juveniles were released in the fall of 2019 but, dismayingly, 'Io, winter storms, and unknown factors took a growing toll on the reintroduced population. Losses accelerated until, by March 2020, only ten crows were left in the wild. By October, only five 'Alalā remained. An alarmed recovery team decided to return these birds to captivity. Biologists spent a month trying to recapture an especially wily male, known as Kia'ikūmokuhāli'i—meaning "guardian of the forest." One of the two 2016 survivors—the other being Mana'olana, meaning "hope"—he had been reintroduced in 2017 and survived in the wild for four years.

Despite the heartbreak of returning such embodiments of hope to captivity, Paul Banko echoed the feelings of conservation biologists worldwide when he stated, after the failed 2016 release: "There is no room for pessimists in endangered species recovery efforts."[47] The seemingly insurmountable threats to the 'Alalā *can* be overcome—*will* be overcome—given enough effort, dedication, funding, and time. Humankind can be as resourceful and benevolent as it is prolific and destructive. A planned 2022 release was postponed until biologists learned more about 'Io—their distribution, daily movements, and territory size—in the hopes that better understanding the predators would lead to more successful 'Alalā reintroductions. Biologists are also considering temporary reintroductions of 'Alalā on Maui, where 'Io don't occur.

New threats certainly will arise, as well. And, indeed, some already have. Reports of rapidly dying 'ōhi'a, beginning in 2010, were a prelude to the daunting realization, by 2013, that a virulent, fungal pathogen was attacking the Big Island's crimson-blossomed trees. The most abundant and widespread native tree in Hawaii, the endemic 'ōhi'a is a keystone species—one upon which countless other organisms depend, and one whose loss would dramatically alter the state's ecosystems. Comprising more than 80 percent of Hawaii's native forests, with over half of the trees occurring on the Big Island, ʻōhiʻa range from sea level to over 9,000 feet (2,800 m) and provide essential shelter and food for endemic birds and other native wildlife.[48] Once an ʻōhiʻa is attacked by the introduced fungal pathogen *Ceratocystis lukuohia* (or the related but less aggressive *C. huliohia*), the tree's ability to move water and sugars is inhibited, its leafy crown turns brown, and the tree dies within days or weeks. This disease has aptly earned the moniker Rapid ʻŌhiʻa Death. Agencies scrambled to address the threat that some have labeled "tree ebola." But despite ongoing efforts to prevent the disease's spread to still-uninfected stands and islands, by 2020, about 175,000 acres of ʻōhiʻa forest on the Big Island showed symptoms of the virulent disease.[49]

The lurking threat posed by other exotic diseases, such as West Nile virus, which decimated American Crows when it first appeared in New York in 1999 and which could still invade Hawaii, is never far from the minds of ʻAlalā supporters.[50] But as conservationists remind those who are concerned with the fate of Hawaii's avian gems, many of the birds are *still here* despite overwhelming odds.[51] And by saving the dawn chorus that reverberates at high elevations, by restoring the lost voice of the ʻAlalā, which contributes so mightily to the forest's mystery and majesty, we will do much to restore a paradise that has been ravaged by deliberate exploitation and woeful indifference.

As difficult as the ʻAlalā's recovery journey has been and as difficult as it remains, there is still hope for the dusky bird that captivated an island, a state, and perhaps one day a nation with its singular voice, superior intelligence, and outsized personality. Habitat restoration continues throughout the ʻAlalā's former range. Control of exotic predators is more sophisticated and widespread. Disease research is exploring and applying innovative technologies. And an ever-growing cadre of fans is investing their support, their hopes, and their hearts in the ʻAlalā's future. Their numbers may not yet match those that ensured the recovery of the Peregrine Falcon, but their passions surely do, and the inimitable ʻAlalā and its beleaguered forest community are no less deserving of our conservation efforts.

Our lives are often enriched and inspired by voices—by songs that resonate with us, by melodies that lift our spirits, by harmonies that comfort us. The voice of the ʻAlalā is not ethereal like that of the Hermit Thrush. It does not cascade in liquid tones like that of the Canyon Wren. But it speaks of place—of Hawai'i's lovely, lush forests—the way the thrush speaks of deep woods and the wren speaks of canyon country. It has inspired wonder, laughter, delight, awe, and even reverence.

When the 'Alalā's voice returns to Hawai'i's forests once again, it will link the past with the present, recall a realm untrammeled by people, and inspire a future that is perchance characterized more by stewardship than by exploitation and disruption. For me, the 'Alalā's unique voice will forever contain echoes of the bold and beautiful Kalāhiki birds, who protected their realm with unwitting dignity and visible ferocity; of Hoapili, whose plaintive cries would resonate with anyone who has ever been left behind; of Hiwahiwa, who inspired with her resourcefulness and charmed with her gentle curiosity; of Lōkahi, who awed as a matriarch despite never laying an egg; of Mālama who met adversaries, seen and unseen, with characteristic boldness; and of Kēhau, who embodied quiet strength and steadfast loyalty. But when the 'Alalā returns to the wild once again, Hawai'i's forests will echo, too, with the ghost birds that others have known and, finally, with the crows of the future, which will one day write the next chapter in the 'Alalā's turbulent, heartbreaking, hopeful story.

PART III

California Condor
Hidden Poison

Do not go gentle into that good night . . .
Rage, rage against the dying of the light.

—Dylan Thomas

CHAPTER SIXTEEN

Lead Shock

S tanding on the edge of one of Peru's, and perhaps South America's, most dangerous roads, I waited all day for a bus that never came. For months now, this unpaved one-lane road—crisscrossed by streams, pummeled by waterfalls, marred by landslides, bordered by precipitous ravines—had been my lifeline to civilization and central to my research. From a desolate, high-elevation pass—at nearly 11,500 feet (3,500 m)—in the wet puna grasslands that served as a gateway to the Manu Biosphere Reserve, the road twisted down through elfin woodlands and montane cloud forest, before reaching the Amazon Basin. Impenetrable vegetation and epiphyte-laden cloud-forest trees bordered the 25-mile stretch of road that bisected my study area and provided its primary access. And more often than not, the road was shrouded in bone-chilling clouds and roiling mist.

I was sent to this remote outpost—a long day's journey from Cusco, Peru, and the most isolated place I had ever worked—by The Peregrine Fund in September 2000. Hired as the organization's South America biologist, I was tasked with studying the natural history of the fiercely elegant Black-and-chestnut Eagle. Inhabiting montane cloud forests in the Andes, from northwestern Venezuela to northern Argentina, this forest eagle, with its chestnut front and black crest of elevated crown feathers, was one of the least-studied tropical raptors in the world. The conservation status of this uncommon eagle was unknown, and the bird was persecuted for its occasional depredations of chickens belonging to farmers who were encroaching on its habitat, so I was tasked with studying its biology and ecology.

For the past few months, I had attempted to observe high-flying eagles and search for their nests in an area whose perilous vertical topography, unstable soils, and impenetrable vegetation made it virtually impossible to pursue the birds on foot—my only means of transport. My field assistant—a young British birder who was funding his travels while building his bird list—and I perched on promontories and walked endless miles along the pothole-filled road, trying to spot eagles before enveloping clouds and torrential rain obscured visibility.

Our Black-and-chestnut Eagle sightings were infrequent, though I enjoyed occasional views of pairs soaring together—shadowing each other's every move in a celestial dance—and once had the good fortune of seeing an adult capture a high-flying bird.[1] Slate-gray Solitary Eagles were far more forthcoming, engaging in highly visible courtship displays and delivering snakes to a streamside nest. And I was treated daily to a vibrant spectacle of extravagantly colored cloud-forest tanagers, electric red-and-green quetzals, iridescent hummingbirds, voluble congregations of fluorescent-orange Andean Cock-of-the-rocks, and a host of other tropical specialties such as oropendolas, barbets, toucans, and motmots. Living out of tents erected on an open-sided wooden platform near a small tourist lodge tucked into the cloud forest, we watched monkeys over breakfast and were besieged by insects during dinner.

Once a month, I returned to Cusco—and civilization—to purchase supplies, do laundry, check in with my supervisor in Idaho, and telephone my family. Now, though, I was embarking on an excursion to search for the Black-and-chestnut Eagle in the highest elevations that it ranged. A ramshackle local bus typically passed our camp, every other day. Because of the narrow road, all traffic moved downslope one day and upslope the next. But our bus never materialized that day. In vain we waited, until it became clear—after eight hours of standing by the road—that we would need another ride.

After eyeing the overflowing *camiones* that lumbered sporadically up the road, hauling goods between the Amazon Basin and Cusco, we eventually waved down a three-quarters full logging truck. Hoisting our backpacks, we clambered onto the slick load, joining four other luckless passengers and an eviscerated swine. Trying to ignore the dizzying drop-offs as the overburdened truck lurched within inches of the crumbling road edge, I chatted with a woman who showed me the carefully wrapped, jewel-bright butterflies she had collected. Slowed by the multiple-point turns needed to get around sharp corners and forced to wait hours for the road to be cleared of a broken-down cargo truck, we reached the dilapidated hut that would shelter us during our eagle surveys after dark.

Our return journey—after three days of enjoying stunning vistas but no eagle sightings—was less perilous but more uncomfortable. Carrying our overloaded backpacks, we walked the 18 miles (29 km) back to camp—clad in oversized ponchos and rubber boots, and battered by torrential downpours. Foregoing dinner and the frigid shower that awaited me at journey's end, I crawled into my damp sleeping bag, nearly whimpering from the pain of blistered feet, and hoped that my forthcoming trip to Cusco would be less fraught.

Despite shivering in a bus whose windows wouldn't close, getting stuck behind a broken-down truck, and being unnerved by a *camión* that had gone off the road and was listing down an embankment, the trip to Cusco passed in relative comfort. I was eagerly anticipating a warm shower, an actual bed, and mad-dash grocery shopping. Unexpectedly, though, the life to which I was only just becoming habituated was suddenly upended by a frantic call from my supervisor.

"I've been trying to get hold of you for almost a month," Rick said with a hint of frustration. "I didn't realize that there literally was no way to get in touch with you when you were at the field site." I tried not to feel reprimanded for circumstances over which I had no control, as he continued, "The field manager for our California Condor project is leaving unexpectedly, and we'd like you to take his place. We've hired a project director, who lives a few hours from the release area, but we need someone on site to manage the day-to-day field operations. We've held the position open for you this past month, but we're anxious to fill it." Stunned, I tried to absorb the news amidst a torrent of emotions. And over the coming days, as I struggled to decide whether or not to accept the job, I imagined a list of pros and cons arrayed on each side of an old-fashioned balancing scale that served as the gateway to my future.

My imaginary scale listed precariously to one side at any given moment then swung wildly to the other. Instead of having to navigate a yearlong absence, I would be reunited with my beloved pets and would be able to talk to my family on a regular basis. I would be working to conserve *California Condors*—one of the most high-profile endangered species in the world. I would escape the endless rain and could take warm showers whenever I wanted. But I would have to abandon working in the tropics—something

I had wanted to do again ever since working with parrots in Guatemala. I would leave behind a cornucopia of avian diversity that, bird by bird, enriched my life on a daily basis and provided me with endless fodder for investigation, enlightenment, and pure joy. I would have to turn my back on a job in which I was heavily invested, and forego potential contributions I might make to the conservation of an elusive tropical raptor. Over several conversations, though, Rick confessed that funding for my eagle work was precarious and encouraged me to accept the condor position, as did the condor-project supervisor, who was anxious for me to start immediately.

But if I opted to work with condors rather than returning to my Montana home as I had planned to do after working in Peru, I would have to leave the mountain country I loved and move to Arizona's canyonlands, where my living situation was uncertain. And yet, this could be the opportunity of a lifetime. The work would be as exhilarating and rewarding as it would be stressful and all-consuming. It was both utterly enticing and a trifle disquieting to know *exactly* what working with California Condors in Arizona would entail. Eight months earlier, I'd had an unexpected opportunity to do just that.

———

Hands on hips, eyes clouded with sadness, biologist Shawn Farry stared down at the dark, feathered body of Condor 116 (One-sixteen). Slumped on a sliver of beach between the swiftly flowing Colorado River and the formidable cliffs that bordered it, 116 lay undisturbed, his wings folded against his body, his right leg frozen in a spasmlike pose. He looked utterly diminished. His impressive stature, his former grace and prowess, his dominant status in Arizona's growing condor flock, the hope so many had invested in his future—all were gone in an instant of discovery on a remote river beach. Condor 116 had once appeared to own the Grand Canyon in which he'd flown. Now he was lost in it. Lost to it.

Hatched in captivity at the San Diego Zoo in May 1995, Condor 116 had been released to the wild two years later from the aptly named Vermilion Cliffs that ring northern Arizona's Paria Plateau. Part of an elaborate effort to return captive-reared California Condors to Arizona's Grand Canyon region, where the birds once roamed, Condor 116 had exceeded all expectations. He had learned to exploit rising columns of air and breaths of wind to fly hundreds of miles in a day. He had learned to find the large animal carcasses on which condors—among the biggest vultures in the world—feed. He had become one of the dominant males in the growing Arizona flock. When he walked up to a carcass, more subordinate condors had backed away like respectful courtiers and let 116 take his fill. Unlike many of his flockmates, he had rarely succumbed to curiosity about the two-legged beasts that gathered at canyon viewpoints. And, at almost five years old, he had nearly completed his transformation from black-headed, smudge-plumaged juvenile, to pink-headed, tuxedo-sharp adult. Recently, he had begun to engage in courtship displays with female condors, exciting fervent hopes in those who watched over him and those who followed his exploits online that he might be among the first to further the celebrated legacy of his species—a legacy that had almost expired, after tens of thousands of years, because of human pressures that pushed the bird to the brink of extinction.

Arizona's condor recovery project—initiated with the release of six captive-reared juveniles in December 1996—was not new to losing condors. Eight of the thirty-five condors that had been released to the wild in Arizona had died or disappeared. All but one of these (that was shot) had been in the wild less than six months. Condor 116, on the other hand, had survived in the wild for three years.

His disappearance, in early February 2000, into one of the Grand Canyon's side canyons initially caused no alarm. Biologists received sporadic signals from his radio transmitter but changes in their location meant that he was moving. After February 9, though, the signal remained stationary. The condor field crew spent weeks trying to pinpoint 116's location amidst complex rock layers and formations. Once biologists finally narrowed it down, they followed a precarious trail down to the river to reach it, but a cliff blocked their passage upriver.

In a last-ditch effort to find 116, Shawn and his field assistant, Kirk, carried an inflatable river raft down the trail, then paddled hard against the Colorado River's powerful current to circumvent the cliff while avoiding treacherous downstream rapids.

At long last, Shawn and Kirk reached 116's undisturbed body, lying on the riverbank at the base of a rock face. Since a cause of death wasn't immediately apparent, and a criminal act, such as shooting the federally protected bird, couldn't be ruled out, Kirk guarded 116 overnight while Sean repeated his hazardous journey—alone and in reverse—to alert US Fish and Wildlife Service and National Park Service law enforcement personnel. A team returned the following morning to document the scene and pack

116 out of the canyon. After x-rays at a veterinary clinic showed no signs of bullet wounds or broken bones, 116 was shipped to the US Fish and Wildlife Service Forensics Laboratory. And the field crew anxiously awaited the necropsy results of a bird that felt more like a family member than a large-winged vulture that had danced with the wind, high in an azure sky.

Weeks later, toxicological tests confirmed what Shawn had suspected—and most feared. Condor 116 had died of lead poisoning after ingesting lead shot or bullet fragments in a carcass—the first such condor death in Arizona. And since the gregarious California Condor rarely feeds alone, the necropsy findings meant that other condors might also have lead coursing through their bodies, threatening their survival.

———

I knew nothing about these events when The Peregrine Fund asked me, shortly after I applied for their South America position, if I could help out with their condor project, which found itself short-staffed at a critical time. Taking a leave of absence from my research job in Montana, I headed to northern Arizona two weeks later for a one-month stint working with condors.

During a whirlwind orientation day following my mid-April 2000 arrival, a condor field crewmember drove me hundreds of miles to show me the reintroduction area, the condors' favorite haunts, and the best topographical high points from which to search for the birds' radio-transmitter signals and to track the condors' long-distance flights. We drove past the arresting Vermilion Cliffs, which border the 25-mile stretch of road between the project's field house and the condor release-site viewing area, and then briefly explored the neighboring Kaibab Plateau. Retracing our route, we visited key viewpoints along Marble Canyon, which encompasses the Colorado River below Glen Canyon Dam and marks the beginning of Grand Canyon. Finally, we drove over an hour south to the east entrance of Grand Canyon National Park. The enormity of my surroundings—the scale and grandeur—eclipsed anything I had experienced. A pastel palette of reds, pinks, mauves, and creams, the towering cliffs and precipitous canyons formed a desert landscape that was as forbidding as it was magnificent—a vast, unknowable land of hidden dangers and hidden beauty.

I tried to absorb place names and directions amidst a steady stream of chatter about individual condors. Every one of the twenty-five condors flying free in northern Arizona at the time had a history—a story—that those who monitored the birds were eager to share. Early in the day, I saw my first condors perched over a mile away on the release-site cliffs. Even at that distance, the birds' size was impressive. North America's largest landbird, the California Condor stands several feet tall, has a wingspan of approximately 9 feet (3 m), and weighs between 17 and 26 pounds (7.7 and 11.8 kg). (Bald Eagles, by comparison, weigh 7 to 14 pounds and Peregrine Falcons weigh 1 to 2 pounds.)

At day's end, I was treated to the unexpected sight of a kenneled condor that was being transported back to the release site. In the wild only a few months, one-year-old Condor 203 had been recaptured after approaching tourists. Cloaked in a coat of dark feathers, his black head and neck covered in a dense gray down that he would retain for another year, 203 still had the ingenuous demeanor of a youngster that had not yet learned that wariness and caution spell survival.

Condor 203 had plenty of company when he was returned later that night to the release pen, which sat atop the Vermilion Cliffs. Because of Condor 116's death, Shawn and his crew were trying to recapture all of Arizona's condors. Early the next morning, I joined another biologist for the hour-and-a-half drive to the release site to help with the ongoing trapping effort. After driving along the base of the Vermilion Cliffs and on a dirt road that traversed the House Rock Valley, which divided the adjacent Paria and Kaibab Plateaus, we turned onto a sandy track that wound its way up through a break in the cliffs to the top of the Paria. Composed largely of sand interspersed with rough segments of jutting rock, the "road" up to the release area was barely passable with four-wheel-drive pickup trucks. After nearly an hour of violent lurching, we arrived "up top," as the release area was dubbed.

After dropping off some equipment at the canvas wall tent that housed anyone monitoring the birds up top, I followed Shawn down a sandy footpath to the cliff rim, where a large chain-link dog kennel served as a walk-in trap. Ensconced nearby, in a camouflaged tent that served as a blind, Shawn would wait until condors walked into the trap to feed on a carcass that had been placed inside as a lure, then pull on a cable to close the entry door, trapping the feeding birds. After helping Shawn bait and set the trap, I tucked myself out of view.

By 0830, Shawn had trapped five condors. Although it had looked easy, Shawn assured me that he sometimes spent days at a time in the cramped blind, unable to catch a single bird. The older condors were extremely wary of the trap, but they found the sight of naive juveniles feeding on a carcass nearly irresistible.

Any thoughts I might have had about captive-reared condors being tamer than their wild counterparts were obliterated as I watched Shawn and another crewmember enter the trap and capture each condor with an enormous, handheld fishing net. The panicked birds flung themselves against the sides of the trap, flapping their giant wings. Once a bird was caught and pinned to the ground by the net, Shawn quickly grabbed its neck, just below the base of its skull, to control the condor's snaky, fleshy head, and prevent it from striking him with its dangerous bill while he extricated the bird from the net and carried it to a transport kennel. Designed to tear flesh and imbued with supreme crushing power, the condor's bill is the bird's primary defensive weapon and the biggest danger to meddling biologists. I became one of those meddling biologists sooner than I expected when Shawn handed me the net so I could catch my first condor. Doing so required careful but decisive movements, and I was relieved to successfully snag one of the fleeing birds without incident.

After transporting each of the loaded kennels to the release pen with Shawn and transferring the birds to the enclosure, I returned to the trapping area to collect my daypack. As I retrieved my gear, Condor 114 suddenly appeared—great wings extended—above the cliff rim, transported upward on an invisible air current. I froze as I watched him float in place and then slice along the top of the cliff. Legs dangling, he tilted his wings, stalled, and came to a graceful landing beside the now-closed trap. Profoundly awed, I turned away to avoid drawing 114's attention, and hiked back to the wall tent, conscious of having been irrevocably transformed by the soul-stirring sight of my first flying condor.

The following morning the field crew convened up top to draw blood from the nineteen trapped condors that were being held in the release pen. Each condor was netted and then carried to a nearby processing area, out of view of the penned birds. There, it was carefully laid onto the lap of an awaiting crewmember who kept the condor's wings folded against its body so the bird wouldn't flap. Another crewmember held the bird's neck and bill to prevent the condor from biting its captors. And a third person held the condor's legs to control its feet.[2]

Rather than having sharp, curved talons like raptors, which kill and carry prey, carrion-feeding condors have blunt talons and long toes like those of a super-sized chicken. A condor's feet provide leverage when the bird stands on a carcass and tears an animal's flesh, but they have little gripping strength and are mainly adapted for walking.[3] Nevertheless, if a condor frees a foot and pushes against its human captor while being carried, it can be difficult to hold on to the struggling bird. So, when carrying a condor, we tucked the massive body under one arm to secure the wings, held the bird's head with the other hand, and tilted the bird's body away from us so the feet couldn't push their way to freedom.

Once a condor was safely secured for processing, a fourth biologist replaced any malfunctioning patagial- (wing) or tail-mounted transmitters, and then drew blood from a vein in the condor's leg. Before inserting a needle, Shawn cleaned the area with an alcohol-soaked cotton ball to wipe away the accumulated white excrement that each condor had deposited on its gray legs. In a behavior shared with storks, New World vultures, including condors, engage in urohidrosis—defecating onto their legs to lower their body temperature by cooling themselves with the fluid as it evaporates.

To my surprise, after we had handled several birds, Shawn expressed unspoken confidence in my abilities when he asked if I wanted to draw blood from one of the condors. I'd extracted blood from the wings of a variety of birds, but as a female biologist, I was unaccustomed to anyone—particularly men—acknowledging my experience so quickly. Invariably, every time I started a new field project, I had to prove my worth from scratch, showing that I could carry the necessary loads, do the required work, and make the appropriate decisions involved in working with wildlife. Shawn's atypical attitude made my month of condor work a halcyon time, when my wildlife experience and biological skillset were immediately taken for granted, rather than being discounted until I had proven myself yet again.

After we had processed and returned the condors to the release pen, Shawn drove the blood samples to the nearest FedEx drop off—four hours away in Flagstaff, Arizona—to send them to a lab that would evaluate their lead content. Several days later, the lab phoned with results. Shawn's stomach dropped when the lab technician asked, "Do you want the worst one first?"

Seconds later, Shawn noted Condor 119's elevated lead level. Blood lead levels are typically measured in micrograms of lead per deciliter (one-tenth of a liter) of blood (μg/dL) or in parts per million (ppm). One ppm is equivalent to 100 μg/dL. In condors, lead concentrations below 20 μg/dL are considered within the normal (or background) range, because low levels of lead occur naturally in the environment. Levels between 20 to 59 µg/dL suggest that a condor has been exposed to lead, 60 to 99 µg/dL means that a condor is clinically affected, and above 100 µg/dL indicate acute lead toxicity.[4] Condor 119 had a blood lead level of 101 µg/dL, clear evidence that she had lead poisoning.

A five-year-old female, released to the wild in 1997 as a two-year-old, Condor 119 had made the longest flight of any reintroduced condor to date, flying over 600 miles (967 km) round-trip from the release site to the Flaming Gorge Reservoir, on the Utah-Wyoming border, in August 1998. On the cusp of adulthood, 119 was one of the recovery program's—and the species'—best hopes for the future. But now, though she showed no overt signs of lead poisoning, her life might be in jeopardy.

No one knew how much lead was fatal to condors. In California, one had survived a blood lead level of nearly 300 µg/dL with treatment. Condor 116, though, had died with a lead level of 320 µg/dL.[5] Fortunately, the condors we'd tested alongside 119 had low lead levels, with only two showing possible exposure.

The next morning, crewmembers again convened up top to draw blood from more captured condors, to take another blood sample from 119 to determine whether her lead levels were rising or falling, and to release the healthy condors. Once freed, the birds preened vigorously, then lined out in a stately aerial procession to the Colorado River for a much-needed bath.

Several days later, we learned that none of the recently captured birds had elevated lead levels. Unfortunately, though, Condor 119's lead levels had risen slightly so we loaded her into a transport kennel, and Shawn took her to the Phoenix Zoo for treatment. As he later noted online, "Driver and passenger were both extremely relieved to arrive at the zoo. Condors, 119 included, typically do not enjoy being in a travel kennel and often make their displeasure apparent by regurgitating. Seven hours, first 100-degree day in Phoenix, windows up, air conditioning on, unhappy condor in the back seat. Let's just say it was a memorable trip."[6]

Upon arriving at the zoo, 119 received another blood test and was x-rayed to determine whether she had visible lead fragments in her digestive tract. Veterinarian Kathy Orr then administered 119's first chelation (pronouced key-LAY-shen) shot. Chelation therapy—a treatment that removes heavy metals from the body—involves injecting a condor in its breast muscle twice daily with calcium disodium EDTA (ethylenediaminetetraacetic acid). This substance binds to the lead (or other heavy metal) molecules and allows them to be excreted from the body. After five days of injections, 119's blood would be retested. If her lead values were still high, she would undergo a second round of chelation therapy; if they decreased, she would be monitored for another week before being returned to the wild.

While 119 underwent treatment at the zoo, I spent a peaceful week up top observing condors amidst the Paria Plateau's orangish sand and otherworldly sandstone outcroppings. Although it was April, my water bottles and I froze overnight at the high-elevation—6,000 feet (1,800 m)—desert camp. Lugging water for the condors through deep sand was no less challenging. Together, the two 5-gallon jugs that I carried on each trip weighed around 80 pounds (36 kg)—or about 70 percent of my body weight. But the numbing cold and challenging physical work were a small price to pay for the joy of watching the captive condors and seeing the wild birds float over the release area, perusing the world below them. As I huddled in the wooden blind that overlooked the release area, I relished one of my favorite pursuits: watching wild animals go about their daily lives, uninfluenced by my presence.

After my quiet time up top, I spent a week at the other end of the condor work continuum, managing condor-tourist interactions amidst the chaos of Grand Canyon National Park's South Rim. The most visited attraction in Arizona, the South Rim hosts over five million visitors per year.[7] A jumble of people and activity, surrounded by wild country, the South Rim is a source of endless fascination to Arizona's condors. Although the Grand Canyon and its environs provide condors with a staggeringly vast wilderness of cliffs for perching, updrafts that facilitate flight, and riverside beaches for bathing, the birds all too often make a direct, one-hour, 50-mile (80 km) flight from the release area to the South Rim's teeming hub of humanity. Here, they buzz like prehistoric pterodactyls over tourists, and lounge on the cliffs below Grand Canyon Village and other viewpoints.

Some have blamed this seemingly aberrant behavior on the condors' captive upbringing. But to biologists, the birds' attraction to the South Rim and other tourist hotspots makes sense when viewed in an evolutionary context. Since its origins in the Pleistocene, the California Condor likely was attracted to aggregations of animals because such herds were the sites of births and deaths, predatory attacks, and unforeseen accidents.[8] Empty landscapes offered little to condors in the way of food. But activity and commotion inevitably meant dead animals upon which condors could feed. The Pleistocene-era mammoths, camels, bison, and horses of northern Arizona have now been replaced by roving human "herds."[9] And nowhere, in modern times, do such large herds gather in the condor's southwestern realm than at Grand Canyon's North and South Rims (and nearby tourist spots such as Zion National Park).

Condors are also drawn to these areas by an abundance of other scavengers. Accessible garbage and human handouts attract large numbers of ravens to these populated areas. Turkey vultures are also ubiquitous. Since sharp-eyed ravens and turkey vultures—which, unlike condors, have a well-developed sense of smell—often are first to locate carrion, condors always investigate these birds' activities.[10] The condors' vigilance is invariably rewarded since the birds find abundant animal carcasses in the park's developed areas, including road-killed elk and deer; mules and bighorn sheep that have fallen from precarious trails and ledges; and a host of other mammals, ranging from beavers to bobcats, that perish in the canyon and its environs.

While at the South Rim, I monitored as many as seventeen condors at once. Soaring over the crowds—legs dangling, heads swiveling as they watched the commotion below them—the condors awed tourists with their graceful flights amidst the timeless setting and ever-changing light of the world's most famous chasm.

Inevitably, the condors' activities prompted a thousand tourist questions, which Shawn and I sought to answer while trying to teach the impressionable young condors "acceptable" behavior. If the birds landed too close to people or on inappropriate perches—lampposts, buildings, or populated observation points—we hazed them by running at them, using noisemakers, or throwing small sticks or handfuls of gravel toward them. Using radio telemetry, we monitored the condors' arrivals, departures, and movements between viewpoints along the cliff rim. Frequently on the move pursuing errant condors, I relished the moments when the birds congregated to loaf or play, and we could simply watch them.

To my astonishment, I quickly discovered that the legendary California Condor was as playful and entertaining as the Hawaiian Crows that had stolen my heart years before. Gregarious condors frequently entwined necks and wrestled in a manner reminiscent of the 'Alalā's foot fighting, with dominant condors pushing subordinates off cliffside perches. Fascinated by any trash that had fallen below the cliff rim, the birds grabbed random items—a baseball hat or soda can—and played keep-away from each other. Furious tugs-of-war ensued over sticks or plastic water bottles. I had never expected to laugh out loud at the antics of vultures.

During lulls in the condors' activities, Shawn and I discussed condor behavior, natural history, endangerment, and recovery. Hungrily absorbing every bit of new information about this intriguing bird, I began to understand how behaviorally complex the condors were and how challenging it was to return them to the wild. During my only break from dawn-to-nightfall condor monitoring over that first hectic South Rim week, I was interviewed for The Peregrine Fund's South America position, which would be based in Peru. A week later, I was offered the job.

On my last day of condor work, I helped Shawn transport a kennel containing Condor 119 to Badger Canyon, one of the many Colorado River side canyons. Finally given a clean bill of health, 119 had proven how ready she was to return to the wild by giving Shawn two nasty bites on his arms when he extricated her from the kennel at the field house so I could attach her new transmitter. Rather than subjecting her to the long ride to the release site, Shawn opted, instead, to release 119 at Badger, a familiar location only a mile from the project's field house.

After we carried her kennel out to a rocky point, Shawn gave me a nod and I opened the door. Condor 119 rushed out, extended her giant wings, and rose into the air. Tears pricked my eyes as I watched her embrace her freedom and drift effortlessly over a condor paradise of wind, rock, and river. Showing no visible sign of weakness from her recent illness or her lengthy stint in a concrete holding cell, 119 danced with the wind. Watching her, I couldn't help but anthropomorphize about what her freedom—her wildness—must feel like after the indignities she had suffered.

Dipping her wings, 119 glided over the river then landed on a boulder at the water's edge. Turning toward the sun, she extended her wings and bathed her feathers in warm light. After watching her sunning, I turned away, believing this would be my last sight of a California Condor. It seemed a fitting end to a month that would forever loom large in my psyche—a month vividly painted in pastel cliffs, rocks, and sand, and animated by the effortless grace and soaring majesty of a regal black-and-white bird that owned the skies like no other.

———

But while I enjoyed summer in Montana and prepared for my new life in Peru, events unfolded in Arizona that forever shattered the recovery program's relative naivete about the threat that lead poisoning posed to the California Condor's future. In mid-June, scarcely a month after Condor

119's return to the wild, Shawn discovered three-year-old Condor 165 dead at the South Rim. After crisscrossing perilous slopes sandwiched between cliff layers, Shawn retrieved 165's body from the talus onto which the bird had tumbled. A necropsy explained Condor 165's sudden death. The hapless bird had *twenty* lead shotgun pellets in his crop and a blood lead level of 300 µg/dL. Once again, the field crew prepared to retrap Arizona's condors.

Days later, a member of the six-person field crew found Condor 191 huddled beside a boulder at the base of the Vermilion Cliffs, in temperatures exceeding 100°F (37.8°C). The two-year-old was emaciated, dehydrated, and heat stressed. Despite the crew's efforts to rehydrate her, 191 collapsed in her kennel in the middle of the night. Treated for shock and stabilized with additional fluids, 191 was transported to the Phoenix Zoo for treatment the next morning. After a blood test showed she was severely anemic—a symptom of lead poisoning—Condor 191 received a blood transfusion from a condor that was being quarantined at the zoo. Heartbreakingly, though, 191 died soon afterward.

Four days later, two-year-old Condor 182 was found dead in the House Rock Valley near the release site. Her deteriorated condition precluded a necropsy, but the timing of her death and the similarity of her premortality behavior to that of Condor 191 suggested lead poisoning had felled her, too. The next day, Condor 150—a four-year-old member of Arizona's first condor release—left the South Rim and headed to the release area. She never completed her journey. And Shawn was never able to recover her body.

As the field crew's desperate efforts to recapture condors bore incremental fruit over the ensuing month, the biologists realized with growing dismay how many of the birds must have fed together on a lead-laced food source.

Shortly after my departure from Arizona, the project had acquired a portable testing machine that could reveal a bird's lead levels in three minutes rather than in the days it had taken when blood had to be sent to a lab. The field tester only revealed exact readings up to 65 µg/dL. If condors had higher blood lead levels, the portable tester flashed "High." During June and July, the tester flashed its ominous "High" for an alarming number of condors, including Condor 119, who had only recently recovered from her first lead poisoning bout.

Condors with high lead values were transported two hours to a veterinary clinic in Page, Arizona, and x-rayed to determine if they had visible lead fragments or lead shot in their digestive tracts. Birds whose x-rays showed no lead were returned to the release pen and chelated twice a day by the field crew, which had now learned to administer the intramuscular injections. Condors with visible lead pellets or fragments trapped in their crops or stomachs were transported to the Phoenix Zoo for treatment, since their lead values were likely to climb as digestive fluids dissolved the metal. Shockingly, x-rays revealed that four condors had shotgun pellets in their digestive tracts.

Follow-up x-rays over the next few days showed that Condors 119 and 136 had passed shotgun pellets that were in each bird's lower intestines (one in 119's and two in 136's). Unfortunately, the situation for a third bird, Condor 133, was more problematic since a pellet was lodged in her gizzard (the muscular portion of a condor's stomach, where digestive juices are more acidic than those of a condor's intestines). When a twenty-four-hour attempt to flush the pellet by repeatedly inserting a tube into 133's esophagus and administering fluids and Metamucil failed, she was transported to the Phoenix Zoo.

There, efforts to extricate the pellet with an endoscope similarly failed. Next, veterinarians attempted to flush out the pellet by placing 133 on an angled table and pumping water through a tube into her gizzard. But still the pellet refused to budge, and veterinarians prepared to operate the following day. The next morning, though, a presurgery x-ray showed that the pellet had finally migrated into 133's intestine, where it would exit her system without additional interference.

The situation was even more dire for Condor 158. His first x-ray revealed six shotgun pellets in his gizzard. By the next morning, three of the pellets had moved into his intestines. Veterinarians removed one of the remaining three pellets using an endoscope, but extricating the final two required surgery. Prior to his surgery, 158's lead level had risen to an astonishing 390 µg/dL, far higher than the level that had killed 116 earlier in the year. (For comparison, blood lead levels of less than 10 µg/dL are thought to cause irreversible harm to human children.)[11]

By the time an exhausted field crew captured the last of the condors in mid-July, nine had been diagnosed with lead poisoning. Each of these had to be netted twice a day and injected with calcium EDTA. Several condors required a second round of chelation. Remarkably, every one of the treated birds returned to the skies, their lives saved by the heroic efforts of biologists and veterinarians.

But condor recovery would forevermore be tainted by the specter of lead. Over the course of two lead-poisoning incidents, five condors—one-fifth of Arizona's population—had died.[12]

As fall slid toward winter, the condor field crew—gutted by their emotional losses, apprehensive about the future, and exhausted by interminably long days and the stress of handling condors—returned to the welcome tasks of tracking the birds' movements and monitoring their daily activities back in the wild.

————

And now, having accepted the position as The Peregrine Fund's condor field manager, I was driving to Arizona to take up the reins of this exceptionally challenging reintroduction effort. I knew I could never fill Shawn's shoes. I was no match for a biologist with years of condor experience, the physical prowess of the ultramarathoner he once was, and a fearlessness that was legendary. But I knew, too, that those whose hearts and minds were captured by the California Condor were inspired to go to unfathomable lengths to protect the birds and keep them flying. And as daunted as I was by the task ahead, I knew that I would do no less.

On my way to Arizona, I stopped by The Peregrine Funds' headquarters in Boise, Idaho, to confer with my new supervisor. The organization's president came out of his office to meet me when summoned by his secretary. Before I could reach out to shake his hand, he paused and looked at me. His unsmiling gaze traveled slowly from my head down to my toes then back up again during what felt like an eternity but surely was measurable in seconds. My face flamed during the long pause that followed his assessment of the new condor field manager.

"A condor's going to fly away with you," he said, then turned his back on me and headed into his office.

I stood for a moment, stunned by his cavalier dismissal. A dismissal that inadvertently played on all of my self-doubts. Then, unconsciously, I lifted my chin, squared my shoulders, stiffened my spine, and followed him into the office.

CHAPTER SEVENTEEN

Hard Lessons

While there was little I could do to resolve my apparent physical inadequacies for keeping condors aloft (and myself on the ground), I tried to address any intellectual deficiencies I might still have by stopping at a bookstore, on my way to Arizona, to purchase a newly published book on California Condors. As I stood looking at the wrinkled, pink condor face on the cover of the prominently displayed book, a young girl and her friend came marching down the aisle toward me. The girl stopped abruptly in front of the book, as riveted as I was by its cover. Then, placing her hands on her hips, she declared to her friend, "California Condors are the ugliest birds in the world." Satisfied with her pronouncement, she marched on, friend in tow, leaving a bemused biologist in their wake.

Impressed as I was that a twelve-year-old readily recognized a California Condor, her words gave me pause. Over the years, I had worked with a treasure trove of avian beauty, including half a dozen elegant raptors, a variety of alluring songbirds, charismatic corvids, a tropical parrot, and a diminutive duck. Was I prepared to work with the ugliest bird in the world? Thoughts of my brief sojourn amidst the castellated splendor of Arizona's canyon country intruded on fleeting images of the striking species that had graced my work life. I pictured a condor opening its great wings and stepping into the wind. I saw ineffable grace. I saw inimitable grandeur. I saw all that couldn't be captured by an image of a bare-headed vulture on a book cover. And I left the bookstore smiling. I couldn't wait to begin working with the world's "ugliest" bird.

But while I delighted in the condors' magnificent flights and intriguing social interactions in the coming weeks, I chafed under the unexpected limitations that were initially placed on my role as field manager. Despite my frustrations, the condors' activities demanded my full attention, and I resigned myself to having to prove my worth, experience, and judgment yet again before being allowed to fulfill the position for which I'd been hired.

A week before my return to Arizona, in December 2000, The Peregrine Fund had experimentally released a pair of adult condors that had spent their lives in captivity. The goal was to try to jumpstart condor reproduction in the wild, which had not occurred since 1984.[1] Previously, the recovery program had released only juvenile condors, but it then had to wait at least six years for them to reach breeding age. Now, I was spending my first days back in Arizona monitoring this pair of eight-year-olds: Condor 74, the male, and Condor 82, the female.

Every California Condor is assigned a sequential "studbook" number—which biologists view as the bird's name—after hatching. The relatively low numbers assigned to these adults meant that the birds were significantly older than the other condors that we were monitoring in Arizona, all of which had studbook numbers in the 100s and 200s. Used as I was to seeing the black-headed juvenile condors and the mottled black-and-pink-headed three- to five-year-olds that had comprised our flock until now, I was struck by the adults' pink heads, deep-red eyes, ivory-colored bills, and elegant plumage. Whereas the younger birds were a uniform, sooty black (with each dark body feather beautifully scalloped in a hint of brown), the aptly named covert feathers covering the base of the flight feathers on the upper side of the mature birds' wings looked like they had been dipped in cream. These pale-tipped feathers created a dividing line between the adults' dark bodies and their silvery secondary (interior) flight feathers.

Despite their mature plumage, however, the adult pair's woefully naive behavior was concerning. The first time I tracked the two birds, I was dismayed to find Condor 82 perched in a field. Although condors often walk, particularly when trying to reach higher ground to launch themselves into flight, our younger birds rarely perched out in the valley unless a carcass was nearby. I was relieved when 82 finally flapped laboriously back to the release site, joining several condors in the air, and I couldn't help but wonder how she must have felt soaring with others of her kind after a life confined to a pen.

Not surprisingly, though, those years of confinement undermined critical survival instincts for these older condors. Whether those life skills could be taught at this late stage was still unknown. Later that day, I found the pair perched on the ground near the dirt road bisecting the House Rock Valley. Despite their years together in captivity, this was their first interaction since being released to the wild nearly two weeks earlier. I watched, enchanted, as they nibbled each other's faces and preened the ruff of spiky feathers that encircled their necks like feathered boas. When the pair began to settle down in a spot where they were vulnerable to coyotes, though, I regretfully intervened, running at them repeatedly to haze them back to the safety of the cliffs.

Over the next week, we kept hazing the pair from inappropriate perches and celebrating small successes, such as Condor 82 finally roosting on a cliff. We also prepared for the forthcoming release of juvenile condors by outfitting each youngster with two large, hourglass-shaped vinyl tags. Each tag was folded over a wing so that the condor's identification number was visible when the bird was perched or flying, and was attached, along with a radio transmitter, with a bolt that passed through a small hole pierced through the bird's patagium—a fleshy area on the wing's leading edge. Each transmitter weighed 2.1 ounces (60 g)—less than 1 percent of the bird's body weight—and rode on top of the condors' tag and wing, its stiff antenna pointing toward the bird's tail.[2] Large birds such as hawks and eagles are usually outfitted with "backpack" transmitters that ride on the bird's back and are secured with straps that loop around each wing and meet at the bird's chest. But because a condor fills its expandable crop with large amounts of carrion when feeding, its chest swells and so it cannot wear a restrictive harness. Instead, biologists developed patagial transmitters that are worn on the bird's wings.

While outfitting the juveniles with their new tracking-related paraphernalia, I held a condor solo for the first time—a slightly terrifying experience, even though I was handling naive youngsters. The condors were an immense and dangerous armful that I found more intimidating than the eagles and smaller raptors I had handled, all of which were relatively docile once measures were taken to avoid being harmed by their perilous talons. Aside from managing a condor's massive, squirming bulk, I had to control the bird's snaky, loose-skinned head, avoid its beak, and ensure that its legs didn't push their way to freedom. Daunting as it was to handle one of the world's largest birds, it wasn't long before I was as comfortable handling condors as I was holding hummingbirds.

Late on a snowy Christmas Eve, I returned to the field house after searching for a missing condor and learned that Condor 82 was still frequenting unsafe areas. A crewmember had hazed her repeatedly that evening, before getting her to perch on a rock outcrop as darkness fell.

I headed into the field early on Christmas Day, ready to continue teaching 82 how to be wild. As I hiked toward her nighttime roost, her transmitter signal boomed from my receiver, telling me she was nearby. But rather than seeing her dark-cloaked form, my eyes were arrested, instead, by the ephemeral imprint of condor wingtips in the fresh snow. The fragile impression left by 82's flight feathers was surrounded by the four-toed, claw-tipped footprints that spelled coyote. Condor 82 had attempted to fly as the predators closed in. Eyes riveted to the story that played out on the ground in front of me, I followed the trajectory of 82's attempted flight and found several small condor feathers. A few steps farther on, two flight feathers lay near scratch marks in the snow where 82 had scrabbled for purchase when coyotes grabbed her wings and tail. Then, 82's struggle intensified, leaving more feathers and displaced snow. And, finally, pitched battle. Amidst the press of canid tracks, deep grooves rent the snow and underlying orange sand where 82 had fought to escape her attackers. Thrashing against the surrounding bushes, her flailing wings and body had left a trail of broken branches. Beyond this heart-wrenching vestige of a last, Herculean struggle for life, a smooth, toboggan-like drag mark disappeared over an embankment.

I broke into a run, heart pounding, adrenaline surging. Leaping over the bank, I rounded a bend then dropped to my knees as I finally saw what I never wanted to find—Condor 82's scavenged body, discarded in a small ravine after being dragged there and fed upon by coyotes. Shocked and saddened by my first condor fatality, I could only vow, as tears swam in my eyes, that I would fight as hard for the lives of the remaining condors as an ill-equipped but valiant Condor 82 had fought for hers. Then, after quietly mourning her loss on a Christmas Day that I would never forget, I set about the sad task of photodocumenting the final moments in the life of a naive, ill-fated condor, and of gathering up her remains.

———

Despite my best intentions, before I could live up to my determination to do whatever it took to keep my charges safe, I was confronted with another painful lesson about the vulnerability of condors that were adapting to life in the wild.

December 29, 2000—release day for the juvenile condors and a big public event for Arizona's condor recovery program—dawned sunny, clear, and cold. I headed out early to the release-site viewing area—dubbed the Ramada for its thatch-covered shelter, which shielded visitors from the sun and storms—to check on the condors that had roosted on the Vermilion Cliffs. I was intercepted by Project Director Chris Parish.

"Adam found Condor 74 dead up on the cliffs this morning," he told me without preamble.

"What? Where? How?" I stammered nonsensically, my mind scrambling for the details that the biologist stationed up top had conveyed to Chris via two-way radio.

"He roosted on top of the cliffs and coyotes got him," Chris responded tersely.

I sat in my truck, my mind reeling, feeling gut-punched and heartsick. As inexperienced in the ways of the wild as his mate had been, 74 had nonetheless seemed less naive and more capable. But a lifetime in captivity had taken its toll. Condors lie down to sleep—head tucked over their back and under a wing, like geese—and can be sound sleepers. If they roost on inaccessible cliff ledges, sleeping like this does them no harm. But newly released birds often struggle to land on sheer cliffs, and 74 had roosted, instead, on the ground near the cliff rim, where he had been as vulnerable to night-prowling predators as his mate had been in the valley.

As I waited for a crewmember to retrieve Condor 74's remains, I vowed, yet again, to do better. From now on, if I couldn't get an inexperienced condor to leave the valley floor, I would either capture it or spend the night nearby to keep it safe. And henceforth, despite some initial resistance from some members of the field crew, I would task whoever was monitoring the birds up top with patrolling the cliff rim before nightfall, checking every condor's radio signal to ensure that all birds were safely perched on the Vermillion Cliffs.[3] We would flush any bird that tried to sleep on the cliff rim, forcing it to move to the cliffs. Each evening, the biologist who was stationed up top would work with a crewmember stationed below the cliffs, communicating via two-way radios to confirm that every bird had found a safe night perch.

I had thought that the condor recovery program was so well established that there was no room to introduce new methods to keep condors safe. Condors 74 and 82 showed me otherwise. In actuality, the program was infinitely more dynamic than I had realized, with each field crew in Arizona,

California, and eventually Mexico, innovating and improving release methodologies as we learned from the very birds that we were working to save.

Tasked with sending 74's body to a pathology lab, I missed my first condor release. But as I headed to town, I met scores of cars arriving at the Ramada, and watched a helicopter, carrying bigwigs, land nearby. The excitement in the air was palpable as nearly 150 condor fans and program cooperators convened at the remote outpost in the House Rock Valley to watch seven captive-reared condors take their first flights. Unexpectedly, though, it was the older, free-flying birds that stole the show. Just before 1100 hours, thirteen condors that had roosted at the Vermilion Cliffs took to the skies. The crowd exclaimed in voluble delight as the black-and-white condors swooped and glided in an indigo sky over carmine cliffs. And no one seemed to mind that release day ended in the same way as poet T. S. Eliot's world, "Not with a bang but a whimper," as the hesitant juveniles cautiously crept out of the release pen to embrace the wild.[4]

Two decades earlier, seeing over a dozen California Condors soaring together was nearly unimaginable, and, for many condor aficionados, the world as they had known it came to an end on April 19, 1987. On that Easter Sunday, the last wild California Condor—a male dubbed AC-9, for Adult Condor number 9—was captured and brought into captivity in a last-ditch effort to save the critically imperiled species.

The California Condor may never have been abundant in North America but, based on paleontological records from the Pleistocene epoch (2.6 million to 11,700 years ago)[5] during which the species originated, the opportunistic bird was widespread, occurring along the Pacific Coast from British Columbia to northern Mexico, across the southern United States to Florida, and as far north as upper New York State.[6] Following the extinction of the Pleistocene megafauna—which comprised large animals such as mammoths, camels, and giant ground sloths—the condor's range contracted, suggesting that the scavenger's feeding opportunities had dwindled. The widespread ceremonial killing of condors by Native Americans in prehistoric and historic times also may have contributed to the condor's shrinking distribution and population.[7]

By the time Spanish explorer Father Antonio de la Ascension became the first European to document California Condors—feeding together on a beached whale in Monterey Bay, California, in 1602—the species occurred only in western North America.[8] Along the Pacific Coast, condors were sustained by a bounty of dead marine mammals and fishes.[9] Two centuries later, members of the Lewis and Clark Expedition observed condors feeding on whales and fish near the mouth of the Columbia River (between present-day Oregon and Washington). Captain Meriwether Lewis dedicated almost three pages of an 1806 journal to a detailed description of a condor that fellow expedition members had wounded and captured alive. Lewis believed it to be the largest bird in North America (it weighed 25 pounds and had a wingspan of 9 feet 2 inches), and proclaimed it "a handsome bird at a little distance."[10]

When commercial harvests reduced the numbers of whales, seals, and other coastal mammals in the 1800s, the proliferation of cattle ranching provided condors with another widespread food source, which supplemented a diet that included deer, elk, and other large animals.[11] Such terrestrial carrion sustained not only Pacific Coast condors but also those that were sighted into the nineteenth century in Arizona, Nevada, Utah, Colorado, Wyoming, Idaho, Montana, and southern Alberta.[12] But the condor's range continued to contract. By the 1940s, condors were largely confined to a wishbone-shaped region in southern California that encompassed the Sierra Nevada and Southern Coast Ranges, between San Francisco and Los Angeles.

The condors' tendency to travel widely in search of food and convene in large numbers at carcasses likely made the birds seem more abundant than they were. But as condors disappeared from places they had long frequented, during the second half of the twentieth century, it became increasingly clear that condor numbers were in steep decline.

Habitat conversion, deterioration, and loss are almost always contributing factors to the endangerment of wildlife and doubtless played a role in condor population declines. Protecting remaining condor habitat in California was a focal point of twentieth century conservationists and led to the establishment of protected areas as early as 1947. Though critical to providing habitat for today's condors, these efforts nevertheless failed to bolster condor numbers, since the birds faced more insidious threats.

According to Lloyd Kiff, a former head of the California Condor recovery team—established in 1973 to address population declines—the condor appeared to be doing well on the Pacific Coast until the 1790s, when Europeans arrived and started shooting it. Throughout the late 1700s and the 1800s, "Virtually every account of the California Condor includes a description of how the birds were killed. Everybody shot condors."[13] The bird's size was an irresistible target to many. Some erroneously viewed the scavengers as a threat to livestock. Others shot condors to use their large, hollow feather quills as receptacles for gold dust. And ornithologists avidly shot the birds for museum collections.[14]

In addition to being shot indiscriminately, untold numbers of condors were unintentional victims of widespread wildlife-poisoning campaigns that targeted unwanted predators such as grizzlies, wolves, and coyotes, and perceived agricultural pests such as ground squirrels and prairie dogs.[15] A 1918 status report stated that California Vultures, as they were then known, were not uncommon in southwestern Idaho before the initiation of efforts to reduce wolf numbers using poisoned carcasses.[16] Condors died not only from feeding on poison-laced carcasses but also from feeding on the animals that succumbed to the strychnine (used in the late 1800s and early 1900s), thallium sulfate (used especially in the late 1920s to late 1940s), Compound 1080 (sodium fluoroacetate, used in the 1950s and 1960s), and cyanide (M-44 capsules) that were widely used as poisons to make western rangelands more hospitable to livestock.[17]

Following biological studies initiated in the 1930s, conservationists highlighted habitat loss, shooting, poisoning, and human disturbance to nesting areas as the primary drivers eroding condor numbers. Yet despite habitat conservation efforts, the reduction of wanton shooting, and the protection of nest sites, condor population declines continued unabated.[18] In 1967, the California Condor received additional protections when it was declared endangered under the Endangered Species Preservation Act of 1966. But by the mid-1970s fewer than sixty condors remained.

As condor numbers declined, hostilities erupted among conservationists, with different factions fighting to enact measures they deemed essential to preserving California's emblematic bird. Swayed by early research suggesting condors were highly sensitive to disturbance, preservationists insisted condors should be left alone, and advocated for increased protection and habitat. Other conservationists, including many biologists, supported the establishment of a captive breeding program to bolster condor numbers. Biologists also wanted to capture condors and outfit them with radio transmitters so they could track the birds' movements, identify important areas and habitats, and determine why condors were dying.

At a time when such research tools were in their infancy, preservationists countered that if condors could not be saved through nonintrusive means, they should be allowed "to die with dignity," unfettered by identifying tags and transmitters. Radio tagging (attaching radio transmitters to animals to track them)—now a wildlife research standard—was characterized as "mutilative biology" by opponents of intensive management.[19] Preservationists were equally horrified about confining condors to cages, arguing that "an animal is first and foremost an expression of its ecosystem; removed from its natural habitat, it literally ceases to be that animal—it is merely a collection of genes in a cage."[20] Captive breeding, they believed, would transform revered condors into "feathered pigs."[21]

Biologists were aware that a critical component of the condor's endangerment was the birds' inability to produce enough offspring to compensate for the species' losses. California Condors don't reach sexual maturity until they are about six years old, and they lay only one egg per breeding attempt. Because of the extended care parent condors give their offspring, successful pairs typically breed every other year. Their numbers are sustained under these conditions because condors, which can live for sixty to seventy years, typically have low adult mortality.[22] But in the 1900s, adult condors were disappearing as fast as juveniles.

In May 1980, amidst dire prognostications for the condor's future and despite objections from preservationists, biologists finally secured federal and state permits to capture a condor for the proposed captive breeding program and to outfit condors with radio transmitters to study the birds' movements and fates. But a month later, during one of the earliest attempts to enter a nest cave and obtain baseline measurements on a half-grown chick, biologists underestimated the stress that such activities might cause it. Having handled a younger nestling without incident a few days earlier, they were unprepared to deal with the sudden onset of the older chick's stress symptoms, and the nestling died moments later. Shock waves reverberated through the scientific and conservation communities, further fueling preservationist's claims that condors were too sensitive to be handled. Capture permits were rescinded and intensive research and conservation activities were terminated.[23] According to long-time recovery-team leader Mike Wallace, "The death of that chick not only brought the whole program to a complete halt, but it also came close to ensuring the extinction of the condor since it prevented the initiation of the captive-breeding program until it was almost too late."[24]

Months of contentious hearings, discussions, and negotiations followed. Finally, two critical discoveries restarted the management activities that many felt were critical to forestalling the condor's extinction. Although yearly condor counts had been undertaken since 1966, population estimates were unreliable because condors travel great distances and often congregate in large numbers.[25] In 1982, a novel effort to census the population by photographing and identifying individuals based on differences in their molt (feather loss and replacement) patterns led to the first definitive count. Shockingly, *only twenty-one condors remained in the wild.*[26] A lone male condor, known as Topa Topa, which had been taken into captivity in 1967, brought the population total to twenty-two birds. Drastic measures were clearly needed to save the species.

Earlier in the year, the inadvertent discovery that California Condors would lay a replacement egg if their first one broke or failed to hatch—a phenomenon known as double clutching—further bolstered arguments for captive breeding. Confident now that females would lay replacements, the recovery team began collecting wild-laid eggs in 1983. Biologists hatched a pair's first egg in captivity and allowed the pair to raise its second egg in the wild, thereby boosting the overall number of nestlings produced. After years of decline, condor numbers inched upward as the captive population began to grow.

But all was not well in the wild. In March 1984, the first-ever radio-tagged (in October 1982) condor was found dead below a tree. X-rays revealed a fragment from a copper-coated lead bullet in the bird's digestive tract, and chemical analyses showed lethal levels of lead in the bird's tissues.[27] The discovery confirmed for the first time that condors could die from ingesting lead bullet fragments.

Worse was in store for condors. During the winter of 1984–1985, six of the fifteen remaining wild condors—40 percent of the population—disappeared, leaving only one established pair. The bird's future looked bleak.

Believing that the only way to keep condors safe was to remove them from the wild, conservationists pushed to capture the last nine birds. Following vehement opposition, the US Fish and Wildlife Service allowed only three condors to be captured in the summer of 1985. In November, though, when biologists trapped AC-3— the female from the only remaining wild pair—to replace a transmitter, they discovered she had dangerously high lead levels. Following this alarming development, the Service authorized the capture of the remaining condors.

The National Audubon Society immediately sued to stop the captures, believing that wild condors were critical to incentivizing habitat protection and to mentoring captive-reared condors once they were released to the wild. As this latest battle was addressed in court, veterinarians fought to save AC-3. Despite their efforts, she died in January 1986. Nearly two decades later, I stood with biologist Jan Hamber by a taxidermy mount of AC-3, on display at California's Santa Barbara Museum of Natural History. As we looked at AC-3's outspread wings, Jan's eyes reflected her loss, and heartbreak resonated in her voice, as she reminisced about her eleven years of monitoring a condor she had come to love.

With AC-3's death, only five wild condors, including a lone female known as AC-8, remained. Surprisingly, AC-8 chose the youngest of the males—AC-9—as her mate. The pair's thin-shelled first egg was inadvertently crushed. Biologists took the pair's second egg into captivity and, after the National Audubon Society lost its lawsuit, captured three more condors, including AC-8, in June 1986.

In February 1987, after capturing AC-5, trapper Pete Bloom glanced up to see AC-9, the last California Condor left in the wild, watching him from an oak tree. Over the next few months, an elusive AC-9 roamed southern California alone. Then, on April 18, Jan Hamber finally spotted him on a remote ranch, north of Los Angeles. Torn between condemning AC-9 to captivity and allowing him to remain free, Jan agonized about alerting the trapping team, but finally drove to a gas station and telephoned Bloom.

Under cover of darkness the next morning, the trapping team placed a calf carcass near AC-9's roost tree and set up a cannon net—an immense folded net attached to projectiles that were inserted into small, cylindrical "cannons." When the cannons were fired remotely, the launched projectiles would drag the attached net in an arc over the target. As the sun rose, the team waited in suspense while ravens and a Golden Eagle arrived at the carcass. AC-9 finally left his perch midmorning, glided toward the carcass and began to feed. Bloom fired the cannons and the net streamed out, ensnaring the world's last wild California Condor. As Bloom somberly extricated the struggling bird and folded his giant wings to put him into a transport kennel, Jan looked at AC-9 and, with a lump in her throat, said "I'm sorry."[28]

Moments later, AC-9 was loaded onto a plane bound for the San Diego Zoo, thereby circumventing an angry mob that was stalking the Los Angeles Zoo, protesting what they viewed as the egregious final nail being hammered into the coffin of a now-doomed species. For the first time in tens of thousands of years, *Gymnogyps californianus*, the bird we call the California Condor, no longer cast its grand shadow over western landscapes.

A year later, in 1988, a tiny celebrity, dubbed Molloko, emerged from the first condor egg conceived in captivity, cementing the success of a captive-breeding program that would help the condor reclaim its place in the firmament.[29] Political opposition to the breeding program all but vanished. According to biologists Noel and Helen Snyder, "Once an ungainly, but cute and apparently happy condor chick became featured on magazine covers and newspaper front pages across the country, the public simply did not relate any longer to stark anti-captivity rhetoric."[30]

Condors bred successfully in captivity, and, by the end of 1992, the condor population had risen from a low of twenty-two birds in 1982 to an astonishing sixty-three captive condors.[31]

Five years after AC-9's capture, following experimental releases in California using the more common Andean Condor to refine reintroduction techniques, the recovery team was ready to reintroduce California Condors to the wild. To address the lead poisoning threat, biologists would provide supplemental food at established feeding stations. On January 14, 1992, a large group of supporters gathered in the Sespe Condor Sanctuary to celebrate the California Condor's much-anticipated return to the wild. The assembled crowd watched avidly as two young condors, Xewe (pronounced GAY-wee)—AC-9's daughter—and Chocuyens (cho-KOY-yens), cautiously embraced their freedom, flapping and hopping along the cliff rim. And the onlookers' hopes for the future soared when Xewe spread her wings and took to the skies to write a new chapter for the endangered condor.

But the wellspring of optimism marshaled by a veritable army of supporters could do little to smooth the condor's path forward. Seven months after the release, a biologist watched in helpless rage as two distant men on a family outing fired a rifle at the large target known as Xewe. Fortunately, the men's skills failed to measure up to their intent and their bullet missed its target, as did two follow-ups fired at the fleeing bird.[32] Sadly Chocuyens was not so lucky, dying in October after ingesting ethylene glycol—a colorless, odorless substance used as an automotive antifreeze—from an unknown source.

Nevertheless, six more condors were released in California in December 1992. Over the ensuing months, three died after colliding with or being electrocuted by power lines. The following year, another condor was killed by a power line. Faced with these devastating events, recovery-team leader Mike Wallace initiated aversive conditioning measures to better prepare reintroduced condors for the challenges posed by an increasingly developed world. Met with initial skepticism, Mike outfitted prerelease pens with mock power poles that gave birds landing on them a mild electric shock. Remarkably, condors soon avoided landing on power poles, and wild condor numbers inched upward.

———————

Despite the challenges of reintroducing condors to a hazard-filled world, biologists soon initiated plans to develop a second, disjunct population of wild condors so that the species would persist if condors in one area were struck by a natural disaster or contracted a virulent disease.[33] Arizona's Grand Canyon—a vast and rugged wilderness with little human development, abundant wildlife, endless cliffs, and innumerable potential nest caves—seemed ideal for a second reintroduction site.[34] But Mike Wallace worried that the area's limited access would pose an insurmountable challenge for biologists tasked with monitoring vulnerable young condors. Nevertheless, while on a reconnaissance flight in the region, Mike spotted what looked like a great alternative. Located just north of the Colorado River corridor and Grand Canyon National Park, the Vermilion Cliffs were remote, unaffected by development, and provided ready access for biologists—a perfect place to establish a second condor population.[35]

Reintroduction plans, public comment opportunities, and highly contentious public meetings followed. Some Arizonans and Utahns, not wanting an endangered species deposited in their backyard, arrived at the first public meeting in Kanab, Utah, heavily armed and bristling with rage at the "Feds'" latest transgression. Most were eventually appeased with assurances that condors would be introduced under a special provision designed to prevent local residents from having their livelihoods circumscribed by the planned reintroduction.

Section 10(j) of the Endangered Species Act allows populations of listed species that are reintroduced outside of their current range, but within their historical range, to be designated "nonessential, experimental populations."[36] To qualify for this designation, such populations must further the conservation of the species but must not be considered essential to the species' survival. The gray wolves reintroduced to Yellowstone National Park are perhaps the best-known 10(j) or nonessential, experimental population in the Lower 48 states. This designation facilitates controversial reintroductions because it allows current land uses—such as ranching, mining, and timber harvesting—to continue on private lands within 10(j)-designated areas, even if the reintroduced species is present.

The 10(j) boundary, within which condors would be considered a nonessential, experimental population, incorporated southern Utah, eastern Nevada, and northern Arizona. Within that designated boundary, landowners could legitimately engage in any *legal* activities even if their actions harmed the species. So, although landowners could not shoot a condor since doing so is illegal under a number of federal laws, they could cut down roost trees used by condors.

Despite such reassurances, Utah's San Juan County filed a lawsuit, in early 1996, preventing the planned spring condor reintroduction in Arizona. After the lawsuit failed, though, The Peregrine Fund, which was managing Arizona's recovery effort, prepared for its first release.

On December 12, 1996, hundreds of condor enthusiasts, reporters, photographers, and dignitaries convened at the Ramada, below the Vermilion Cliffs. After a rousing speech by Secretary of the Interior Bruce Babbitt, the joyful crowd joined in a countdown for the return of the condor to northern Arizona. As six young condors tentatively emerged from their enclosure and extended their wings, the wind stirred their flight feathers. Watching them from just over a mile away, the crowd roared in delight as an updraft launched Condor 150 skyward and into the history books as the first California Condor to fly free in Arizona in over seventy years.

The following May, The Peregrine Fund released another nine condors, and Arizona's reintroduction program was well under way.

———

And now, after rejoining the recovery effort in December 2000, I was driving from one end of the 10(j) area to the other, looking for a young condor that had gone missing months earlier. Even among condors—masters of the skies that can soar for hours on a breath of wind and travel vast distances with hardly a flap of their wings—Condor 176 stood out as a bird that was driven to fly. Long after other members of her 1998 release cohort had settled down for the night, 176 remained airborne, flying back-and-forth over the cliffs until darkness forced her to part with the sky. Within six months of her release to the wild, she had explored parts of four states, skirting Las Vegas, Nevada (a 200-mile [320 km] round-trip flight), thrilling visitors at Mesa Verde National Park in Colorado (400 miles [640 km] round trip), and exploring wild corners of northern Arizona and southern Utah.[37] First in her cohort to find a nonprovisioned (not supplied by biologists) carcass, 176 was an independent loner that rarely stayed in one place for long.

In the early fall of 2000, she disappeared altogether. Unsure if her transmitter batteries had failed, we worried that she might have succumbed to lead poisoning, a power line, or even hunger, given that she wasn't even two years old. We searched for her constantly—scanning for her transmitter signals from trucks and small planes—to no avail. Then, unexpectedly, in mid-January 2001, we received a late-afternoon call from a tourist who had seen a condor with the number 76 on its wing two days earlier, at the faraway outpost of Toroweap, in Grand Canyon National Park.

Throwing food, water, camping gear, warm clothes, a gas can, condor trapping equipment, a transport kennel, and several calf carcasses into a four-wheel drive truck, I left the field house with a visiting friend for a harrowing night journey that included over 60 miles of slick, barely passable, snowy and muddy dirt roads. Bleary with fatigue when we finally arrived, well after midnight, at Toroweap's euphemistically dubbed "undeveloped" campground, I strained to hear a transmitter signal amidst my receiver's static. Whenever and wherever I drove in condor country, I connected the receiver via a coaxial cable to a radiolike antenna—known as an omni—that rode on my truck's roof. I programmed the receiver to scan through all the condors' transmitter frequencies (like channels on a radio) if I was looking for condors generally, or I entered a particular frequency into the receiver if I was looking for a specific bird. If I received a signal with the omni—indicating that a condor was within range—I jumped out of the truck and attached my receiver to a Yagi antenna to determine the bird's direction, so I could locate it.

Blasted by a biting wind as I exited my truck at Toroweap, I pointed my Yagi toward the cliff rim, hoping that 176 had opted to linger in this wild place. As snow swirled around me, I heard a faint "blip, blip, blip" from my receiver and broke into a broad smile. Moving to the cliff edge, I turned the Yagi vertically and pointed it over the rim toward the cliff face. "BLIP, BLIP, BLIP," my receiver shouted into the night. Condor 176 was there below me, roosting on the cliff. I had found our elusive bird at last.

I arose before dawn the next morning, eager for my first sight of the mythical 176. Huddled on the clifftop, I waited for the sun to rise over one of the most dramatic vistas I had ever seen. The sheer cliff that I had cavalierly leaned out over the night before dropped precipitously for 3,000 feet (900 m) and was mirrored by an equally dramatic cliff on the other side of the Colorado River, which flowed like a ragged-edged ribbon in the distant trough formed by the gorge. Not a single human trace disrupted the monumental vista. When the sun finally began to warm the cliff rim,

176 took to the skies and soared into view with the cliffs, the canyon, and the cobalt-blue sky as her backdrop. Captivated, I watched her for hours, marveling at her independence and wildness. As transient as the sweeping shadow that she cast across the sandstone cliffs, 176 captured my heart and my imagination that day, epitomizing a truly wild condor, despite the trappings she wore on her wings.

————

Over the ensuing weeks, winter storms battered northern Arizona's high-elevation country. While driving to Toroweap again, to check on the still-elusive 176, I was blocked by several feet of snow. Unable to advance, I tried to turn around but got stuck. Grabbing a shovel from the back of my truck, I worked up a sweat digging myself out, only to get stuck again several hours later, after darkness had cloaked the landscape in a moonless shroud. Again, I dug myself out, but I had to sleep in my truck since my way home was now impassable. The following morning, a faulty starter sidelined my vehicle. Fortunately, I was rescued by someone who worked for one of our condor-project partners, but I spent the next few days digging our trucks out of snowdrifts, rescuing stranded crewmembers, and retrieving abandoned vehicles.

More storms thrashed us in early February, with violent winds tearing over the Paria Plateau and nearly destroying our observation blind. The condors huddled tightly in their cliffside roosts. Several of the newly released juveniles also spent time loafing on the hack-box portion of the release pen. Swirling snow and unrelenting winds hampered visibility, but I became concerned when we didn't see juvenile Condor 228 on February 7. Although I received her signal toward the top of the hack box at twilight, I could only spot two other juveniles roosting there. Not wanting to flush the youngsters from their safe roost by taking a closer look, I could only hope that 228 was tucked in with them.

The following morning, unable to shake the disquieting feeling that something was wrong, I set out at dawn to find 228. Despite receiving her signal toward the hack box, I still couldn't see her. Suddenly, though, I remembered an incident from the day before that hadn't seemed significant at the time, but now resonated with a sickening import. From the blind, I had watched three older condors clustered on the hack box, pecking under branches that helped break up the box's flat profile. When a fourth condor displaced them as it arrived to investigate, I'd seen what looked like several small, black feathers drift over the release-pen netting. Moments later a coyote had appeared, flushing the condors, and I became absorbed in hazing the predator away with a noisemaker.

Now, the image of condors pecking at something on the hack box returned to me in sharp relief and I raced toward the release pen, already knowing what I would find. Condor 228's stiff body was tucked under a branch. Had she been trapped inadvertently while roosting? Had she succumbed to exposure in the bitter cold? My mind swirling with questions, I methodically photographed the scene, took copious notes, then gently extricated her body. Hours later, I tearfully handed 228 over to a crewmember, who would send her to a pathology lab.

After trying to distract myself with mindless chores at our tent, I returned to the observation blind, burdened with sadness and guilt. Warmed by intermittent sunshine, I watched the other newly released condors feeding aggressively at a calf carcass that I had laid out near the cliff rim the night before. Hours later, as fantastical rock outcroppings cast lengthening shadows over the Paria Plateau and condors left the feeding area to settle on the cliff, I watched our youngest condor—newly released Condor 235—trying to grab a few extra bites before bedtime. Unfortunately, Condor 187—176's fierce brother—dominated the carcass, pecking viciously at 235 whenever she approached. Nonplussed, 235 lay down nearby, waiting for her turn even though her dark crop bulged visibly from the day's feast. The sun dropped behind the neighboring Kaibab Plateau and 235 tucked her head under a wing and fell asleep. The departure of 187 failed to rouse her, as did a cold wind that swept over the clifftop. I watched tensely, willing 235 to wake up, knowing how vulnerable she was to prowling coyotes, but she slept on. It was time to teach the young condor a valuable life lesson.

Creeping out of the blind, I ran down the sandy slope toward the carcass. Sneaking up behind 235, I yelled abruptly and reached out as if to grab her. Awakened with a terrified start, 235 scrambled into flight and shot off the cliff. Using my two-way radio, I called the crewmember stationed in the valley below to tell her that I had hazed 235, who was now flying over the cliff. Moments later, I received word that 235 had landed on a cliff ledge and was settling down. She never roosted unsafely again. Gratified that my efforts were keeping some condors safe, I nonetheless spent an uneasy night wondering how I had failed 228.

The answer, when it came, was shocking. The necropsy report listed Condor 228's cause of death as "poor body condition." She had been severely underweight, so the bitter cold might have contributed to her physiological stress. Heartbroken and appalled, I scoured the daily data sheets on which crewmembers recorded every bird's activities—including whether or not it had fed. We had cheered 228's success when she was first to arrive at a carcass in January and had recorded her eating regularly. But, clearly, she had not been getting sufficient food.

Again, I vowed to do better and, in the coming weeks, adopted a number of changes. In the winter, the calf carcasses that we laid out after dark sometimes froze overnight, making it difficult for the inexperienced and less-developed juvenile condors to access the meat the following morning. Henceforward, when temperatures dropped below freezing, we kept the carcasses with us in our wall tent overnight—piled around our small wood stove—to keep them thawed, and placed them out early in the morning, before light. We also provided more carcasses at each feeding, and sliced them open so that the hungry juveniles that usually arrived first at the carcasses would not have to wait for a powerful adult to tear open the hide and provide access, and so would not be excluded from feeding by dominant older birds.

In the aftermath of 228's death, we retrapped the juveniles to weigh them and assess their body condition. (The keel, or central breast bone, to which the large flight muscles attach, feels sharp in underweight birds because of a loss of breast muscle and associated fat.) Eventually, we assured ourselves that the remaining juveniles were healthy. In the interim, the condor flock provided much-needed comic relief, which helped allevi-ate my sadness and the tedium of long, cold trapping days. Walking like ponderous turkeys toward the opened release-pen door to access the bait we had secured inside, condor after condor stopped short upon spying the snaky ropes that we used to open and close the door. Cocking its head, each bird reached out with its bill and tugged on a rope. Then tugged again. Invariably, other condors hustled in to investigate, muscling out their com-padres as they vied for the best pulling spot. Vigorous tugs-of-war ensued, sometimes with multiple condors arrayed along a rope. As I huddled in the cramped darkness of the trapping blind, stifling laughter, and holding the door steady, I couldn't help but delight in the condor's capacity to entertain as well as inspire. And I fervently hoped that the lessons we were wittingly or unwittingly teaching each other would help condors and caretakers alike navigate less turbulent skies in the months to come.

CHAPTER EIGHTEEN

Growing Up

Bundled against winter's cold, I positioned my spotting scope and began searching for condors. Morning sunlight had not yet touched the prone forms scattered among favorite perches on a west-facing portion of the Vermilion Cliffs that we'd dubbed "Africa" because its rocky outline bore such a striking similarity to the African continent. When at the release site, condors invariably roosted together in this area lying alone on a ledge or clumped together with other condors. As the birds began to stand up, stretch, and preen, I noted each bird's wing tag number and circled the related studbook number on my data sheet. Field crewmembers stationed at the South Rim and roving on the Kaibab Plateau and along the Colorado River corridor did the same with the condors they sighted, so we could verify that every member of our flock had been accounted for when we compiled our data at the end of each day.

When sunlight finally illuminated Africa's cliff ledges, the condors extended their wings and faced the sun, bathing their feathers in golden light. A widespread behavior in both New and Old World vultures, sunning is also seen in storks, pelicans, cormorants, and other birds. Waterbirds, such as anhingas and cormorants, typically sun to dry their wings, as do condors and other vultures after bathing.[1] But condors also sun when their wings are dry, particularly in the morning when sunlight first touches them or when a glimmer of sun pierces overcast skies.

Some vulture species are able to reduce their nocturnal body temperature by several degrees to conserve energy.[2] By embracing the sun's morning rays with their extended wings while sunning, condors may raise their body temperatures back to normal daytime levels before taking flight. Sunning may also help condors keep their flight feathers in optimum condition. Over time, air pressure bends a soaring condor's flight feathers. It can take hours for them to return to their normal positions, reducing flight efficiency if the bird takes off again soon after landing. Because the sun's heat warms and softens the keratin feather shafts when a bird spreads its wings, sunlit feathers resume their shape more quickly than do the shaded feathers of closed wings, which may explain why condors often sun after landing late in the day.[3]

While the sun moved imperceptibly across the sky, I identified condors and noted their activities. As I focused on six-year-old Condor 123—a dominant male—he extended his wings as if to sun but pointed the tips downward. Facing female Condor 119, he rocked from one foot to another. Lowering his head to flaunt the back of his colorful mauve and pink neck, 123 engaged in a timeless, ponderous-but-graceful dance. Excitedly, I noted that 123 was displaying to 119, a relatively new activity initiated by these young-adult condors about five months earlier, in September 2000.

After spending their lifetimes feeding and preening, soaring and bathing, playing and sleeping, our adult condors had discovered a new purpose, which they embraced with abandon. Despite the penetrating chill on that mid-February day, I watched a veritable dance fest on the cliffs. Even two of the four-year-old males got in on the act, though they did so indiscriminately, courting a submissive three-year-old male that seemed befuddled by their attention.

In addition to displaying to 119, 123 repeatedly hopped onto her back, as a precursor to mating. While early mating attempts consisted of 123 standing on a disgruntled-looking 119, his more recent attempts were more coordinated. Female Condor 127 frequently disrupted the proceedings, vying unrelentingly for 123's attentions. Fueled by hormones, 123 also displayed to 127 and once mounted Condor 133, though that particular amorous effort was foiled when male Condor 114 walked up to the pair and tugged on 123's tail.

Such comic moments sometimes made it difficult to take our condor's nascent reproductive activities seriously, but I watched their antics with growing excitement, aware that I was witnessing history in the making. It had been fifteen years since condors had bred in the wild, and reintroduced condors had never done so. Biologists in California and Arizona had waited for years for their charges to reach breeding age, wondering whether the birds would instinctively initiate breeding activities since they had no wild mentors to mimic. Fortunately, the birds' courtship displays confirmed that such behaviors were innate, as did other pair-bonding behaviors. We had recently begun to witness synchronized flights, in which one condor shadowed the other's every move so closely that the two seemed to move as one. Arcing across fall and winter skies, these tandem flights were grace and power, speed and synchrony.

Our adults were also showing an intriguing interest in caves that pocked the Vermilion Cliffs, despite having ignored these concavities for years. As the sun descended toward evening, I watched as 119 and 123 took turns disappearing into a small cave near the cliff rim. While standing guard outside the cave that 119 had disappeared into, 123 pushed around small stones on the cliff ledge with his bill. A precursor to a behavior known as nest grooming, 123's actions would one day clear debris from the floor of a cave or tree cavity, creating a smooth surface on which his mate could lay her egg.[4]

In the coming days, 119, 123, and tagalong 127 expanded their new interest in caves to the Colorado River corridor. Disappearing for longer and longer periods into caves in Marble Canyon, 119 and 123 instinctively followed a path that condor fans hoped would ultimately lead to the hatching of a wild nestling. But although 123 displayed most often to 119 and mated with her exclusively, I couldn't help but be dismayed in the coming weeks that our nascent condor "pair" behaved more like a "trio," with 127 unwilling to relinquish Condor 123, despite the efforts of other males to attract her attention.

————

As February drew to a close, our condors began traveling more widely again—foregoing their winter habit of spending day after day at the Vermilion Cliffs. When Arizona condor project director Chris Parish checked for the birds' radio-transmitter signals at the fog-enshrouded South Rim, he was astounded to hear a frequency he had seldom heard. Doubting his ears, Chris checked and rechecked his list of frequencies before calling to tell me that our elusive Condor 176 was finally back in our midst. To our delight, she had met up with several other condors, ending her self-imposed months of solitude.

The condors' signals soon faded as the birds headed away from the South Rim during a break in the weather, but, miles away, a crewmember stationed at Badger Canyon began receiving 176's signal for the first time since joining the project five months earlier. Soon, three specks materialized in the skies over the Colorado River, taking the shape of condors as 176 and two older males glided up the river corridor before settling down to roost at Badger Canyon.

The next morning, I joined other crewmembers in rushing out to Badger to view 176 before she disappeared again into the firmament that was more home to her than any terrestrial haunts appeared to be. A few hours later, another beaming crewmember reported that 176 had just arrived at the Vermilion Cliffs and was battling for a place at a carcass.

Eager to replace her one nonfunctioning transmitter and assess her health before she disappeared again, we spent much of the next week trying to trap our itinerant condor. As I watched 176 bounce on top of the release-pen netting like a child high-stepping across a trampoline, tug on the rope attached to the trap door, and walk around with a puff of down stuck to the top of her bill, it was hard to imagine that she was anything but a goofy, playful two-year-old condor. And yet she refused to walk through the trap door to reach the baited carcass as the other condors did.

Finally, not wanting to risk her leaving the release site in search of food, we gave up on trying to trap her, and placed several carcasses on the cliff rim while the condors slept. As I watched 176 at the carcasses in the days that followed, I was struck by her wildness. While all the condors glanced around warily when feeding, 176 did so with a frequency and intensity that spoke of her many months alone in the wild. Heart in my throat, I watched her line out toward the Kaibab after several days of feeding. To my relief, though, she circled over the Kaibab and rejoined the other condors. After so many months of flying solo, she seemed content, at last, to accompany her flockmates on their sky journeys and settle with them on beaches, cliffs, and carcasses.

––––––––

While 176 was finally consorting with others of her kind, I was doing the opposite and gaining a little personal space. In early March, The Peregrine Fund rented a two-room hogan in the village of Vermilion Cliffs, where I could live with my border collie and three indoor cats. After months of living in a crowded field house, I was thrilled to have a little place of my own, with a small hummingbird garden and dramatic views of the Vermilion Cliffs.

As March days warmed and lengthened, Condors 119, 123, and 127 seemed equally eager to carve out a bit of space for themselves, focusing with increasing intensity on a potential nest cave in Badger Canyon. In addition to disappearing into the cave for long periods, the trio nest groomed, pushing around debris on the entrance ledge with their bills. Soon, 119's attachment to the cave superseded all else. On March 13, she remained at Badger while 123 and 127 returned to the release site. I fervently hoped that her "independence" from 123 portended future nest-sharing duties, but 119's speleological fervor meant that 123 and 127 spent more time together. In 119's absence, 123 displayed to 127. When 119 returned to the Vermilion Cliffs to feed, 123 and 127 stayed behind, roosting together in the river corridor.

One of the behavioral cues that would tell us if a condor had laid an egg included the bird remaining in the nest while its mate left to forage. If the foraging condor then relieved the cave-attending condor, this parental switch would suggest that the birds were taking turns incubating. But although 119 spent significant amounts of time in the Badger cave, she didn't seem to be alternating nest duties with 123, and 127's presence confounded our expectations for a nesting pair's behavior.

Sunday, March 25, 2001, began, as did most of our condor-project days, with an early start as crewmembers departed to monitor birds at the release site, in the river corridor, on the Kaibab Plateau, and at the South Rim. My duties that day were to monitor 119, 123, and 127's activities at Badger Canyon while keeping tabs on wayward juvenile Condor 198. Recaptured for approaching people too closely soon after his 1999 release, 198 had spent nearly a year back in captivity before being rereleased.

After observing young condors, monitoring their weights, and studying their parentage, it had become increasingly clear to me that juveniles that were submissive, slightly underweight, or the offspring of certain parent condors often behaved concerningly. These youngsters sometimes failed to integrate well with other condors, had a tendency to approach people, and endangered themselves by perching in areas with no ready escape route. While such "problem" condors were often more cautious after being given a brief "time-out" in our new flight pen on the Vermilion Cliffs, repeat offenders invariably benefitted from additional growing-up time in captivity. Upon their return to the wild, these problem birds usually were warier and responded better when we hazed them away from people and dangerous situations. Nevertheless, the first forays of rereleased birds into developed areas were always fraught, and we monitored their interactions with people and structures closely.

After checking on Condor 198, who was roosting alone on the cliffs beyond the town of Marble Canyon, I headed to nearby Badger Canyon to monitor the adults. Seated on a point of rock protruding over the canyon, I focused my spotting scope on our trio's cave. A sudden movement caught my eye as a blue-gray streak winged over the river. As the peregrine disappeared into a cliffside aerie, my mind's eye flashed back to Wyoming cliffs and the five falcons that had initiated me into the bird realm. Watching falcons and condors from my current precarious perch, I couldn't help but reflect on how far I had come from the days when I had been daunted by the small cliffs that cured me of my fear of heights and launched me into an unimagined future.

Now, I watched 119 disappear into her cave while 123 and 127 loafed nearby. The blue-green Colorado River glinted hundreds of feet below me as it churned through its confluence with Badger Canyon. As I noted the birds' activities, 198 suddenly appeared, heading straight for the cave and the loafing adults. Dodging the peregrine that pursued him as he glided past her aerie, 198 landed with a flourish next to 123 and 127. After several days spent alone, he had found fellow condors. The adults seemed less than pleased. Chaos ensued.

As 119 charged out of her cave toward 198, 123 unexpectedly attacked her. He seemed to be exhibiting displacement behavior by targeting his presumed mate rather than the intruder. Condor 127 showed no such confusion, lunging at 198 with her bill. Fleeing 127's attack in a flurry of wingbeats, 198 was pursued by the three adults. Apparently oblivious to what had precipitated their unprecedented aggression, 198 circled and landed on the nest cliff. The adults settled nearby. But when 198 moved closer to the cave, they immediately attacked, chasing him away from the cliff. In a more measured approach, often seen in territorial birds of prey, 123 then "escorted" 198 downstream, flying behind and to the side of him until the younger bird left the area.

Circling back upstream, 123 landed at the cave and walked in. Nine minutes later, he emerged, pushing something around with his bill. I refocused my scope, thinking he must be nest grooming, but saw, to my surprise, that the object consuming 123's attention was smooth, elliptical, and looked like . . . an egg. Frantically, I tried to increase the magnification on my scope for a closer look, but it was zoomed in all the way. Scarcely breathing, I stared at the cream-colored object, trying to remember if it had been in view before. Had I just failed to notice a smooth, white rock? Or was it the first condor egg laid in the wild since 1986?[5] I needed to be absolutely sure. I struggled to contain my excitement, wanting to celebrate what would be a gigantic leap forward in the recovery of the critically endangered California Condor. But I could not afford to be wrong about this. Reining in my stampeding emotions, I stared through the scope, calmly recording what I was seeing in my field notes.

For almost an hour, I stared at the mysterious object. Meanwhile, 119 left the area and 123 entered the cave again then perched in the entrance. Finally, 127 walked along the cliff ledge and stopped at the possible egg. Reaching down, she placed her beak into its hollowed-out back end. And then I knew. Elation warred with disappointment. I was indeed witness to a historic moment in the condor's epic recovery saga. But the egg was broken.

Condors often fail to hatch their first egg. Sometimes it is infertile. Sometimes a member of the pair breaks it accidentally or intentionally. But this egg being broken in no way diminished the extraordinary achievement of these captive-reared birds, which had survived in the wild for four years, overcome lead poisoning, found themselves a cave, and produced their first egg.

I sat, for a time, surrounded by sun and wind, suspended over cliffs and flowing water, and absorbed the moment. Euphoria mingled with gratitude. Over the years, countless biologists and conservationists had contributed to the condor's recovery, but I'd had the unbelievably good luck to be in the right spot at the right time to document the first confirmed egg laid in the wild by condors in fifteen years. As my smile widened and my affection surged for the remarkable birds I'd watched over for months, I reached for my newly acquired cell phone and began to let the world know that the California Condor was on track to recapture the skies over which it had long reigned.

————

The next few days were a blur of media interviews interspersed with my usual condor-monitoring and crew-management duties. I paced nervously around my house while being interviewed on National Public Radio and was delighted to see news of 119's egg appear in national and international media outlets in the coming days.[6] The news resonated with a public that had followed the condor's story for years, that had been mesmerized (as I had been in high school) by video footage of zookeepers using condor arm puppets to raise the first condors hatched in captivity, and that had watched enthusiastically as the legendary birds were first returned to the wild. At year's end, *Discover Magazine* included our condor egg in its list of the 100 top science discoveries of 2001.

An unexpected breakup of our trio followed the egg discovery, and my anthropomorphically inclined heart broke a little for 119, who appeared to lose her "mate" in subsequent days to 127. Condors 123 and 127 had been inseparable for several days before the egg appeared. After it was broken, the duo nest groomed and spent the rest of the day together. Their attachment to each other faded in the coming weeks as the season's courtship activities waned, but, unbeknownst to us, the bond they had formed foreshadowed events and milestones to come.

———

Though our adult condors garnered the public's attention in the spring of 2001, the remainder of our flock commandeered the bulk of the field crew's energies. As the weather warmed, our twenty-five condors ventured far and wide, visiting favorite locations and exploring new areas. Standing beside the Colorado River, my receiver beeping at my side as the antenna I held aloft received one radio-transmitter signal after another, I marveled at what appeared to be condor highways in the sky. Invisible to me, one high-flying group of condors that had left the release site and headed to the river mingled with birds that were leaving Badger Canyon. Some birds headed downriver toward the South Rim, while others headed upstream to Navajo Bridge—a dramatic span that crosses the Colorado River near the town of Marble Canyon. Still other condors headed westward over the Kaibab, where they mingled with birds that were leaving the release site and heading north to Zion National Park. If the condors were flying high in the sky, I could receive their signals as far as 70 miles (113 km) away from an unobstructed viewpoint. If the condors dropped into the canyon, my receiver might only pick up their signals if the birds were within a mile or so.[7]

But while I loved picturing the condor's flyways and movements, and delighted in the intrigue of which birds joined together and which went where, I also tracked the moving signals with trepidation. Would our youngest birds land too close to people after leaving the release site for the first time? Would our older condors indulge in riotous (and, from our perspective, inappropriate) play with the fascinating items comprising an unattended river camp? Although the latter activity was an early habit that the condors had largely abandoned, they sometimes couldn't resist tugging on tent guy wires, dragging around slithery sleeping bags, tearing apart foam mattresses, and playing with camp trash—as long as no people were around to frighten off the unruly birds.

Such behaviors had caused considerable consternation within the recovery program as biologists debated whether these actions were a product of the natural curiosity of juvenile condors, which typically explored the world with their bills but now lacked parental oversight and wariness, or whether these behaviors resulted from inappropriate captive-rearing techniques. Though answers remained elusive, animal keepers continually refined rearing protocols, ensuring that nestlings were raised by parent condors when possible, and that juveniles—whether raised by parents or human puppets—had suitable condor mentors once they fledged.

Having worked with other playful and intelligent birds—Common Ravens and Hawaiian Crows—that test novel items with their bills to gain exposure to what is important, I viewed such exploratory behavior in condors as natural curiosity that only seemed anomalous because the parentless youngsters found themselves in a world heavily impacted by people. Curiosity and playful exploration likely had adaptive value in the condor's evolutionary past. Indeed, as corvid expert Bernd Heinrich observed while raising a brood of ravens, the birds' *curiosity* exposed them to items in their environment. His juvenile ravens never passed up anything novel and their proclivity for exploring such objects—by manipulating them with their bills—ultimately contributed to their food-finding abilities. Heinrich's ravens played enthusiastically with the "toys" that he provided to them, tearing apart milk cartons and drumming on plastic bottles. And, as Heinrich noted, "There are countless newspaper accounts of ravens doing damage to roof shingles, parked automobiles, delicate airplane wings. Young ravens, like other youngsters, are open to exploring a variety of stimuli and to learning what they mean. In the wild, this play behavior is a survival mechanism, but in an urban environment their exploratory behavior can be a source of human annoyance"[8]

Unfortunately, whereas people might tolerate a raven pecking at roof shingles, enormous condors tugging at a backcountry bathroom's loose siding in Grand Canyon National Park caused serious alarm. And for some rafters, who were accustomed to leaving riverside camps unattended, the condors' playful but destructive antics caused apoplectic rage. We could only haze the condors when we caught them in the act, and hope that their curiosity declined either with age (as it did with ravens) or when they were distracted with raising their own young.

In the meantime, though, when transmitter signals indicated that a large group of condors was clustered together, I invariably looked for a carcass or a source of entertainment. After locating eleven condors on the Badger Canyon beach, I quickly surveyed the scene and breathed a sigh of relief when I saw no people, no boats, and no tents. Binoculars glued to my eyes, I sat back to enjoy the show. Two condors played tug-of-war with a branch that had washed ashore. Others walked up and down a drift log, jostling for prime positions. Several paddled in an eddy at the river's edge then sent water flying skyward as they bathed. Suddenly, Condor 196 emerged from a mid-beach condor scrum holding an irresistible prize aloft in her bill: a plastic, one-gallon jug. She took off at a run, the other condors in hot pursuit. But with a sudden smack, 196 rammed face-first into a boulder. Momentarily stunned, she sat back on her legs, still holding the jug that had obscured her view. Regaining her balance, 196 scrambled to her feet and ran on, as the other condors closed in.

Minutes later, one-year-old Condor 227 commandeered the jug and flung it skyward with his bill. When it landed unexpectedly on his back, 227 leaped in mock fright and turned on his "attacker." Seeming captivated by this new twist on the game, 227 again flung the opaque monster skyward, whereupon it fell back down and again landed on his back. Unnoticed on my distant clifftop perch, I laughed out loud, as captivated by the condors' activities as the birds were by their frenzied play.[9]

Like the play behavior of crows and ravens, condor play often seems purposeless. But it likely has survival value, helping the birds to build strength, develop food acquisition and manipulation skills, and interact appropriately with fellow condors and other wildlife. Playing tug-of-war and dragging around objects with their bills helps condors develop the musculature and coordination for tearing open carcasses. When adult Condor 114—who I once watched drag around a piece of plywood for hours—arrived at a carcass, he had the strength to tear open the tough skin that foiled awaiting juveniles. And while punting around and playing keep-away with an empty water jug seemed like pure fun—akin to ravens sledding down snowy slopes on their backs—such behavior likely contributed to the condors' dexterity and helped refine each individual's place in the flock's dominance hierarchy.

The condors' playful antics are evidence of and likely contribute to the behavioral flexibility of the species. Rather than responding instinctually to a set of cues, condors must evaluate their flockmates' behavior and choices, and react to them while playing. The awareness that is developed through these encounters may later benefit condors when they encounter complex and unpredictable feeding situations. When a condor lands at a carcass, it must avoid predators and compete for food with other scavengers, making decisions that help it survive. Scavengers face different threats and competitors depending on where and on what they are feeding. A condor feeding on an elk carcass in the forest must be alert to different potential predators and competitors than a condor consuming fish offal on a river beach. The behavioral flexibility required of avian scavengers to function successfully in the complex and variable social environments found at carcasses may help explain why many of these birds are intelligent and quick to learn new tasks.[10]

The condor's notable but unsung intelligence was underscored for me one evening when biologist Kris Lightner returned to the field house and described Condor 127's solo play with an empty plastic bottle on the Badger beach. Exhibiting World Cup dexterity, 127 repeatedly punted the bottle down the beach with her bill and chased after it. As she pursued it, the bottle moved closer and closer to the river's edge until the water suddenly caught it and whisked it downstream. In vain, 127 ran along the shoreline, eyes fixed on the bobbing bottle. Just when it seemed lost to her, the current swept the bottle into an eddy several feet from the riverbank. Without hesitation, 127 waded into the water and reached out to retrieve her toy. But it floated just out of reach. Looking around, 127 spotted a partially submerged boulder beyond the bottle. Extending her wings, she hopped out onto the boulder. Now, by stretching her neck, she could just reach the bottle. Nudging it with her beak, she gave it a push shoreward. As it floated toward the beach, she hopped into the water and gave it another nudge. And then another. As the bottle touched solid ground, 127 resumed her frenzied play, punting it down the beach.

"I know it's anthropomorphizing," Kris remarked, "but 127 seemed crushed when she lost her bottle and she seemed so excited when she was able to retrieve it and could play with it again."[11] As entertained as I was by Kris's story, I was equally struck by 127's intelligence. No level of rote, trial-by-error learning had enabled 127 to retrieve her toy. Instead, she had engaged in multiple steps to enact an outcome that she had envisioned, figuring out how to reach the bottle, then pushing her prize shoreward (and, counterintuitively, away from her) step-by-careful-step, so she could gain control of it again.

Although I relished the condors' intelligence, playfulness, and curiosity, these captivating qualities sometimes put my charges at risk in a world that had been transformed by humanity in an evolutionary blink-of-the-eye. As a result, the steps that juvenile condors took in their journeys of exploration and discovery were fraught with anticipation for our field crew and were often followed by controlled harassment (such as flushing birds from inappropriate perches), then careful monitoring to ensure that lessons taught had been lessons learned.

The juveniles we had released in December had made their first exploratory flights away from the Vermilion Cliffs when tiny flowers painted the House Rock Valley purple. But it wasn't until mid-May that all the one-year-olds—Condors 223, 224, 227, 234, and 235—first streamed after the older birds en masse and followed them to the river corridor. While 234 and 235 had cautiously completed their 50-mile round-trip journey and returned to the release site to roost, the other three youngsters had spent their first night away from "home" perched on cliffs bordering the Colorado River. Meanwhile, a nervous field crew had anxiously wondered how these juveniles would handle their first people encounters.

Condors 223 and 227 had exhibited stellar behavior in the following days, exploring the river corridor and avoiding people. I'd delighted in watching them cautiously approach the Colorado River and then jump back in surprise as the water unexpectedly moved. I couldn't imagine how vast the river must appear to young condors whose only prior water sources had been the water tub we filled for them and the rainwater that accumulated in clifftop potholes. Regrettably, rather than join the two males in exploring empty river beaches, Condor 224 had been drawn to activity at the nearby Lees Ferry Campground. Two crewmembers had hazed 224 to a safer area, but not before a tourist threw her a piece of chicken. When 224 had returned to the campground the following day, we'd netted the unfortunate bird and returned her to the release pen.[12]

We had awaited our youngsters' first journeys to the South Rim with particular trepidation. Now armed with cell phones, our crew could better coordinate its efforts. Only four days after first leaving the release site, Condor 234 flew to this tourist mecca for the first time. Two days later, with characteristic independence, 234 left the adult condors that he had accompanied to the park and returned to the Vermilion Cliffs, achieving a milestone that was notable for its lack of noteworthy interactions. Three weeks later, as I stood on the cliff rim, checking transmitter signals to determine which condors were in and around the South Rim, I heard juvenile Condors 223 and 227's signals for the first time. And judging from the volume of the beeps that emanated from my receiver, they were approaching fast.

Moments later, I watched 223 glide over hordes of tourists congregating along the cliff rim. Looking skyward, I almost laughed out loud as I saw what looked like a wide-eyed, giddy-with-excitement juvenile condor speed over my head. Instead of exhibiting the dreamy, floating quality of an adult condor's flight, 223's flight looked hyperactive, edgy, and barely controlled. Legs dangling to slow his speed, head swiveling to and fro, shiny black eyes taking in all the commotion below him, 223 careened along, looking as excited as a child on his first visit to an amusement park. Nevertheless, Condors 223 and 227 exhibited exemplary behavior. Although each bird had to be hazed from occasional "inappropriate" perches—several of which were on stone buildings that blended in with the cliffs—they responded well to our harassment and were soon perching safely below the cliff rim.

But while our youngsters were successfully navigating the South Rim, some of our older condors took a step backward when they discovered a new form of entertainment across the canyon. When I tracked a group of condors to Grand Canyon's less-chaotic-but-still-crowded North Rim, I expected them to be buzzing over tourists and perching on cliffs. Instead, I followed their signals to a grassy area abutting the cliff rim, beyond the tourist facilities. And, once again, I almost laughed aloud at the sight that met my eyes, though in reality it was no laughing matter.

Looking like trench-coat-wearing hoodlums discussing something nefarious, nine condors huddled around the rim of a massive metal water tank that the park had installed so fire-fighting helicopters could ferry buckets to a nearby forest fire. Condor feathers floated in the water and the birds dipped their heads in for periodic drinks. As I ran toward the group, flushing them into the air and away from the area, I couldn't help but shake my head at the condors' capacity for discovering new venues and entertainment. Despite having the Colorado River at their disposal, this water—in close proximity to the tourist herds and with a ready escape route into the canyon—had proven irresistible.

Over the next few days, we repeatedly flushed the obstinate birds from the North Rim water tank, alerting the fire crews whenever condors were in the area. But despite these precautions, during one tense moment, I watched sibling Condors 176 and 187 glide into the area just as a helicopter began lifting logs to transport them to the park's trail crew. The two birds circled under the helicopter, investigating the rope attached to the logs. Heart in my throat, I watched anxiously until the helicopter finally moved away and the condors dropped into the canyon. To our relief, park personnel soon solved the water-tank problem by installing a large, removable cover. And the condors' discovery of several carcasses in the canyon eventually drew the flock away from the entertainment at the North Rim.

Although the older condors gave us no more trouble, a particularly challenging youngster gave us some tense moments before the onset of fall began to curtail the flock's far-flung journeys. Rather than being released to the wild in 1999 as planned, Condor 210 (Two-ten) had been kept in captivity for an additional year of growing-up time because she was unwary of people and unwilling to interact with other condors. Released to the wild in December 2000, two-year-old 210 was our most challenging juvenile from the outset. After spending her first two weeks post-release alone at the base of the Vermilion Cliffs, where we repeatedly hazed her to safe locations, 210 finally returned to the cliff rim and fed alongside her flockmates. But her limited interactions with other condors were notable for such a gregarious species.

After making short forays away from the release site in early April, Condor 210 began traveling more widely and discovered a new favorite place: Navajo Bridge. Enjoyed by thousands of tourists each year, the historic site consists of two parallel, steel-arched bridges that cross the Colorado River near Marble Canyon. The original structure—built in 1929—was decommissioned and serves as a walking bridge. Its vehicular replacement was built in 1995. With the river to bathe in, beaches to play on, and people to observe from lofty cliff perches, Condor 210 had little incentive to leave this newfound haven. Discovering her first carcass further cemented the area's appeal, and seventeen days passed before 210 returned to the release site. Given her proclivity for landing on bridge railings and other inappropriate spots, we watched over 210 from daybreak to nightfall each day, hazing her when necessary and answering innumerable questions from curious tourists.

When the one-year-olds with which she had been released to the wild began making regular journeys to the South Rim, 210 stayed close to home, maintaining her solitary ways. As the summer progressed, though, she began interacting with the other youngsters. After watching her allo-preen and cuddle up to Condor 227 at the release site, biologist Chris Crowe wrote in his field notes that it was "good to see 210 be social and have a buddy."

But after making encouraging steps forward, 210 took a step backward. After embarking on a flight with nine other condors, 210 was either waylaid by a looming storm or became attracted to the activity at a scenic overlook near the release site. When Chris Crowe rushed up to the overlook, he found 210 standing beside a parked car. Perhaps she was seeking shelter from the storm by what she may have viewed as a large boulder, but her behavior was alarming. Chris hazed 210 to a nearby ravine and she roosted on a cliff ledge.

Over the next few days, 210 loafed in the ravine and periodically hiked to the nearest high point to get better lift for takeoff. Unfortunately, that high point was the scenic overlook, so we had to haze her back into the small canyon whenever she reached it. When she finally soared back to the release site three days later, we breathed a collective sigh of relief.

But 210's adventures were not over. After a successful sojourn on the Kaibab with our youngest condor, 235, whose independence, wariness, and ability to find carcasses made her a stellar companion, 210 left the release site with two adult condors. The adults arrived at the South Rim. Condor 210 did not. Instead, she was drawn to the irresistible chaos of the North Rim. And there the fun really began.

Touching down near the park's forested housing area, adjacent to the command-and-control center for fire-fighting operations, 210 perched in one inappropriate place after another. Each time biologist Kris Lightner hazed 210 to move her to a safer spot, the young condor fled on foot or with short flights and landed on another terrible perch. Several firefighters tried to corner the miscreant condor, but wily 210 eluded them, sprinting away at breakneck speed and leading Kris on a breathless chase. Aghast at the sight of 210 hopping up onto the windowsill of a maintenance building, Kris ran inside, grabbed a broom from an astonished worker, and thrust it through the open window, pushing 210 off the ledge.

As Kris emerged from the building, she saw 210 stop suddenly, mid-run, as she spied something on the ground. A *toy*! A small, bright-orange ball lay in her path and 210 couldn't resist pushing it around with her bill. But Kris bore down on her, and 210 was forced to relinquish her newfound toy. Soon after, to Kris's relief, 210 took flight and landed high in a pine on the outskirts of the North Rim campground, where she roosted.

In darkness the next morning, Kris and I somberly returned to the North Rim to capture our troublesome 210. Several hours after dawn 210 left her perch and glided toward some ravens that were hopping around the nearly empty campground. Kris and I quickly headed her way with our capture net. But as soon as 210 spotted us, she launched herself skyward, landing in a pine on the canyon rim. Moments later, 210 took flight, then headed back to the release site.

In the ensuing months, 210 explored the Kaibab and the Colorado River corridor, spending more and more time with other condors. To our astonishment, she exhibited exemplary behavior, roosting in remote areas far from people. Although we celebrated these encouraging steps, we nevertheless spent the summer and fall on tenterhooks, awaiting 210's first trip to the South Rim. To our relief, we waited in vain. Our adventures with 210 at that particular condor haven would have to wait until the coming of a new year.

———

Despite 210's welcome absence, the South Rim remained popular with the other condors as summer edged toward fall. After scanning through transmitter frequencies to check for new arrivals, I sat on a stone wall adjacent to the clifftop walkway in Grand Canyon Village to update my data sheet and catch my breath. It had been a typical South Rim day. I had arrived in Grand Canyon Village just after dawn, checked frequencies to record which birds were present, then kept an eye on those I could see while they stretched, preened, sunned, and, finally, took flight.

The questions had begun the moment I raised my antenna to determine which condors were in the area. Weeks earlier, Chris Crowe had tallied the number of tourists that asked him questions during one day of condor monitoring at the South Rim. He counted 107—a fairly typical number. As the sun rose in the sky, I told Grand Canyon visitors how big the condor's wingspan was (about 9 to 9½ feet [3 m]); how far it was from the South Rim to the release site (about 50 miles [80 km], one way—a condor could make the round-trip flight in two hours); how fast they could fly (radio-tagged condors had been tracked by airplanes traveling at around 40 to 60 mph [70 to 95 kph]); how they had become endangered; and what the numbers on their wings meant.[13] Upon learning about the individual condors they were observing, tourists often felt a connection with a bird they had never expected to see, but whose exploits they now could follow on the field notes that I posted on The Peregrine Fund's website.

In addition to answering questions, I also spent time explaining my actions to tourists who yelled at me for harassing condors.

"Leave them alone! They're beautiful!" shouted an outraged group, as I ran at a juvenile, waving my arms to flush it when it perched on the cliff rim adjacent to a footpath. After making sure the youngster landed on a distant perch, I informed those who had been haranguing me that I was involved with the reintroduction program and was teaching condors to keep their distance, since the birds could be shot or otherwise harmed if they approached people too closely. Once I had explained that our hazing taught the condors to select safer perches, tourists invariably forgave me for my obnoxious behavior and expressed gratitude for my work.

At day's end, I sat now, watching my charges settle down to sleep on cliff ledges. As the sun sank below the horizon, casting a muted glow over towering rock walls, colorful domes, and lofty pinnacles, a reverent hush permeated the canyon. Scattered visitors watched the sunset along the rim, gazing quietly at the immense abyss or speaking softly to companions. Lost in my own thoughts, I absorbed the canyon's tranquility, mesmerized by the ever-changing light that continuously transformed the iconic vista.

Nudged out of my reverie, I glanced up and smiled when a middle-aged woman asked if I was monitoring condors.

"Someone told me they were here, in the canyon, and I couldn't believe it," she said. "Forty-five years ago," she continued after a pause, "when I was five years old, my father put his hands on my head and pointed my gaze up to a condor perched in a eucalyptus tree in California. 'That's something you'll never see again in your life,' he told me."

As I pointed out juvenile condors 223 and 227 sleeping together on a cliff ledge and saw tears sparkle in her eyes, I couldn't help thinking how privileged I was—how privileged we all were—that her father's prophesy hadn't come to pass, and these great birds were still among us. In the months to come, though, I was reminded that their survival was still precarious and the species' very existence was still tenuous.

CHAPTER NINETEEN

Shot

Energized by coffee and anticipation, I impatiently scanned skies through which perfect puffs of clouds drifted and a weak winter sun gleamed. Internally, I hopped up and down like a little girl awaiting Santa. Externally, I tried to maintain a modicum of calm, though my constant smile and animated conversation belied my excitement. After what felt like an interminable wait, I heard a plane's engines. Turning my binoculars toward the sound, I glimpsed an orange-and-white Twin Otter US Forest Service plane as it appeared over the Vermilion Cliffs escarpment. Arriving on fixed wings, our field crew's hopes for the coming year moved steadily toward us.

Minutes later, the plane landed at the Marble Canyon airstrip, to the cheers of our assembled field crew from The Peregrine Fund and assorted personnel from the US Fish and Wildlife Service, National Park Service, Bureau of Land Management, and Arizona Game and Fish Department. As the cargo doors opened, a hush fell over the crowd. My eyes feasted on a veritable sea of gray transport kennels, containing ten juvenile condors and one condor that was being given a second chance after having been recaptured.

Quietly, the assembled team transferred the kennels to awaiting vehicles. Our view of the condors was obscured by the dark cloth covering each kennel's door to minimize the birds' stress. Soon, a convoy of trucks wound its way up to the release site. Upon arriving, two people carried each kennel down the sandy trail to our flight pen. Eager for the first glimpse of our new charges, I could hardly bear the anticipation as we moved the kennels into the pen. Finally, it was time to introduce the birds to their new home.

As I opened the first kennel, Condor 243 hopped hesitantly into his new life. Condor 249 followed next, but this youngster came out fighting. As he stepped out of his kennel, he turned and leaped toward crewmember Kevin Fairhurst, who had just opened the door, biting and clinging briefly to the string on Kevin's sweatshirt hood before dropping to the ground and racing off. Some of the juveniles were cautious and fearful. Others came out hissing and fighting. All were shiny and dark, noble and beautiful. They looked freshly laundered, each feather crisply perfect compared to those of our timeworn adults.

Looking at the youngsters, as one hopped to a high perch and stretched out its wings, I felt elated and a trifle choky, contemplating their futures. Which of these sleek-feathered, fuzzy-headed youngsters would make it? Which would not? And what lay in store for these juveniles that carried their species' future on their untested wings?

———

Just over a month later, as I sat huddled in our flight-pen blind, watching snow fall on the young birds that had already stolen my heart, I wondered about their future for the thousandth time. And I thought about the year that was coming to a close on this New Year's Eve. We had celebrated the first documented condor egg laid in the wild by captive-reared birds. And to my delight, Condor 119, who had laid the famous egg, had increasingly been consorting with Condor 122. The two were now flying to and from the South Rim together and engaging in glorious, synchronized courtship flights.

Though she still traveled widely, our independent Condor 176 was more often in our midst and, in August, had received the project's first solar-powered satellite transmitter. Similar in size and shape to a conventional transmitter, 176's new accoutrement beamed its signals up to receiving satellites, which sent data down to a processing center that transmitted her coordinates (taken every few minutes) to our computers.

Several of our other birds had made impressive journeys. Most recently, Condor 198, who had disrupted our first nesting attempt, had visited southern Arizona. Alarmed by reports of 198 perching on a highway guardrail and hissing at passing cars, I had sped to the Parker Dam area, south of Lake Havasu City, intending to recapture him. Upon finding the wayward bird, however, I was reassured by his wary behavior, and realized that the "highway" he had encountered was a state park road, running between low cliffs and the Colorado River. Chancing upon the road as he'd hopped upslope to get more height for takeoff, 198 had been frightened by a passing car and, after trying to defend himself with a threatening hiss, had hightailed it to the safety of the cliffs.

I had realized earlier that year, when juvenile Condor 203 perched by a road that provided him with an open flyway during his first visit to a heavily forested area, that context was important when evaluating condor behavior. What seemed like unwary, "bad" behavior to a casual observer often consisted of a condor making the best possible choice it could when confronted with a novel situation.

At year's end, twenty-five condors flew free in Arizona. We had lost one condor in 2001 (our unfortunate 228), but the majority of our youngsters were now successful, independent birds. Eleven more juveniles would bolster our flock's numbers in 2002. As snow accumulated on the backs of these youngsters, pressed wing-to-wing on their favorite perches, I relished the quiet and my chance to watch them. What others might have found tedious—watching birds rest—I found captivating since it gave me insights into the condors' lives and personalities. And it was only by spending long periods watching inactivity that one was rewarded with seeing the unforgettable: like the time a tiny mouse darted between the feet of dozing Condor 133, startling her into comically panicked motion; or the time Condor 235 landed on the water tub, lost her balance, and fell face-first into the water.

Watching the birds up top also gave me a much-needed break from the endless office duties that supplemented my field work—reviewing the crew's data sheets and producing summaries of our birds' daily activities and roost locations, making the crew schedule, writing semimonthly field notes for The Peregrine Fund's website, training new employees, and contributing to scientific and management articles. While watching these condors, I could still field the countless daily phone calls from crewmembers, who gave me constant updates on the locations and activities of the condors they were tracking, and asked advice on which birds to follow or how to handle unexpected situations.

Now, warmed by a propane heater that made long hours in the frigid blind more comfortable, I watched in awe as every one of the eleven penned condors suddenly shifted on its perch and extended its great wings. I hadn't even noticed the glimmer of sun that had pierced through the dense afternoon clouds. But clearly the condors had, and they responded as one to the sun's tenuous warmth. My two-way radio clicked and crewmember Courtney Harris—stationed at the Ramada—gushed, "Every condor I see on the cliffs is sunning!" Minutes later, twenty of the free-flying condors shook the snow off their backs and took to the skies. It seemed a fitting final image to cap a wonderfully successful year.

———————

Headlamp piercing an inky gloom, I steeled my muscles and managed to lift the garbage can a foot off the ground. It was no ordinary container, but one to which we had riveted backpack shoulder and waist straps so that we could transport the calf carcasses that we fed our condors. Given my small stature, I often had to be creative in using what strength I had to accomplish seemingly impossible tasks. Earlier that summer, when stranded in the desert with a flat tire, I had discovered that, although my arms couldn't lift the heavy spare tire onto the awaiting wheel studs of the truck that I had jacked up, if I placed the tire on my feet and lifted my toes, I could raise the tire just enough to guide it onto the studs with my hands. Now, I hoisted the garbage can, which was filled with a 60-pound carcass, onto the footrest of our all-terrain vehicle (ATV). Heaving the can again, I raised it onto the ATV's seat, where it was high enough for me to slip my arms into the awaiting straps. Bracing my legs, I pulled the load onto my back and set off with a slight stagger, pushing through darkness and deep sand to the cliff rim.

Once there, I dropped to my knees and wriggled out of the pack. Dragging the calf out of its plastic bag, I laid it out on a rocky outcrop and opened the carcass with a sharp knife to make it easily accessible to juvenile condors the next day. I thought back wryly on the days when cutting sheep heads had seemed appalling. Though I was no less saddened by this animal's fate, I had become more comfortable than I'd ever imagined I could be with cutting up dead animals.[1]

After carrying another carcass and 12 gallons of water to the cliff rim, and refilling the condors' water tub, I met up with crewmember Dave McGraw, who had just provisioned the penned juveniles. He had dropped his carcass through a chute so that the youngsters wouldn't associate people with food and accessed their water tub via a large side door in their pen.

Our work completed, we drove wearily back to the field house. I struggled to stay awake by focusing on Dave's taillights. Because darkness fell early in February, we could begin our nighttime carcass drops earlier than we could in summer, when I sometimes finished my workday well after midnight. Even so, the need for sleep almost overwhelmed me as I carried my equipment into the field house. Our arrival was met with the grim faces of our fellow biologists. They had just received a report from someone who had seen two young condors, 223 and 235, feeding on a coyote carcass a few miles from our release site, earlier in the day. A dead coyote almost invariably meant a shot coyote. Had our youngsters ingested lead? And if so, how much, and how quickly might they succumb to it?

Time seemed to stand still as we discussed our options and agonized over what-ifs. Moments later, I laid out a plan, profoundly grateful for this particular crew's willingness to put their own needs aside to keep condors alive. Kevin Fairhurst and Marta Curti would drive the hour-and-a-half to our wall tent up top, set up our condor trap and blind, and move the carcasses I had laid out earlier into the trap. Meanwhile, the rest of us headed out to the House Rock Valley to find and collect what turned out to be two coyote carcasses, to prevent other condors from feeding on them, and to x-ray them to check for lead fragments. By 0230 hours, the trap was set. Two hours later, I headed back up top to join Kevin in the trapping blind.

Several hours after daylight, 223 and 235 returned to the release site and soared over our trap, heads swiveling as they eyed the carcasses. For the next hour, each time one bird entered the trap, the other ran out. We waited, not wanting to scare one away while trapping the other. Finally, though, both birds entered the trap and began to feed. Moments later, we had them in hand.

As Kevin and Marta held them, I drew blood from each condor's leg. After two interminable three-minute waits, our tester relayed the birds' negligible lead levels. To be absolutely certain that neither bird's digestive system contained lead fragments, we then drove the youngsters two hours, to the Page Animal Hospital, and x-rayed them. Fortunately, the x-rays showed no bullet fragments. X-rays of the meager coyote remains showed a possible fragment in one of the animal's muzzles. Later that evening, we transported 223 and 235 back to our flight pen, where we would observe them and retest their blood in a few days' time.

But our ordeal was far from over. Discussions with the crew and a perusal of our data sheets revealed that thirteen additional condors had been near the coyotes. And only one day remained before the highly anticipated release of our first group of juveniles. If we couldn't trap the other condors that might have fed on the coyotes in time, we would have to cancel the release—foiling scores of people's plans—since we couldn't trap when newly released juveniles were taking their first flights.

Astonishingly, though, we managed to capture ten of the thirteen birds the next day. All were lead-free. The release could proceed as planned. Sobered that a possible lead-poisoning incident had occurred so close to "home," we nonetheless breathed a collective sigh of relief that we had reacted quickly, taken every precaution, and avoided a potential disaster. And now, it was time to give the awaiting juveniles their chance to fly free in our imperfect world.

———————

The following morning, I crouched in the small, enclosed viewing box attached to our release pen, watching, through one-way glass, the birds that we had been readying for release for more than two months. We had taught them to avoid power poles by electrifying a mock power pole in their flight pen, and we had seen every youngster progress from landing on the pole (and receiving a mild shock) to avoiding it altogether.[2] We had initially ensured that daily food was available to help the juveniles navigate the transition from being fed by their parents or human puppets at the captive breeding facilities to feeding on calf carcasses at the release site. Then we had gradually increased the number of days between feedings so the birds would get hungry and learn to feed aggressively as a group at carcasses, rather than feeding tentatively, one at a time, as they did initially. We had watched the juveniles' interactions and behavior, and had chosen a subgroup of the most dominant condors for this release. Those that were more submissive to other flock members or competed less aggressively at carcasses would be released once they had more time to mature.

I had been met with significant resistance from my Idaho supervisor when I first proposed sequentially releasing small groups of juvenile condors that we deemed ready, rather than releasing all the youngsters at once on a predetermined date, as had always been done in the past. By doing multiple smaller releases rather than one big, yearly release, we risked reduced public interest (and possibly the critical financial support on which the recovery project depended). But, eventually, my arguments won out that condors mature at different rates and our juveniles were more likely to survive and less likely to be recaptured for behavioral problems if we released sequential subgroups of the most dominant juveniles—those that could better compete with the older condors and better cope with people challenges. Releasing smaller groups of condors also meant that the field crew could more easily monitor and protect each young bird. Ultimately, this release strategy prevailed and was adopted program-wide.

Now, I looked beyond the release pen and out over the valley, where a crowd of about 150 people had gathered to watch condors embrace their freedom on a cold, sunlit day. My two-way radio clicked and Chris Parish's voice came through.

"Sophie, we're ready to start a countdown whenever you are."

"10-4," I whispered back, using the universal acknowledgment of a radio transmission. With butterflies dancing in my stomach and a heart full of hope for the future, I turned the crank on a nearby winch as the countdown finished. The large wire gate that formed the front of the release pen inched upward.[3] I couldn't hear the cheers, over a mile away, down in the valley. Instead, I heard a profound stillness, punctuated by the symphonic sound of wind whistling through wings, as adult condors sped over the cliffs, their heads tilting to look down at the penned juveniles.

I waited in suspense for what felt like hours. But only six minutes after I raised the gate, intrepid male Condor 252 walked toward the cliff rim. Moments later, animated by the comings and goings of the free-flying condors, Condor 248, the only female in the release group, rocketed out of the pen and into flight. A smile nearly split my face. Condor 248 was the younger sister of Condor 176—our greatest flier. How appropriate that she should be the first of our youngsters to take flight. As she made her first wobbly landing on a rock pinnacle, Condor 240—her brother—followed suit. Not for the first time, I was struck by the similarities I saw among sibling condors.[4] A short while later, Condors 243 and 258 hopped out of the pen in quick succession. Twenty-eight minutes after I opened the gate, the last juvenile condor, regal Condor 246, took to the skies, capping off a successful release. Cars soon trickled away from the viewing area and our field crew began the work of keeping track of the new flock members and making sure naive condors perched safely for their first night in the wild.[5]

———

As the weather warmed and days lengthened, the juveniles flexed their wings, taking short flights, while the older birds started going to the South Rim. Their visits came over a month earlier than usual, and on April 2, we were blindsided by the event we had been dreading for a year: Condor 210 arrived at the South Rim. Landing on Lookout Studio—a historic building on the canyon rim, constructed to look like part of the cliff—210 let curiosity overwhelm caution, and in the process fulfilled our worst fears.

Had we been present to haze her when she landed on the outer patio of Lookout Studio, 210 wouldn't have approached the tourists that had gathered there, and much of her subsequent behavior might have been avoided.

Park volunteers were not authorized to haze condors, and were told instead to bring the tourists inside and close the patio doors. While curious tourists peered out the windows, 210 wandered around the patio, even nibbling on the door through which the people had disappeared. Finally, a gift-shop employee, frustrated by the standoff, overrode park rules and chased off our recalcitrant condor. After approaching more people, 210 eventually settled on the cliffs bordering her new play area.

Belatedly alerted to her presence, several members of our crew convened at the South Rim at daybreak the next morning, hoping to reverse the harm caused by 210's unsupervised introduction to the condor's summer mecca. Upon arriving, I found her perched on a cliff, overlooking Grand Canyon's Bright Angel Trail. From her vantage point, 210 could watch people disappear into a tunnel, which ran through a cliff segment on the trail, and reemerge seconds later on the other side, a sight that seemed to fascinate her.

From my viewpoint on the cliff rim, I looked down at 210 with inestimable sadness, wondering if she would be the first of our condors to fail in the wild. Over the last year and a half, we had hazed her off the valley floor and taught her how to choose safe roost spots. We had chased her off the Navajo Bridge railings, away from the House Rock Valley scenic overlook, and out of the North Rim housing area. We had seen her become more social, and we had cheered each incremental step she made toward becoming a successful free-flying condor. She was our nemesis bird, but, despite the heart-racing stress she routinely caused us, she had become a great favorite with the field crew and members of the public. Nevertheless, we could not afford to have her approach people and negatively influence the newly released juveniles. I resigned myself to retrapping her.

Condor 210, of course, had other plans. As she had done repeatedly over the previous year, just when we thought she had reached the nadir of bad behavior, 210 turned over a new leaf. Unable to get close enough to net her, we hazed her from several marginal perches the first day, fewer bad perches the next two days, and almost none at all after that. As each day passed, 210 learned which perches solicited a rush by loud, two-legged predators. Harassed unrelentingly if she landed anywhere near people, her inclination to do so diminished. Within days she began focusing on what captivated the rest of the condor flock: superior flying conditions, abundant toys, and, above all, a bounty of natural food. I felt like cheering when 210 returned to the release site on April 9, after days of exhibiting acceptable behavior.

But 210 had found a new favorite spot and returned to the South Rim two days later. To our delight, though, she improved with each subsequent visit, and I felt like a proud parent when she was the first to discover an elk carcass at the end of May. In the coming months, 210 bounced between the release site and the South Rim like a feathered yo-yo, but she was a model bird, focusing on the doings of condors rather than tourists.

————

Precocial Condor 248—176's younger sister—was the first of our juveniles to visit the South Rim metropolis, arriving the same day 210 did. Absorbed with 210, we paid scant attention to 248. Fortunately, she behaved better than perhaps any other first-time condor visitor. Meanwhile, Condor 240—248's brother—seemed equally intent on bolstering his family's reputation for being wide-ranging fliers. In the months following his February release, 240 often flew long after sunset and after the other condors had gone to sleep, gliding over the cliffs in the waning light. On one of his first journeys from the release site, he accompanied 176 to the Zion National Park area in southern Utah—one of her favorite destinations and an area infrequently visited by other condors at the time. Impressed with 240's long flight and adventurous spirit, we were hardly surprised when he disappeared, the way 176 had once done, on his third trip to the Zion area in mid-August.

But when we failed to find him, after days of ground searches and one overflight, we became seriously worried. One of his transmitters had stopped functioning weeks earlier, so we tried to reassure ourselves by speculating that his second one had failed, too. Finally, though, during a second aerial search on August 25, Chris Parish received a faint signal from 240 when flying over a remote ranch northwest of Zion. For a moment, I was elated upon hearing the news, before realizing that if 240 had been flying, we would have found him on our ground searches. "I am sick with worry about him," I wrote in a personal notebook that night, before succumbing to a fitful sleep.

Having received permission from the landowners to access their ranch, I drove up there at first light on August 27. Soon after arriving, I heard the drone of a plane. Seconds later, my two-way radio crackled.

"Sophie, do you copy?"

"10-4. I copy," I responded.

"Okay, we see you and we're almost above you."

Because we hadn't found 240's signal during our ground searches, Chris had relocated the signal from the air and would direct my movements along ranch roads until I was close enough to hear 240's transmitter. After driving through a locked gate that I'd opened with the landowner's key, I sped over a rough road leading to a distant water tank. Racing along below the plane that shadowed me from above, I felt for a brief, surreal moment like I was acting out a part in a James Bond movie.

"Are you getting his signal?" Chris asked as I reached the end of the track.

"Negative," I responded.

"Okay, we'll try another road. We have another option." As the plane circled overhead, I backtracked, and then took a different road leading west. Suddenly, I heard the almost imperceptible "blip, blip" of 240's transmitter.

"I've got it!" I called up to the plane, before jumping out of my truck to determine the signal's direction with my Yagi.

"10-4," Chris responded. "Good luck! We'll head back to the airport and fly back this afternoon if we don't hear from you."

Moments later, I was enveloped in quiet as the plane's engine faded into the distance. I drove on a bit farther then readied my pack for a prolonged hike into unknown territory. As I left my truck, I took a bearing with my compass (GPS units not yet being part of our field equipment) toward where I was headed, in case I needed help finding my vehicle on my return journey. Climbing steadily, I hiked up a steep, timbered ridge, heading toward 240's barely audible signal. Eventually, I reached a heavily grazed meadow. A Golden Eagle soared over me, its head swiveling to pierce me with its fierce gaze. Condor 240's signal was booming now and, as I approached a stand of dying aspens, I scanned the ghostly branches, hoping I would see his bright eyes peering down at me. But I searched the treetops in vain and soon resigned myself to scanning the ground. It wasn't long before I found him.

Condor 240's desiccated body lay beside a log, amidst scattered woody debris. His tail was pitched upward and his wings were outspread, as though he'd dropped from the sky mid-flight. Intently, I stared at those great wings—then shivered involuntarily. Not a feather was out of place. But 240 no longer wore his identifying tags and transmitters. Somebody had removed them.

Eyes wide, I looked around, searching for his transmitter, searching for someone who might be watching me. Moving my Yagi in a slow circle, I fiddled with my receiver's controls, trying in vain to narrow down the location of 240's still-booming transmitter. There was no discernible difference in the signal's volume in any direction. I had to be within feet of the transmitter, but I saw no sign of it. Detaching my Yagi from the coaxial cable, I held out the end of the coax and moved in another circle. This time, the signal was slightly stronger on my right side. I stared at the ground. Small branches littered the dirt, and, as I looked at them, my mind racing, their placement struck me as being slightly unnatural, as though they had been laid down intentionally rather than falling haphazardly.

Pulling my camera out of my pack, I photographed the sticks and the bare ground under them. Then, I moved the woody debris aside and dug into the dirt with my fingers. Within seconds, I felt 240's buried vinyl identification tags and their attached transmitters. Drawn unwittingly into a situation that suddenly felt intensely threatening, I again glanced around uneasily, searching for a hidden figure amidst the scattered trees. After reassuring myself that I was alone, I left the buried transmitters as I had found them, methodically photographed the crime scene, and recorded everything I had seen and done in my notes. I then hiked to the top of a nearby ridge, hoping to get cell coverage. In luck, I phoned Chris, so he could alert the authorities. Sitting on the ridgetop, arms wrapped around my knees, I gazed at distant gray cliffs and a sparkling reservoir, prepared for the hours-long wait for Chris and the US Fish and Wildlife Service law-enforcement agent.

As images of 240 filled my thoughts, a movement suddenly caught my eye. A rider was approaching on horseback. A rider who probably didn't know I had permission to be where I was. A rider who might have shot a critically endangered California Condor. And buried the evidence. Adrenaline surging, I scrambled behind some bushes and dropped to the ground, heart in my throat, as he came fully into view, a hat pulled low over his eyes, a shotgun lying across his lap.

Heart racing, I pressed myself into the dirt, trying to make myself invisible, as my mind flashed back to the only other time I had felt such intense fear. I'd been a teenage girl, out for a morning run, when two beer-drinking men passed me in their low-slung sedan. Braking, they'd turned around then passed me again, leering. Once again, they'd turned then pulled over, waiting for me to approach. I'd ducked into the yard of the only house for miles around and cowered behind garbage cans, as the men drove into the yard, got out of the car, and searched for me, before deciding I'd gone into the house. Then, as now, I'd felt an unreal sense that I was outside my body, watching events that were unfolding around me. Events over which I had no control.

The rider neared. His horse's footfalls kicked up small circles of dust. And then they moved past me. Fervently, I hoped the man wouldn't see me, hoped that he wouldn't have a wide-ranging dog that would discover my prostrate form. After an eternity, the horse's hoofbeats faded, and I began to breathe again. Hours later, I hid once more as a moving wave of grazing sheep approached then surrounded me, jumping spookily as they caught sight of me. But the flock and its unseen minder eventually moved on, and my pulse again returned to something approximating normal.

Darkness dropped silently around us as we somberly loaded 240's body into Chris's pack, and Special Agent Bell thanked me for my careful treatment of the crime scene as she repacked her equipment.

"So, time to find our way back to our vehicles in the pitch dark," Chris laughed, injecting a much-needed moment of levity.

"I took a compass bearing." I said, spinning the dial 180 degrees to navigate back to my truck.

"Great. Lead us out of here then."

As we stumbled down the mountain, fighting our way through sudden tangles of brush in a moonless darkness that hid any landmarks, I fervently hoped that the compass skills I had learned when searching for goshawk nests in Idaho's forests years before would serve me well now. An hour later, we stumbled onto the road 20 feet from my truck, and I smiled for the first time that day.

Four days later, as I was writing up the fatality report for Condor 240, I received a call from a crewmember.

"Sophie, I have a mortality signal on 186," Roger stated tersely.

I felt a surge of dismay before reason reasserted itself. Some of our transmitters had erroneously been kicking into mortality mode of late. (Mortality signals were tripped by twenty-four hours of inactivity and consisted of a more rapid series of beeps than ordinary signals.) Each time we had investigated, we had found a live condor. Nevertheless, I told Roger to drop everything and search for Condor 186.

A few hours later, Roger called back to tell me he was having trouble finding 186, and I gave him some tips on how to narrow down the location when receiving a strong signal in a forest. Within an hour, he called again.

"Sophie," he said breathlessly, "I found 186 and he's dead."

Minutes later, I was on my way to the Kaibab with crewmember Kris Lightner, but it wasn't until I saw 186 lying on the forest floor that the terrible news really sunk in. As we hugged each other, Kris whispered, "He was so beautiful."

I looked down at 186's pink and yellow neck, his elegant black feathering. Only four years old, 186 had been acquiring his adult colors early.

Condor 186 had had a particularly troubled past. Released in November 1998, he had been recaptured the following March, after landing in an RV park. He was rereleased a month later, then briefly disappeared into the vastness of Grand Canyon, before resurfacing at a remote airstrip where he was corralled by airport personnel. Having shown little awareness of his own safety, 186 was promptly returned to captivity. Eighteen months later, 186 was rereleased again, in December 2000. We trapped him a few months later to replace a transmitter, and even though he almost overwhelmed me with his ferocity as I held him, 186 was still insufficiently wary around people and was recaptured at the South Rim in May 2001.

We gave 186 a final chance in December 2001—rereleasing him a fourth time—and, at long last, he was a bird transformed. In the months that followed, 186 stayed away from people and focused on finding carcasses, the way a wild condor should. He had been on his way, finally, to becoming one of our best success stories.

And now in August 2002, he lay dead, near the remains of a deer carcass, deep in the Kaibab National Forest. As I photographed the scene while waiting for law enforcement, I noticed that several feathers on his left wing looked like they had been cut with scissors. Several other sheared feathers lay nearby. Years earlier, while working with Rough-legged Hawks, I had learned that such feather shearing was a reliable indicator that a bird had been shot.

Weeks later, though, we received appalling details from his necropsy. Condor 186 had been perched up in a tree when he was shot with an *arrow* that had pierced through a leg and wing. It had not killed him outright. Instead, 186 had been left to bleed to death on the forest floor.[6] Despite a reward of $22,000 for information on 186's death, the perpetrator was never found.

———

Aside from earning the field crew's devotion, our condors—with their massive size, their tag numbers marking them as unique individuals, their awe-inspiring grace in flight, and their captivating behavior—garnered a nationwide following online. Following his release in February 2002, Condor 258 attracted particular attention. Perhaps the big number "8" that he wore on each of his wings stood out to people. Our condor's tags typically consisted of the last two digits of an individual's studbook number. However, when two condors had the same last two numbers (as with Condors 158 and 258), we used only the last digit for the younger bird's tag number.[7] Perhaps people found 258's bright eyes and fuzzy gray head endearing. Regardless, we received frequent inquiries about "Condor #8." And as our youngest condor and a stellar juvenile, 258 was a favorite with the field crew.

In early May 2002, a parade of delighted tourists watched 258 at Navajo Bridge. While I answered their many questions, 258 put on a show, soaring over and under the bridge, eliciting awed exclamations. Suddenly, an intense wind gust caught him off guard. Still an inexperienced flier, 258 was driven headfirst into a steel bridge support and landed shakily on a nearby ledge. Concern rippled among the observers. Many remained for hours, fixated on a subdued 258. Long after the last tourist left, I watched the hapless youngster fly to a safe roost and settle down for the night.

A week later, I received a phone call from Texas. It was a long-haul trucker named Ben. Reminding me that we had met at Navajo Bridge, he asked about Number 8.

"I just can't get that bird out of my mind," Ben explained. "I had to know if he was okay." Happily, I was able to reassure him that 258 showed no ill-effects from his collision. After our chat, Ben began following 258's progress online and wrote us a gracious letter applauding our recovery work. In my next online Notes from the Field installment, I noted that, "With biologists and birders, kids and conservationists, tourists and tour guides, and a caring citizen driving a big rig around the country pulling for him, surely 258 has every chance at success!"[8]

And indeed he did. Condor 258 was a model youngster, behaving appropriately in people areas, interacting well with other condors, and adeptly finding food and feeding alongside older birds. So it was all the more devastating that 258's many fans and the field crew that watched over him weren't by his side to protect him from the person who pulled over on a remote forest road, pointed his gun out the window, and leveled it at our bright-eyed youngster on October 25, 2002.

Still reeling from the deaths of 240 and 186 in August, we were devastated by the senseless killing of a young condor that could be counted on to do everything right. Again, tears flowed as our field crew assembled at the crime scene to mourn the loss of another member of our small condor family. Gutted, angry, heartbroken, we nevertheless regrouped and redoubled our efforts to inform Kaibab hunters of our condors' presence and protected status. Beginning in 2003, the Arizona Game and Fish Department included information about condors in their hunting regulation handbook and with the permits they mailed out to hunters. Although 258's killer was never found, no more condors were shot in Arizona during my tenure as field manager, and many local hunters became avid condor supporters. Nevertheless, shooting remains a serious threat to condors.

At least thirteen condors were confirmed to have been shot between 1992, when condor reintroductions began, and 2022.[9] The first shooting death of a reintroduced condor in Arizona occurred, quite shockingly, in Grand Canyon National Park when a university student inexplicably shot female Condor 124 multiple times.[10] But perhaps the most devastating shooting in the condor recovery program occurred a few months after Condor 258's death.

Years earlier, when only four wild condors still floated over California's chaparral-clad hills, AC-8 had been the only remaining female. Captured in June 1986, AC-8 became a critical member of the captive breeding program, producing nine offspring during her fourteen years in captivity. After 1995, though, AC-8—who was revered as the recovery program's matriarch—stopped producing eggs. With condor reintroductions well underway, supporters called for her release to the wild, hoping she would serve as a mentor for young condors. Despite concerns for her safety, biologists finally returned AC-8 to the skies, amidst tremendous fanfare, in the Sespe Condor Sanctuary, on April 4, 2000.

AC-8 easily adapted back into life in the wild, an astonishing feat for an animal that had been confined to a pen for so long. Soon, she was revisiting old haunts and being followed to them by young condors that had never ventured to these areas before.[11] AC-8 unwittingly stitched together two eras: the eons in which California Condors ruled North America's skies before the species was driven toward extinction, and recent decades in which, until AC-8's release, only captive-reared condors existed in the wild.

But the dangers that had felled so many of AC-8's fellow condors remained, and when biologists trapped her to replace a transmitter in November 2001, a blood test revealed alarming lead levels. X-rays showed a metal fragment in her digestive tract, and, as her lead levels soared, AC-8 stopped eating and drinking. Veterinarians worked frantically to save her—chelating, force feeding, and hydrating her. Miraculously, she eventually recovered.

With more trepidation and less elation, biologists again returned AC-8 to the wild, in December 2002. As she bolted out of her transport kennel and rocketed into the sky, no one could doubt that allowing AC-8 to fly free—despite the hazards—was the right decision.

But less than two months later, on February 8, a twenty-nine-year-old out hunting pigs with his dad, raised his rifle and shot the world's most-beloved condor, leaving her hanging in a tree.[12]

An icon that transcended the recovery program, AC-8 was mourned the world over, and eulogized in every form of media. Her ancestors had flown over herds of mammoths, gathered at carcasses killed by saber-toothed cats, and perched in trees whose leaves were browsed by giant ground sloths. She was a painful reminder of how irrevocably we humans have transformed our world, and of the perils to which we have subjected the wildlife that share our habitats. She left a legacy of more than twenty condors in Arizona alone, but the ways that she touched the lives of those who read about her or followed her exploits or merely saw her perched in a tree—as I once did on a visit to California—were incalculable.

Sadly, shooting poses an ongoing threat to condors. Three more Arizona condors were shot (two in 2014 and one in 2015); and in the summer of 2018, Condor 526—daughter of the famed AC-9—was shot on private land in California. Although 526's death underscored the danger that illegally fired bullets pose to wildlife the world over, her life highlighted the more insidious threat posed to condors by lead ammunition, since she was treated for lead poisoning five times during her truncated six-year life.[13]

We had our own revelation about the ways lead ammunition threatens condors back in 2002, when the US Fish and Wildlife Service relayed its findings from the investigation into Condor 240's suspicious death. To our profound astonishment, 240 had not been shot, as we had assumed. Instead, he had been found dead by a ranch hand who knew that condors were protected, and panicked, convinced that he would be blamed and suffer severe consequences—perhaps even deportation—if anyone learned of the bird's death. Carefully removing the transmitters, he had buried the evidence. But unbeknownst to him, that evidence had remained track-able, and its burial had only made the supposed perpetrator appear guilty. Nevertheless, the necropsy conducted by the forensics team on 240's desiccated body revealed unequivocally that our young adventurer—like so many condors past and present—had died not by being shot with lead but by ingesting it.

CHAPTER TWENTY

A Sisyphean Predicament

How many times had we been here before, I thought wearily, as I readied our trap so we could test our condors' lead levels again. Hunting season was underway on the Kaibab Plateau, and more and more of our birds were feeding on gut piles (visceral remains) left by hunters that had field dressed the deer they'd killed. The Kaibab had proven to be a reliably bountiful cafeteria for our condors.

Given early concerns about whether captive-reared condors would find sufficient food in the wild, the number of carcasses that our condors found in the Grand Canyon area was as gratifying as it was surprising.[1] Road-killed elk and deer were common, as were free-range cattle that died with disturbing regularity while roaming the Kaibab National Forest. Cattle were hit by vehicles and mired in the quicksandlike mud of drying waterholes. A massive bull that we dubbed The Refrigerator died after falling on a steep slope. In August, a new crewmember had hiked deep into the Kaibab forest following multiple condors that she suspected had found a carcass. And indeed they had. But, as the condors' signals boomed from her receiver, nothing prepared Meghan for the sight of twenty-four bovine legs reaching for the sky. The bloated bodies of an enormous bull, three cows, a yearling, and a calf were arranged in a circle around a tree. Disconcerted by the creepy scene, Meghan left the area to call me and I soon joined her in the forest. A local livestock officer later confirmed my suspicions that the unfortunate animals had been killed when a lightning bolt struck the tree under which they had gathered for shelter. The rest of the condor flock soon abandoned other venues and streamed into the

Kaibab forest, where they feasted for weeks on what we dubbed the "six-cow strike."[2]

Now, though, with the Kaibab painted green and gold with aspen stands scattered amidst a panorama of dark conifers, mule deer had become the pièce de résistance for our insatiable condors. Worryingly, the birds had found three headless deer within a week. Unethical hunters had shot these animals, removed their trophy heads, then discarded their bodies, illegally wasting the meat and subjecting untold numbers of ravens, eagles, condors, and scavenging mammals to lead-filled carcasses. In addition, innumerable gut piles from legally killed elk and deer awaited our sharp-eyed birds.

Hunting provides a food bonanza for California Condors. In 2000, hunters killed more than 106,000 game animals in the eight California counties comprising the condor's historical range. Annually, these hunters left behind over 36,000 large animal carcasses and gut piles, including more than 8,000 deer gut piles, offal from more than 17,000 feral pigs, and the bodies of nearly 11,000 coyotes.[3] Hunters on the Kaibab Plateau routinely shot deer, elk, bison, and coyotes, leaving approximately 700 gut piles each hunting season.[4]

We had hoped that the prevalence of National Park Service land—where hunting was illegal—in the condor's Arizona and Utah range would reduce our condors' chances of ingesting lead from hunter-killed carcasses. But the birds ranged over hundreds of square miles and, during the fall hunting season, preferentially patrolled the Kaibab and other heavily hunted areas.[5] Because our concerns about lead intensified during this time period, we routinely trapped our condors to test their blood and replace any failed transmitters each November. Accordingly, at dawn on November 5, I opened our trap for our first day of fall trapping. Heavily bundled against the cold, I waited for condors to show up. Three hours later, not a single bird had appeared.

My luck soon changed, though, and my spirits lifted when I heard the unmistakable sound of wind singing though condor wings and large prehistoric forms began patrolling the sky above me. Minutes later, condors landed on the trap, cocking their heads as they eyed the bait. After nearly two hours of unremitting suspense, I pulled the trap door shut, capturing the first eight of our thirty condors. And now began the gratifying but arduous task of replacing transmitters, and drawing and testing blood.

After pressing the sensor slide, onto which I'd placed a droplet of Condor 187's blood, into our field lead tester, I sat back on my heels with the assembled crew to wait the 180 seconds until the tester revealed its results. Numbers flashed on the screen sequentially as the countdown began: 180, 179, 178, 177 . . . No idle chatter broke the silence. All eyes were riveted on the flashing numbers. An eternity passed . . . 127, 126, 125 . . . Somebody chewed a cuticle . . . 63, 62, 61 . . . Someone else rubbed their eyes, as though unable to bear the forthcoming verdict. No one seemed to breathe . . . 4, 3, 2, 1, 0. And then a painful pause before a low 14.7 µg/dL flashed on the screen.[6] Smiles broke out on every face and several people huffed out breaths of relief. Two crewmembers carried the kennel in which Condor 187 awaited his verdict back to the trapping area and opened the door. Running to the cliff edge, 187 opened his wings and stepped into the wind.

Our work was just beginning, though. After netting dominant Condor 158 and restraining his hefty bulk while we replaced a transmitter and drew blood, we assembled in front of the field tester, as we would do repeatedly in the days ahead. Again, the countdown plodded through 180 seconds then paused for an agonizing moment before rendering its verdict. This time, "High" flashed on the screen, indicating a value over 65 µg/dL. Our hearts sank. Condor 158, who had barely survived after ingesting six shotgun pellets in 2000, would require an x-ray. And then, if he showed no discernible lead fragments in his digestive system, at least one round of chelation.

After multiple weeks, we finally managed to trap all of our birds. Fourteen more condors joined 158 in the flight pen for monitoring or treatment. A disproportionate number of the birds that had spent time on the Kaibab in the preceding weeks had high lead levels.[7] And with each passing day, we became increasingly loath to release birds that hadn't been poisoned, in case they ingested lead fragments in what remained of the fall hunting season.[8]

Given that California Condors had died with visible lead pellets and fragments in their digestive tracts, that we most often documented elevated lead levels during or after the fall hunting season, and that we had recorded high lead levels in our condors after seeing them feed on hunter-killed carcasses, it seemed increasingly clear to those of us who worked with condors that ingesting lead ammunition posed an existential threat to the species and undermined its recovery. But in 2002, the threat that ingested lead *bullet* fragments posed to condors and other scavengers was not universally acknowledged—and was vociferously disputed by much of the hunting community, even in the face of mounting scientific evidence.

That lead killed birds was well documented. Although it had been known since the late 1800s that waterfowl ingested spent lead shotgun pellets (also known as lead shot) that settled in staggering quantities into wetlands during the hunting season, it wasn't until 1959 that a watershed study by waterfowl researcher Frank Bellrose detailed the extent and seriousness of the threat that ingesting lead shot posed to these birds.[9] In the mid-1970s, the US Fish and Wildlife Service proposed the use of nontoxic steel shot for waterfowl hunting, estimating that approximately two million waterfowl died of lead poisoning each year.[10] Bald Eagles and other scavengers also died after consuming lead-laced waterbirds.[11] Despite intense controversy, the Service banned the use of lead shot for waterfowl hunting nationwide in 1991. However, lead shot continued to be used for upland birds, and lead bullets were universally used for big game and other animals—with little apparent concern about potential consequences for those that fed on the remains of these hunter-killed animals.

———

As the California Condor population declined precipitously in the 1980s, an intensive research effort sought to determine what was killing the birds. While some speculated that food scarcities and other factors might be affecting the condors' ability to successfully raise young, condors appeared to be reproducing as expected. However, starting in 1982, the condor population declined at over 26 percent per year, over a four-year period, dropping from a probable twenty-three wild birds to five.[12] The known causes of mortality—poisoning, shooting, habitat loss, DDT contamination, and a host of minor factors—couldn't explain the population's freefall.

Outfitting condors with radio transmitters so the birds could be tracked finally allowed researchers to examine why condors were dying. Ironically, biologists were initially stymied because few of the transmitter-carrying condors died. Indeed, in the winter of 1984 to 1985, five of the seven condors without transmitters died, whereas only one of the eight with transmitters died. (This at least helped allay fears that carrying transmitters might increase condors' chances of dying.) Of the fifteen condors that died between 1982 and 1986, the bodies of only four were recovered.[13] But three of these birds transformed our understanding of why this magnificent vulture—which some had dubbed a Pleistocene relict, "with one foot and even one wing in the grave"—was perishing at an unsustainable rate.[14]

Necropsies of the recovered condors revealed that the birds had died of lead poisoning (in separate events).[15] Two of the three had visible lead particles in their digestive tracts, confirming that condors could die from ingesting lead bullet fragments. Here at last was the smoking gun that explained why numbers of this opportunistic scavenger had steadily decreased since Europeans first arrived in North America. Prior to this discovery, lead poisoning had not been recognized as a significant threat, even though biologists were aware that condors often fed on hunter-killed animals.[16]

Evidence that condors were dying after ingesting lead ammunition continued to accumulate during the 1980s onward.[17] And a growing understanding of the prevalence of lead in the condor's range and beyond further underscored the magnitude of the threat. In the mid-1980s, more than a third (36 percent) of 162 Golden Eagles captured in the condor's historical range had elevated lead levels, with the highest levels seen during the fall and winter.[18]

In the early 2000s, researchers from The Peregrine Fund gained important insights into the prevalence of lead in hunter-killed animals and the way avian scavengers might be ingesting it. Over a two-year period, researchers collected and x-rayed the remains of deer killed by cooperating, licensed hunters during fall hunts in Wyoming and California. The x-rays revealed in graphic detail the extent to which lead bullets will fragment when fired into an animal. (Subsequent ballistics studies bolstered and expanded on these findings.)[19] Every one of the whole or eviscerated deer killed with lead-based bullets contained visible lead fragments, which glowed like misshapen stars in the x-rays. Three-quarters of the deer had 100 or more visible fragments each. Ninety percent of the offal remains (or gut piles) also contained a constellation of bullet fragments.

The lead fragments ranged in size from a fraction of an inch (0.2 inch, 5 mm) to fine, dustlike particles that could not be seen without a magnifying lens. The fragments scattered as far as 6 inches (15 cm) from where the bullet passed through the animal.[20] Suddenly, we had a visual illustration of how a group of condors clustered at a gut pile or deer carcass might ingest tiny, easily absorbed, particles of lead. And every hunter on our field crew gained an unwelcome insight into what *they* might be ingesting when they feasted on the deer they had killed during the hunting season.

That condors were indeed ingesting such bullet fragments was confirmed by later scientific studies that identified lead ammunition as the *source* of the lead found in the blood of condors and other birds. Lead exists in different forms, known as isotopes, which have similar chemical properties but slightly different structures (differing in their number of neutrons). The relative abundance (or ratio) of different lead isotopes in industrial products such as lead paint, gasoline, and ammunition differ because the lead used in making these products comes from different sources. These lead-isotope ratios are as identifiable as fingerprints.

In the late 1990s, researchers in Canada used lead-isotope ratios to identify the products to which lead-poisoned birds had been exposed.[21] Herring Gulls that the researchers examined had lead-isotope ratios that were consistent with leaded gasoline.[22] The waterfowl, loons, and eagles that they examined had lead-isotope ratios that matched those of lead shot that the researchers had recovered from nearby lake sediments and purchased locally for comparison. These birds had ingested lead shot (or lead fishing sinkers in the case of loons).[23]

In 2006, researchers in California used similar techniques to identify the source of the lead found in condors. They measured and compared lead levels and lead-isotope ratios of reintroduced condors living in the wild in California to those of captive condors that had not yet been released. They also measured lead-isotope ratios in condor dietary items and lead ammunition samples. Lead levels in captive condors were low, and the lead was isotopically similar to background environmental lead in California. Most free-flying condors, on the other hand, had higher blood lead levels, and the lead was similar in its isotopic composition to lead ammunition, suggesting that the birds had ingested lead bullet fragments while in the wild.[24]

Subsequent research expanded on these findings. Lead that is present in a condor's blood is deposited in the bird's feathers during feather growth. By examining sequential subsections of condor flight feathers, researchers realized that they could determine the frequency, severity, and source of a bird's lead exposure during the feather's two-to-four-month growth period.[25]

Further analyses provided additional confirmation that lead ammunition is the source of lead poisoning in condors. After biologists observed a condor feeding on a feral pig in central California, they recovered the pig carcass and x-rayed it. X-rays revealed a snowstorm of lead fragments in the pig's head and two intact bullets in its chest and gut. The intact bullets were retrieved and their isotopic composition analyzed. Upon being captured, the condor had high blood lead levels. Later analysis showed that the lead-isotope ratios matched those of the ammunition retrieved from the pig.[26]

Before the condor was returned to the wild, researchers clipped sequential feather vane segments from one of its growing flight feathers and analyzed their lead content. Lead levels in the feather portion that grew before the condor fed on the pig carcass were low and matched the isotopic signature of feathers from captive condors. Subsequent samples later in the feather's growth showed high lead levels and an isotopic signature that matched that of the ammunition recovered from the pig. Finally, lead levels in the most recently grown feather vane segments declined as chelation therapy removed lead from the bird's body. Not only could researchers now determine the lead source to which a condor had been exposed, but, when they fit a timeline to actively growing feathers, they could also establish a date range for when a condor had ingested lead.[27] In essence, researchers now had the ability to capture a bird's lead-exposure history over the time it took for a flight feather to grow.

The single most significant threat to reintroduced condors, lead poisoning has occurred at unsustainably high rates.[28] Between 1992 and 2009, roughly two-thirds of the adult condors and a quarter of the juveniles that died did so of lead toxicosis (poisoning).[29] That number would have been infinitely higher were it not for intensive oversight by biologists and lifesaving chelation therapy. A young condor in California that survived in the wild for only six months was exposed to lead four times prior to her death (from lead toxicosis).[30] Between 1996 and 2005, twenty-eight condors in Arizona received a total of sixty-six rounds of chelation (each round consisting of an injection in their breast muscle twice a day for five days). Several condors were treated as many as six times (meaning that each of those condors was netted, restrained, and received injections sixty times).[31] One condor in California was known to have had lead poisoning thirteen times as of 2009.[32]

After 2002, when Arizona condors began to patrol the Kaibab regularly during the fall hunting season, lead levels and the *proportion* of condors that ingested lead ammunition increased significantly. In 2006 alone, 95 percent of Arizona's condors (more than fifty individuals) were exposed to lead, and 70 percent received chelation.[33] Although the treatment has saved countless condor lives, the long-term effects of chelation and repeated acute lead exposure are unknown. What is clear is that without ongoing management—monitoring condors with telemetry to detect potential lead exposures, trapping birds, testing their blood levels, treating lead-poisoned condors with chelation therapy, and removing ingested lead fragments through purging or surgery—condor populations would decline and again face extinction.[34] For now, intensive oversight and treatment are keeping condors alive, but following detailed analyses, scientists concluded in 2012 that "condors are chronically lead-poisoned" and "the lead poisoning rates in condors are of epidemic proportions."[35]

California Condors may be especially prone to lead-poisoning given their proclivity for consuming large animal carcasses, which often have been shot, but condors are not alone in suffering from lead toxicosis. More than 130 animal species, including mammals, a wide variety of birds, and even amphibians and reptiles, have suffered adverse effects from consuming lead shot, lead bullet fragments, or animal prey that contains lead ammunition.[36] As many as fifteen million Mourning Doves may die each year, after ingesting lead shot that is sprayed over fields by dove hunters, reaching densities of more than 400,000 pellets per acre in some areas.[37]

The vulnerability of Bald Eagles to lead poisoning—a concern for over half a century—helped precipitate the federal ban on lead shot for hunting waterfowl, because eagles regularly fed on waterbirds that had ingested or been injured with lead shot.[38] Like condors, Bald and Golden Eagles may also ingest lead shot or bullet fragments when feeding on hunter-killed animals or gut piles. More than half of the Golden Eagles captured in a Montana study during fall migrations in 2006 and 2007 had elevated blood lead levels, and three-quarters of Bald and Golden Eagles in a Wyoming study did so. A more recent long-term study evaluating the lead exposure of eagles of both species from 38 US states found that nearly half of the birds had chronic lead poisoning (determined by analyzing bone). Up to a third had suffered recent acute lead poisoning (based on analyses of liver, blood, and feathers). Lead levels in eagles were significantly higher during the hunting season (fall and winter) than in the nonhunting season.[39] The same was true for lead levels in Common Ravens, which also feed on the remains of hunter-killed animals. About half the ravens trapped in and around Grand Teton National Park had elevated lead levels during the hunting season, whereas only 3 percent of them did so during the nonhunting season. Moreover, raven blood lead levels were about five times higher during the hunting season than during other seasons.[40]

The impacts of lead can be insidious, felling an eagle, raven, or condor days or months after it has fed on a deer carcass. Lead impacts can also be dramatic. Between 1999 and 2008, more than 1,500 Trumpeter and Tundra Swans died from ingesting lead shot that had accumulated in and around Judson Lake, which straddles the US-Canada border in the Pacific Northwest. To dissuade more swans from settling in this perilous place, biologists were driven to haze the elegant birds away from it using noise makers, laser lights, and airboats.[41]

Regrettably, such dramatic measures are not being employed to prevent human consumers from ingesting lead ammunition fragments from hunter-killed animals, despite a growing recognition that people, too, are consuming such lead. After Arizona condor project director Chris Parish, a lifelong hunter, told a gathering of Peregrine Fund board members that he had found metal fragments in processed venison, a North Dakota physician initiated a study to investigate this concern. In 2007, William Cornatzer and several colleagues examined randomly selected ground-venison packages from the thousands that the Hunters for the Hungry program had donated to North Dakota food banks, and found that 59 of the 100 packages they x-rayed contained one or more visible metal fragments. Generous hunters sharing their bounty with the needy were inadvertently subjecting them to unacceptably high levels of lead.[42]

Following Cornatzer's announcement about his findings, North Dakota's Health and Human Services, Agriculture, and Game and Fish departments advised food banks to discard any remaining venison donations. Officials in Minnesota made similar recommendations after laboratory tests revealed lead in *their* hunter-killed, game-meat donations.[43] Wisconsin soon followed suit after conducting its own studies.[44] Roughly 3 percent of all US households seek emergency assistance from food pantries, churches, and food banks annually. A 2008 study of hunter donation programs, which operate in all fifty states and four Canadian provinces, found that about half the total number of organizations that use such donated meals provide approximately nine million meals annually to needy people, including children.[45]

The Consumer Product Safety Commission regularly publicizes and issues recalls for children's toys containing lead, as it did in 2004 for 150 million pieces of metal toy jewelry sold widely in vending machines.[46] Yet lead ammunition—likely the greatest source of lead knowingly discharged into the nation's environment—is largely unregulated.[47] The roughly 13.7 million US hunters (sixteen years and older) and the 1.8 million US children (six to fifteen years old) who hunt, along with their families, may be consuming lead-laced game meat with little awareness and few precautions about doing so.[48]

Studies have shown that people who eat wild game shot with lead bullets have higher lead levels than those who do not.[49] Yet many hunters who receive warnings about the danger of consuming game killed with lead ammunition are highly skeptical. They believe that no science exists to support such claims and view such warnings as antihunting initiatives—spearheaded by environmentalists—that threaten their Second Amendment right to bear arms. Upon learning about the North Dakota venison study, for example, the executive director of Farmers and Hunters Feeding the Hungry—an organization that coordinates the butchering and distribution of hunter-harvested deer in twenty-eight states—said, "We haven't seen a health problem in the 500 years since humans have been killing deer with firearms." A bullet manufacturer stated, "This boils down to an anti-hunting initiative. It's as simple as that. . . . [It's] an issue to divide hunters and thin our ranks."[50]

Lead is a neurotoxin. The effects of acute lead exposure—from large doses of lead over a short period of days to months—can be dramatic, causing seizures, lethargy, blindness, paralysis of the lungs and intestinal tract, anemia, mental impairment, and even death. More often, though, the effects of lead are subtle and nonspecific, particularly when doses are small and occur over months or years. All of the body's systems may be affected, but the most profound effects are seen in the nervous, digestive, and circulatory systems. Nerves require calcium to transmit the signals that allow our bodies to function. Many of the health problems that arise from ingesting lead occur because the body mistakes lead for calcium, and incorporates the metal into the nervous system and vital tissues. After lead is ingested, it moves from the blood into soft tissues, and is eventually sequestered in bone, where it remains a potential long-term lead source (particularly during bone loss due to aging or osteoporosis).[51]

No level of lead is considered safe for human beings, and toxic lead effects are largely irreversible. Chronic effects include high blood pressure (hypertension and cardiovascular disease), decreased kidney function, reproductive issues (including impotence, decreased fertility, miscarriage, premature birth and stillbirth), developmental problems (learning disabilities, lowered IQ, stunted growth, decreased brain volume), and increased aggression. Lead can interfere with smooth muscle contractions (peristalsis), which are essential to digesting food, causing stomach and abdominal pain.[52] Or, as sometimes occurs in lead-poisoned condors, lead can paralyze the digestive tract, so that food becomes stuck and rots in the bird's crop while the bird starves to death.[53]

Lead disrupts the activities of certain enzymes and the formation of new red blood cells in bone marrow, leading to anemia. Lead can cause muscle and joint pain, as well as peripheral neuropathy or weakness, numbness, and pain from nerve damage in the hands and feet. The human "wrist drop" this causes is heartbreakingly manifested in the "droop wing" of birds. The dragging wings, inability to stand, and drooping heads of lead-poisoned eagles are all-too-familiar sights at wildlife rehabilitation centers across the country.

Lead exposure may be related to cognitive decline (and dementia) in older adults.[54] And lead also affects longevity. In one study, adults with higher blood lead levels (20–29 µg/dL) were 46 percent more likely to die from all causes and 68 percent more likely to die of cancer compared to those with levels less than 10 µg/dL.[55] In another study, lead levels as low as 2 µg/dL led to an increased risk of death from strokes and heart attacks.[56]

Developing fetuses and young children are particularly vulnerable to lead's toxic effects. Even blood lead levels as low as 2 µg/dL can cause significant, permanent effects in children, impairing both their physical growth and their intellectual development. Decades of studies have shown that even low maternal and infant lead exposure leads to lower IQ levels in children.[57] And although a few IQ points may seem of little consequence, one study estimated that the slight (2 to 5 point) increase in US children's IQ scores following the decline in blood lead levels from the phaseout of leaded gasoline, would lead to IQ-related increases in future income of between $110 to $319 billion for the 1998 birth cohort of 3.8 million children.[58] Follow-up studies of lead-exposed children have shown that they have poor reading abilities; poor test scores; a higher frequency of delinquent, antisocial, and aggressive behavior; and lower high-school graduation rates.[59]

Lead-exposed children suffer irreversible changes to their brains, and numerous studies have found an association between early lead exposure and later criminal behavior.[60] Researchers examined criminal records for a cohort of children, whose blood lead levels were measured repeatedly from before birth (in utero) through the age of 6.5 years, and found that each 5 µg/dL increase in lead levels led to an increased number of total arrests and arrests for violent crimes.[61] Other studies have shown an association between atmospheric lead levels—resulting from leaded gasoline—and violent crime rates.[62] Homicide rates in US counties with the highest air lead levels were four times higher than those counties with the lowest air lead levels. Areas with higher sales of leaded gasoline also had higher rates of violent crimes.[63] The reduction in childhood lead exposure that followed the phaseout of leaded gasoline, in the late 1970s and early 1980s, is thought to have contributed significantly to widespread declines in violent crime in the 1990s. Indeed, ambient lead from gasoline may explain as much as 90 percent of the rise and fall of violent crime in the second half of the twentieth century.[64]

Although atmospheric lead levels declined when leaded gasoline was banned, lead remains a serious threat for those who consume game killed with lead bullets. Whether potential links exist between the ingestion of lead ammunition and violent crime today—including mass shootings—is not known. Nor is it known whether ingesting lead ammunition contributes to increased aggression that might be linked to today's political violence and rising extremism.

But when hunters and gun advocates claim that no one has ever had a problem from ingesting game killed with lead, they almost certainly are unaware that their possible digestive pains, impaired sexual performance, headaches, high blood pressure, kidney issues, or their child's aggression, ADHD (attention deficit hyperactivity disorder), or trouble in school may have something to do with ingesting lead.[65] Nevertheless, scientists and medical doctors in North America and Europe have expressed their consensus that the inadvertent ingestion of lead ammunition poses a significant threat to people and wildlife alike.[66] And despite the continued protestations of hunters and hunter-advocacy groups that there is no scientific basis for such claims, between 1975 and 2016, nearly 600 peer-reviewed scientific papers addressed the environmental, wildlife, and human-health impacts of lead ammunition.[67]

Fortunately, there is a simple way to avoid the problems caused by lead ammunition. As I told countless hunters on the Kaibab Plateau when I spoke to them about the existential threat that lead ammunition poses to condors, this is not a *hunting* issue; hunting is beneficial to condors, given how much food it provides them. Rather, it is an *ammunition* issue. And a wide range of lead-free bullets—typically composed of copper or brass (an alloy of copper and zinc)—is now available. While most lead-based bullets lose 30 percent or more of their original mass by fragmenting when penetrating an animal, lead-free bullets retain most of their mass and produce virtually no fragments.[68] They are as powerful, lethal (not more likely to wound an animal), and accurate as lead bullets. And as of 2013, nonlead bullets of varying calibers and grains were available for hunting every type of game.[69]

Nevertheless, although the costs of comparable lead and nonlead ammunition is similar, some brands of lead-core ammunition are significantly cheaper than high-quality, lead-free ammunition, fueling continued opposition to nonlead bullets from cost-conscious hunters.[70] Increased market demand likely will lead to cost reductions, as well as a greater availability of desired bullets and more resources to help hunters navigate the transition to nonlead.[71] However, given widespread hunter resistance and the vociferous opposition of firearm and shooting-sports advocacy groups, such changes may be contingent on the adoption of new regulations requiring the use of nontoxic bullets.[72]

———

Back in 2002, with lead ammunition's danger to people still unconfirmed by extensive scientific research, the prospect of any restrictions being imposed on the use of lead bullets was inconceivable. But after holding more than half of our condors in our flight pen to treat the birds for lead poisoning and keep them safe during the hunting season, it was finally time to return them to Arizona's stormy skies.

On December 6, 2003, half of our field crew headed up top to release the condors while the rest of us gathered at the Ramada to celebrate our birds' return to the sky. Seconds after receiving notice from up top that a condor had been freed, we watched it rocket into view, then lurch skyward, as though whisked aloft by a puppeteer, when it hit the updraft along the Vermilion Cliffs. Eager to join their newly freed flockmates, the condors that we hadn't been holding launched into the sky to greet the released birds. Swooping, diving, black-and-white wings slicing through the air, the birds coursed over the cliffs, their seeming exuberance matched by my soaring happiness as I watched almost as many California Condors flying together as once existed.

Apparently, the birds had been more repressed by their enforced incarceration than we had realized. As soon as they began touching down on cliff ledges, male Condor 158 spread his wings, lowered his head, and began to display to our inveterate traveler, Condor 176. My emotions swirled as chaotically as the condors had flown over the cliffs moments before. My romantic soul delighted in the courtship display and hoped that the coming year would bring us our first condor chick. But worry about the condors' future clouded my horizon as dusk crept over the carmine cliffs. Would all the lead the condors had ingested over the years foil their chances of reproducing successfully? What impact would so many rounds of chelation have on them? Would chronic lead exposure threaten their very lives in the months to come? As the condors settled down to roost and the year drew to its inevitable close, I looked to the future with equal measures of hope and trepidation for the great birds that electrified all who bore witness to their freewheeling, celestial majesty.

CHAPTER TWENTY-ONE

No Tags

An extra minute of daylight, the warmth of sunshine on a winter's day, two giant wings encircling a chosen female in an age-old dance. The inception of the condor's breeding season never failed to ignite an ember of optimism that soon flamed into speculation and excited prognostications about the future, no matter how challenging recent events had been. Those who work with endangered wildlife are invariably prisoners of hope that keep fighting for their charges in the face of inevitable obstacles, disappointments, and setbacks.[1]

Predictably, the path to captive-reared condors raising young in the wild was a rollercoaster. The thrill of documenting Condor 119's first egg in 2001 in Arizona came with the disappointment that it was broken. Events in California followed a similar trajectory. Since California's oldest reintroduced condors were a year older than those in Arizona, our field crew avidly followed news of the California birds, hoping that successes for them would forecast progress for own birds, as well as notching milestones for the species.[2]

A California pair had disappeared repeatedly into a remote canyon in 2000, but they abandoned the area before biologists could confirm whether the female had laid an egg. In 2001, California struggled, as we did in Arizona, with a trio trying to breed, rather than a pair—a drawback of small populations with limited mate choices. In California, however, the two females each laid an egg in a cave in Los Padres National Forest, and the three "parents" took turns incubating. When biologists rappelled to the cave to determine if the eggs were viable, they found one fertile-but-dead egg and a second egg that was developing poorly. Hoping to save the lone ailing embryo in captivity, the team collected both eggs and left behind a dummy egg to encourage the adults to keep incubating.[3]

Having learned that nesting condors tolerated some disturbance in captivity, biologists felt that strategic interference could facilitate breeding in the wild. Giving reintroduced condors opportunities to experience incubation and chick-hatching was critical. Breeding in the wild was not only elemental to condor recovery but was also a critical component of the US Fish and Wildlife Service's management goals. For the condor to be considered "recovered," or diverted from the path to extinction, the Service called for the establishment of two geographically distinct, wild populations of at least 150 condors each, and one captive population. Each wild population had to contain at least fifteen breeding pairs, be reproductively self-sustaining (meaning reintroductions were not needed to sustain population numbers), and have a positive population growth rate.[4]

When the California trio's "egg" was due to hatch (had it been real), biologists rappelled down to the nest and replaced the dummy egg with a ready-to-hatch egg laid by captive condors. In late June, the first condor chick to hatch in the wild since 1984 struggled out of its tough-shelled egg. Although the hatching had followed biologists' intervention, photos of Condor 111 gently touching the wobbly nestling with her bill were captivating. And those tender moments made the chick's eventual fate even more heartbreaking.

One of the most perilous periods for first-time condor nesters is when a parent that was absent during hatching returns to the nest and finds a chick instead of an egg. Condors typically incubate for two to four days before being relieved by their mate (although some birds may stay away longer). Because it can take three days for a hatching condor to break out of its egg, both parents usually experience the transformation from egg to nestling. But the three-parent situation in California, in 2001, complicated an already fraught stage, since the two birds that accompanied each other while the third incubated were less anxious to return to the nest.

Condor 111 had been in the cave for more than a week when male Condor 100 and female Condor 108 finally returned. Condor 111 flew out to join the errant pair, and 108 went to the cave. But rather than finding an egg, she found a tiny intruder instead. After attacking the little invader, she flung it out of the cave.[5]

Having witnessed similar behavior by inexperienced breeders in captivity, biologists consoled themselves by focusing on the progress that had been made: three captive-reared condors had now experienced incubation, and one had experienced hatching. Meanwhile, the recovery team had learned that breaking up trios by temporarily recapturing one bird might allow a pair to cement its bond and increase its chance of nesting successfully.

Hopes were high in 2002 that captive-reared California Condors would finally raise their first nestlings in the wild. California confirmed its first egg of the year in mid-February. Two more pairs soon followed suit, laying their own creamy-white, approximately 4.25-inch-long (108 mm) eggs.

Events in Arizona were equally exciting. Courtship displays, which had begun the previous October, reached a crescendo by January. Watching a chaos of extended wings, lowered heads, and rocking bodies, crewmember Kevin Fairhurst recorded twenty-three courtship displays, involving eight different condors, during one day at the Ramada. A not-entirely-discriminating Condor 123 displayed to six different females in one day but seemed particularly focused on Condors 127 (part of 2001's trio) and 133, a member of Arizona's first condor release.

Our egg-laying Condor 119 and Condor 122 had been inseparable since the previous fall and began investigating potential nest caves in early January. Near the end of February, when 122 returned to the release area alone for the first time in nearly a month, we had our first inkling that 119 might be incubating. To our delight, rather than selecting an obscure nest site in a remote location, the pair chose an enormous cave in a prominent South Rim butte called the Battleship.

Given 123's Cassanova-like tendencies, we worried that we would have only one breeding condor pair. But when we trapped condors in mid-February, 133 had elevated lead levels and had to remain in our flight pen. At long last, Condor 127 had 123's exclusive attention! Soon the pair began searching for nest sites and settled on a small cave in Dana Butte, across the drainage from 119 and 122.

On March 3, when both females left their South Rim nests and returned to the release site while their respective mates remained in the caves, we finally celebrated with abandon. Our seven-year-old condors were incubating their first eggs. As a bonus, both nest caves were visible (through a spotting scope) to tourists and volunteer nest watchers, who could witness conservation history in the making without disturbing its star participants.

Nevertheless, since we couldn't see into the depths of either cave, we anxiously watched our pairs' comings and goings, searching for subtle changes in their behavior as expected hatch dates neared. While we waited, California announced the hatching of its first chick, followed soon after with two more hatched eggs.

Meanwhile, in Arizona, we waited and waited. The roughly fifty-seven-day incubation period passed without clear signs that the eggs had hatched. Our pairs had been averaging two to four days between incubation switches—one mate taking over incubation duties from its mate. Once a chick emerges from its shell, both parents are quicker to return to feed and brood their nestling than they are to incubate their egg. We therefore expected to see our condors switch nest duties at least daily and possibly more frequently once their egg hatched, but no such change in attentiveness occurred.

Our hopes soared on May 3—two weeks after our earliest expected hatch date—when 119 returned to the nest twice in the same day. Surely the egg had hatched. The next morning, though, the pair stayed away from the nest for more than two hours. And that evening, our lingering hopes were crushed when both birds roosted outside their cave.

Across the drainage, Condors 123 and 127 continued their diligent every-two-to-four-day nest switches. And another month dragged by. Finally, near the end of May, the pair abandoned their three-month dedication to an egg that clearly wouldn't hatch. Disappointed as we were that our condors' nesting attempts had ended in failure, we tried to focus on our birds' progress. And we looked to California to achieve the goal the recovery program so fervently awaited: the fledging of the first wild chick by captive-reared parents.

With three growing condor chicks, success in California seemed almost certain. But in mid-September, a month before the oldest chick was due to leave the nest, its father, Condor 100, disappeared. He was never seen again. Biologists hoped the female would compensate for her missing mate. And for a time, she did. But on October 2, observers couldn't see the chick in the cave. Two days later, when biologist Greg Austin hiked to the nest area for a closer look, he spotted a pile of feathers below the cave. When he saw the flies, he knew it was too late.[6]

Because the chick was partially decomposed, necropsy results were inconclusive. Nevertheless, biologists suspected lead poisoning since both parents had fed on nonprovisioned carcasses and the female had elevated lead levels when the team trapped her following the chick's death. If the parent condors had indeed fed on a carcass containing lead, they likely regurgitated tainted meat to their hapless offspring.

Two weeks later, when biologist Allan Mee couldn't spot the second chick in its cave, he scanned nearby ledges, wondering if colleagues had missed the chick fledging. Suddenly, though, to Allan's dismay, Condor 98 moved out of the cave's shadows, dragging the body of his nearly full-grown nestling.[7]

Upon learning the results of the nestling's necropsy, our collective sadness over the chick's unexpected demise turned to shock. X-rays of the chick's digestive tract revealed shards of glass and plastic, metal washers and screws, electrical parts, and no fewer than *twelve* bottle caps. Suddenly, what had seemed harmless when condors played with litter now revealed itself to have potentially deadly consequences if adults regurgitated ingested trash into the awaiting bills of their hungry nestlings.

California Condors are not alone in their proclivity for ingesting trash and other items (like wool, wood, and fibrous materials). Although condors are closely related taxonomically to other New World vultures, the group is not closely related to the Old World vultures seen in documentaries feeding on lion-downed prey.[8] Nevertheless, as a result of what is known as convergent evolution, vultures worldwide share traits that are indicative not of shared ancestry but of their similar lives as soaring scavengers. For example, California Condors have large wings and featherless heads and necks like Old World Griffon Vultures (of the *Gyps* genera). Also like the *Gyps* vultures, condors feed primarily on soft tissues—such as muscles and internal organs (viscera)—of large mammals, and share a tendency to ingest trash and regurgitate it to their nestlings.[9] Why they do so is not clear.

Since these vultures feed on animals that have few small bones, the birds may seek out bone fragments to meet their calcium needs. With fewer bone-crunching predators on the landscape thanks to habitat loss and human persecution, bone fragments are likely scarcer than they once were. But human trash resembling bits of bone is ubiquitous, possibly leading the vultures to ingest white plastic, ceramics, and other garbage. The dearth of calcium in their muscle-and-guts diet may be a particular concern for eggshell-producing females and growing chicks.[10] In the 1970s, researchers successfully addressed nestling wing deformities and the presence of trash in South African Cape Vulture nests when they provided bone fragments at feeding stations.[11] However, other researchers have found that these vultures seek out bones and trash when food is scarce because natural habitats have been converted to farmland and other land uses.[12]

Condors and other vultures may also seek out trash to help them produce and eject pellets. Raptors that eat feathers, fur, and small bones regularly cast pellets to rid their digestive tracts of these indigestible items, as do Turkey and Black Vultures. Since condors and *Gyps* vultures consume more digestible materials, they cast pellets more sporadically. When these larger, long-necked vultures have small, undigested items lodged in their stomachs, they may consume additional objects—including roughage that binds the items together—to provide the necessary bulk for esophagus muscles to contract and cast a pellet. Because trash is so prevalent in our landscapes, vultures may inadvertently pick up trash, rather than natural fibrous materials, when trying to cast pellets. No adult condors have been found with abundant trash in their digestive tracts when necropsied, but nestlings may be less adept at regurgitating pellets and indigestible materials. In addition, cave-bound nestlings may have no access to fibrous materials that bind to ingested items and help with pellet formation.[13]

Finally, condors may ingest trash because of the same neophilia—love of novelty—shown by other opportunistic scavengers, such as ravens, which tend to pick up new objects, particularly when they are young and are learning which items may serve as food.[14] Condors are particularly inquisitive, and they explore the world with their bills. Their neophilia may be adaptive because of the uncertainty the birds face in finding widely scattered food sources.[15] However, a behavior that once had survival benefits could now be life threatening—as it was for California's 2002 chick.

Unaware at the time how serious an impediment trash ingestion would be to successful condor reproduction, we now pinned our sights on California's third chick. In mid-October, it appeared active, healthy, and well fed. But on October 21, a stunned Allan Mee watched female Condor 112 drag her lifeless chick onto a ledge outside the nest cave. When biologists made the sad trek to recover the body, male Condor 107 defended it by flaring his wings, opening his bill, and hissing at them.[16]

The unthinkable had happened. Despite having three condor nestlings, California's breeding season had come to a devastating end, and the successful fledging of a condor chick seemed as elusive a goal as ever.

––––––––

The onset of breeding activities in 2003 pressed a reset button, and the disappointments from the previous year morphed, once again, into hopefulness. Our expectations were soon rewarded by four condor eggs, one in California and three in Arizona. Biologists in California worked diligently to avoid the loss of another chick by increasing the availability of supplemental food to parent condors, scattering bone fragments around feeding areas, and climbing into the nest to remove any accumulated trash. Even at that early stage, when an egg glowed like a beacon in the cave's dark recesses, biologists recovered a bottle cap packed with glass shards and calf hair, and the wadding from a shotgun shell mixed with calf hair. The presence of hair from supplemental carcasses provided definitive evidence that the adult condors had ingested these materials and then regurgitated them, thereby discounting the possibility that ravens were depositing such trash into condor caves.[17]

In Arizona, meanwhile, new breeders selected a cave in the Vermilion Cliffs, south of the release site. To our consternation, though this breeding effort was carried out not by a pair of condors but by a *foursome*. Our quad, as we dubbed them, consisted of two males and two females that were constantly together and set their sights on a single cave. Knowing how unsuccessful a trio had been, even the most optimistic among us suspected this arrangement would fail. And indeed, in late February, before we could trap two of the birds to try to force a pairing, one of the females laid an egg that was destroyed within days, either by birds moving into and out of the narrow cave or because of squabbles over incubating the lone prize (something that had occurred historically with some pairs).[18]

Fortunately, our two established pairs each produced an egg at Grand Canyon in early March. Condors 119 and 122 again nested in the Battleship cave. Condors 123 and 127 moved westward, selecting a cave—not visible from the cliff rim—in the neighboring Salt Creek drainage. Viewing the nest required a grueling 24-mile (39 km) round-trip hike into a portion of the Grand Canyon dubbed The Inferno for its semicircular, flaming-red cliffs, which evoked Dante's concentric circles of hell. The name just as aptly described conditions in the drainage—even in March. Summer temperatures routinely topped 110°F (43°C) and could climb to as high as 130°F (54°C). And, with little available water, hiking to the area could be perilous. Indeed, *Backpacker* magazine dubbed the hike into and out of the canyon, which included climbing 3,800 vertical feet (1,160 m) in just under 5 miles (8 km), one of the ten most dangerous hikes in America.[19] As a result, once we had confirmed the nest cave location, we listened to the birds' transmitter signals from an overlook above the Salt Creek drainage to monitor the pair's comings and goings during the incubation stage.

As the eggs' projected hatch dates approached, we observed our condors' activities with equal measures of anxiety and anticipation. Surely, this would be the year that Arizona's condors hatched their first chick. The second half of April was hectic, with newly released juveniles making their first South Rim visits, and as many as twenty-two condors—as many as once existed—in the area on most days. Amidst the chaos came the disquieting realization that 119 and 122 seemed less attentive to their nest cave than expected. Over a two-day period, the pair loafed together on the cliffs below Grand Canyon Village. Several days later, Condor 122 took two overly long incubation breaks. Like fretful grandparents, we assiduously monitored the pair's activities, relieved each time one of them returned to the cave and disappeared inside.

But on May 9, our hopes were dashed. Both birds spent most of the day away from their cave, loafing and playing with other condors. And though they returned to it that night, they failed to do so the next day. In a heart-breaking repeat of 2002, the Battleship nest had failed.

The failure in 2002, though disappointing, had not surprised us, since first-time breeders are often unsuccessful. But the failure in 2003 was disheartening indeed. To help us determine why the pair had failed again, park personnel helicoptered to the top of the Battleship then made the perilous climb into the cave. Dangling like spiderlings on ropes roughly 200 feet (61 m) from the cliff rim and 400 feet (122 m) above the ground, the climbers snagged a knotted rope around a rock in the cave's entrance to navigate a large overhang and pull themselves inside. To their amazement, the cave was enormous, perhaps 40 feet (12 m) deep and 8 to 10 feet (2.4 to 3 m) wide in places. Amidst the cave's sandy substrate, they found condor eggshell fragments, suggesting that the nesting attempt had failed during the hatching stage.

We could only speculate on what might have happened. Perhaps the pair had fought over the egg, destroying it in the process. Or maybe the chick had been unable to extricate itself from its shell. To hatch successfully, the embryo had to be positioned with its head tucked under its right wing and its bill pointed toward the air cell at the egg's wider end, so that it could breathe when it broke through the internal membrane and began tapping at the external shell with a temporary projection (the "egg" tooth) on its bill. Cracking open the egg was a lengthy, fraught process, and inexperienced parents sometimes failed to provide needed assistance.

Along with eggshell fragments, the climbing team found several bottle tops and some cloth, dashing hopes that Arizona's condors would be immune to the trash problems that had plagued California's birds. But there was an unexpected consolation prize for the findings inside the cave. While sifting the powdery substance at the back of the cave to search for trash, one of the climbers discovered several long bones. Subsequent analysis by paleontological expert Jim Mead revealed that the bones were those of adult and juvenile California Condors that had died during the Pleistocene.[20]

That our captive-reared condors had selected a nest cave chosen by their ancestors, thousands of years earlier, was remarkable. The bones also added to fossil evidence suggesting that condors had occupied a wetter, cooler Grand Canyon region during the Pleistocene. Condors first appeared in the area's fossil records more than 43,000 years ago, when juniper woodlands and scattered desert plants dominated the canyon's Inner Gorge, and grasslands and montane forests comprised the surrounding plateau.[21] Condors shared the skies with Black Vultures (rather than today's abundant Turkey Vultures, which were scarce) and an enormous vulturelike bird known as Merriam's Teratorn, which had a wingspan of about 12 feet (3.7 m). The scavengers fed on a host of large animals that included Columbian mammoths, shrub-oxen, large single-humped camels, donkeylike horses, Harrington's mountain goats, and enormous ground sloths that stood close to 6 feet (2 m) tall and weighed nearly 400 pounds (180 kg).[22]

The diverse and abundant megaherbivores—and the carnivores that preyed on them—thrived in this region for many thousands of years until the ending of the last Ice Age. Beginning around 12,000 years ago, as glaciers retreated from the higher mountains of the Colorado Plateau, Grand Canyon's climate warmed and dried, while seasonal temperature fluctuations became more extreme. The changing climate and associated vegetation changes from woodlands to deserts coincided with the extinction of the Pleistocene megafauna, though evidence for the disappearance of these animals also points to unsustainable persecution by human hunters, whose signs first appear in the region's archeological record 11,000 to 12,000 years ago.[23]

Fossil evidence suggests that condors disappeared from the Grand Canyon region along with the Pleistocene megafauna. Thus far, no condor bones younger than 9,750 years old have been found in Grand Canyon. Avian paleontologist Steven Emslie, who explored many of the area's caves, believes that condors died out when they lost their megafaunal food source.[24] Nevertheless, abundant animals—such as bison, pronghorn, elk, and deer—on which today's condors feed, remained, so perhaps human persecution played a greater role in the bird's fate than has been recognized. To date, only a small fraction of the area's caves has been explored. Finding more recent condor bones would bolster the possibility that a vestigial population continued to cast its shadows over the area's iconic cliffs into historic times.

Condors indisputably occurred in Arizona in small numbers in the 1800s and early 1900s. Emslie argues that condors from California recolonized the interior West after settlers introduced horses, cattle, and sheep in the 1700s, which provided scavengers with a novel food source.[25] Several observers recorded California Condors in Arizona in the late 1800s. Renowned early ornithologist Elliott Coues, who discovered and named the lovely Grace's Warbler after his young sister, claimed the condor was a resident of southern Arizona, having documented several individuals at Fort Yuma (on the California-Arizona border) in 1865.[26] Miles Noyes shot one of two condors that flew over his camp at Pearce Ferry, on the western edge of today's Grand Canyon National Park, in March 1881. He marveled at its size, claiming that its wingspan equaled more than three lengths of his 1876 Winchester rifle.[27] And rancher Jack Alwinkle, who knew the species after having lived in California, shot a condor in southeastern Arizona's Santa Catalina Mountains in 1890.[28]

Ornithologist Edouard Jacot, in 1924, was the last to record a condor in Arizona, prior to the bird's reintroduction in 1996. He observed it feeding on a carcass with Golden Eagles near Williams, just south of Grand Canyon.[29]

Opponents to the condor's Arizona reintroduction sometimes argued that the bird was not native to the Grand Canyon region, since its presence was unconfirmed between the late Pleistocene and the late 1800s. Nevertheless, the Pleistocene condor bones in the Battleship cave provided an irrefutable connection between the condors of old and those of the present, and I couldn't help but be delighted that condors across the ages had viewed the same cave as an optimal site in which to raise their young.

———

For now, in 2003, our immediate hopes for a new generation of condors rested entirely on Condors 123 and 127. Beginning on May 5, the pair began switching incubation duties almost daily. Since they had done this for several days the previous year before returning to longer incubation stints, we viewed this encouraging development with restrained optimism. But the trend continued, and, after three weeks of daily nest switches, our unconfirmed suspicions morphed into unconfirmed belief and then, eventually, into outright euphoria.

The pair soon seemed to be on a constant food-finding mission. When not at the cave, each parent spent virtually all of its waking hours searching for carcasses whose meat it would consume then later regurgitate into its nestling's awaiting bill. On some days, 123 and 127 switched nest duty multiple times, returning to their cave whenever they finished eating. By mid-June, when the chick was about six weeks old and could thermoregulate (regulate its own body temperature), the pair began to leave it for longer periods, flying far and wide in search of sustenance for their growing nestling. On one occasion, 123 flew to the release area from the South Rim, fed on a calf carcass, returned to Salt Creek to feed his chick, then flew back to the Vermilion Cliffs to roost. He had flown at least 150 miles (241 km) in the course of just one day in the six-month-long nestling period. After eating the next morning, he again returned to the cave to feed his chick. While one parent ranged farther afield searching for food, the other usually spent time in or near the nest, making shorter-distance food-finding forays. In July, both parents began spending nights away from the cave.

Absorbed as we were in our condor management duties, the summer sped by. We navigated our newly released juveniles' first interactions with people, trapped all our condors to inoculate them with a specially developed vaccine against West Nile virus—a frightening new threat to North American birds—and treated several birds for lead poisoning after they fed on a shot coyote.[30] It was mid-August before we finally braved Salt Creek's fiery temperatures to view the newest addition to our thirty-five-member condor flock for the first time. By then, monsoon rains were ushering in cooler temperatures. Nevertheless, we planned to begin the trek to Salt Creek in the evening to avoid the still-blistering-hot summer heat. Ironically, upon arriving at the South Rim, I was met with rain, hail, and cold temperatures. Nonetheless, after shouldering my heavy backpack, I set off down the cliff-side trail with a park biologist.

I had hiked into the canyon only a few times and had paid for my first descent with shins so sore I could barely walk the next day. I'd made each journey in sweltering heat and accompanied by gaggles of breathless tourists, so I was unprepared for the tranquil magic of the canyon at night. Sheer cliffs loomed over us, casting great shadows on the waterlogged trail. After the stormy skies cleared, a tapestry of bright stars illuminated a velvet darkness. The first Mexican Spotted Owl I'd ever heard hooted softly. After completing a long descent through geologic time—as demarcated by the various rock layers through which the trail passed—we reached the Tonto Plateau, midway to the Colorado River.[31] We hiked the next 7 miles, along the Tonto, to an intermittent musical accompaniment of high-pitched clicks. Unseen by us, spotted bats foraged for moths in the darkness, emitting echolocation calls that, unlike those of most bats, are discernible to human ears. Inhabiting semiarid regions in western North America, the spotted bat resembles a miniature Holstein cow, with a white front and three prominent white patches on its black back. Among North America's most distinctive but least-seen mammals, the bat has enormous, translucent pink ears that it curls around its head when resting.

Arriving at the Salt Creek drainage just past midnight, after a five-hour hike, we set up our camp and crawled into our sleeping bags. My alarm seemed to go off moments later. The sun had arisen, but the narrow drainage's magnificent Redwall Limestone cliffs were still in shadow as we hastily breakfasted and then trekked to the top of a steep scree slope for a better view of the nest cave. Although we, on the field crew, were certain we *had* a chick given the parent condors' behavior, members of the public, the press, and even the condor recovery program would not believe Arizona had a nestling condor until someone visually confirmed it.

I had prepared myself for long hours of watching and waiting for a nestling to emerge from the cave's depths, but when I looked through the spotting scope, I drew in a sharp breath as I beheld what looked like a full-grown chick in the cave's entrance. Keenly aware of the moment's import, I stared avidly at the first wild-hatched condor nestling in recorded Arizona history. Eye riveted to the scope, I watched it stretch both wings over its back, and I realized that it was less developed than it had at first appeared to be.

While we had monitored 123 and 127's activities from the cliff rim, their nestling had undergone several transformations. After hatching, chicks are covered in white down and have a featherless, pinkish head and neck, and grayish legs. Within twenty to thirty days, a dark-grayish down begins to push out the white natal down. At around two months of age, nestling condors begin growing black body and flight feathers. This juvenal plumage is nearly complete by five to five-and-a-half months, but the nestling's flight and underwing feathers may still be growing when the chick is ready to leave the nest at around six months.[32]

Young condors are adult sized when they fledge. Our nestling was nearly full grown at not quite four months old, but its flight feathers were only partially developed. The chick sported puffs of down on its legs and body, and its tail feathers were barely visible. Bright, dark eyes peered from a gray head covered in short, wooly down as the chick surveyed its realm.

Scarcely noticing the intense heat that enveloped the drainage as the rising sun chased away cool shadows, we huddled in the shade of umbrellas and watched the chick rest and preen. Cocking its head, it watched a passing butterfly and, later, a gliding Turkey Vulture before retreating into the cave's cooler depths, midafternoon. Several hours later, we hiked down to camp, weary but elated.

The chick came into view after we arrived at our observation point the following morning, but quickly retreated into its cave. Midmorning, we began receiving 127's signal. Minutes later, she landed with a flourish and disappeared into the cave to feed her nestling. Emerging soon after, she wiped her bill, preened, then spread her wings and headed northwest. Energized, the nestling rushed to the front of the cave and watched her go, then bounded around like an erratic rabbit, flapping its unwieldy wings.

Later that evening, we reluctantly left Salt Creek's otherworldly peace, slept at the Havasupai Gardens Campground, and then made the punishing ascent to the canyon rim the next morning to share news of our sighting with the world—and deal with the ensuing media firestorm.

———

Several weeks later, the Salt Creek nestling was assigned its studbook number—or, what was for us, its much-anticipated name. Henceforth, it would be known as Condor 305. Beginning in September, we watched over our developing nestling during every daylight hour, monitoring its feedings, activity levels, and health. We took turns hiking to Salt Creek alone at night, remained in the drainage for three or four days, then hiked out on the evening of our last day. For safety in the remote backcountry, we staggered our visits so that two of us were always camped out at Salt Creek, watching over 305. Although we couldn't ensure our nestling's success by watching its activities so closely, we were on hand to assist if the need arose, and we would document its departure from the nest if it fledged.

I came to love the night hikes into Salt Creek, feeling a perverse delight in setting out on my journey into the canyon when tourists were leaving the area in droves, and relishing the ensuing solitude and the canyon's timeless tranquility. And I embraced every opportunity to monitor 305. Immersing myself in the daily life of a condor nestling, I delighted in its curiosity and bursts of activity—wing flapping and sprinting around its cave—which fueled my hopes for the future of the species.

But even though 305 radiated energy and health, my faith that condor populations would continue to grow and the species would be saved faltered, for the first time, when we received devastating news from California. Their nestling—Condor 308—had appeared as healthy as 305, but when a team climbed into the nest to remove potential trash, administer a West Nile virus vaccine, and outfit the nestling with radio transmitters, they found a poorly developed chick. Although 308 was feisty, its feather growth was severely retarded, it was underweight, and a rasping noise accompanied its breathing. Worse, its crop appeared to contain hard objects. Reluctantly, the team returned with a helicopter to transport the nestling to the Los Angeles Zoo for treatment. To reduce the nestling's stress during transport, it was given oxygen, wrapped in a dark cloth to obscure its view, and draped in cold towels to prevent overheating.

At the zoo, veterinarians took x-rays and administered fluids and antibiotics. When the chick regurgitated its crop contents, a shocked veterinary team found four aluminum can rings, a bottle cap, part of a plastic shotgun shell, two pieces of rubber, three pieces of melted metal, an electrical connector, and two pieces of bark. And still the chick's raspy breathing continued. Upon conducting an exploratory surgery, the vets discovered that one of the ingested objects had punctured the nestling's digestive and respiratory tracts. California's hope for a wild-fledged chick was euthanized on the operating table.

As I watched Condor 305 run around, flapping its wings energetically and hopping up onto a large rock at the cave entrance, I couldn't imagine that its health might be compromised by having ingested trash or lead bullet fragments. But California had lost all four of its nestlings and I scarcely dared to dream that 305 might have a different fate. And if condors couldn't reproduce successfully in the wild, what hope was there for the species' future?

————

The inner canyon's cottonwoods turned gold and temperatures moderated, as summer gave way to fall. And still 305 received meals from its diligent parents, watched passing raptors, and dashed erratically around its cave, flapping its unwieldy wings with an enthusiasm that fanned my smoldering optimism. In October, I watched, enchanted, as Condor 127 landed at the nest cave, fed her nestling, then cuddled up to it for a few hours, while the two allopreened. Shortly after 127 left, 123 arrived and fed 305 its second meal of the day. He, too, then cuddled up to the nestling and gently preened it before returning to the skies. Parental care by California Condors is among the most extreme of any bird species. It includes nearly two months of incubation, six months of caring for the nestling, and as long as a year of continued feeding post-fledging.[33] But such statistics hardly captured the dedication with which 123 and 127 searched for carcasses, patrolled the skies to keep away intruders, and interacted with their nestling. Energized by the food and attention, 305 rushed around the cave following its parents' departure, flapping ferociously.

With each passing day, 305 seemed more constrained by the confines of its cave. Cringing inwardly, I watched the nestling clamber repeatedly onto a narrow ledge adjacent to the cave and, chest pressed against the cliff, attempt to flap its giant wings. The four-hundred-foot drop might be deadly if the youngster wasn't yet ready to fly. Each team that emerged from the canyon, disappointed that 305 hadn't fledged, regaled the rest of the crew with accounts of breath-catching moments when 305 balanced on ledges or spread its wings as though about to take flight.

Late in the day on November 3, I hiked into the canyon again, enjoying the cooler temperatures and delighting in the unseen bats that flitted around me as darkness extinguished a radiant sunset. Lost in thought, I walked briskly along the trail as the miles unfurled behind me. Despite my heavy backpack, I'd become accustomed to the 12-mile (19 km) journey and joined a Grand Canyon biologist at Salt Creek after less than four-and-a-half hours.

Any hopes that I might see 305 spread its wings and leave its cave on this trip evaporated on my second day in the canyon. The nestling appeared not only to have given up on trying to clamber out of its cave, but also to have forsaken any activity altogether. Huddled against the cold, I watched 305 doze. And I struggled to stay alert myself as an unusually quiet morning gave way to afternoon.

To my relief, 305 roused itself at last and, for the first time that day, flapped, ran, jumped up on rocks, and balanced on ledges. As though making amends for its prior lassitude, 305 was active longer than usual. Finally, though, the youngster began to wind down, and I settled more comfortably into my camp chair, ready for a note-taking break. Eye still glued to my spotting scope, I watched 305 crane its neck and look fixedly at an indeterminate spot out beyond its cave. Suddenly, with no preamble, the young condor leaped toward it—and into the void.

"It's fledging!" I yelped, jumping to my feet. Following its trajectory with my binoculars, I watched 305 plummet earthward like a paraglider in a semi-controlled free fall. Flailing wings half extended, 305 struggled to slow its precipitous descent. Finally, after what seemed like an eternity, the fledgling touched down surprisingly gently.

As though turned to stone, 305 stood motionless at the base of the cliff. Adrenaline surging, horror mingling with euphoria, I stared at the young condor, willing it to move. Minutes passed while I tried to determine whether it had sustained injuries. Then, 305 stretched its neck toward a banana yucca, clamped on to it with its bill, and gave the plant a good tug. The breath I hadn't realized I was holding whooshed out in a burst of laughter. Condor 305 was doing what all young condors do: exploring the world with its bill, tugging and pulling and checking things out. Our newly fledged condor was just fine.

While 305 rested, I reflected on the magnitude of what I had just witnessed. Before now, only three condor nestlings in recorded history had ever been seen taking their first flight.[34] Moreover, I had just observed the first fledging of a wild condor since the inception of the recovery program. I had watched over and held every condor in Arizona. I had observed many of the reintroduced condors in California. But 305 was unique. For the first time since AC-9's capture in 1987, I was looking at a condor untouched by human hands, a condor free of the trappings that we had inflicted on the dwindling species. Staring avidly at 305's dark, tagless body against the dramatic red cliffs, I relished the moment, unable to stop smiling.

An hour later, Condor 127 glided into the Salt Creek drainage. She and 123 had been absent for the past three days. Touching down at the cave, she walked out of view. She reappeared seconds later, head swiveling as she searched for her chick. Far below 127, the hungry fledgling solicited food by wing-begging in the universal way that juvenile birds do when their parents land beside them. Nevertheless, what looks like small, shivering wing motions when enacted by a sparrow looked like ungainly flopping as 305 frantically worked its massive wings to get its parent's attention. Although she must have spotted 305, 127 went back into the cave several times, as if to reassure herself that her chick wasn't tucked into a hidden corner. Finally, gliding out of the cave, she dropped to the ground and fed 305 its first meal as a free-flying condor.

It would be several months of short, hesitant flights and hops from one cliff ledge to another before 305 followed its parents out of Salt Creek and soared over Grand Canyon's castellated cliffs. But as condors initiated another round of nesting in early 2004, Condor 305 made its debut at the South Rim, treating countless visitors to a glimpse of the California Condor as it had been for tens of thousands of years and as, someday, it might be again.

CHAPTER TWENTY-TWO

Condor in the Coal Mine

When biologist Jan Hamber, who monitored condor nests in the 1980s, could finally bring herself to describe her involvement in the capture of AC-9—the last wild condor—on Easter Sunday, April 19, 1987, she wrote in her diary: "It is still painful for me to think about those huge wings held captive in a cage . . . and know that never again will I thrill to the sight of that majestic bird, soaring for miles so effortlessly on the wind currents." Sitting in a distant vehicle, Jan had watched AC-9 fly toward the baited trap site and couldn't help herself from screaming ineffectually for him to get out harm's way. She saw a puff of smoke as the cannon-driven net went over him. "For all the effort that had been expended, for all the plans that had been laid, for all the hopes that had been imagined," Jan wrote, "I just knew that I had been part of a historic moment: the purposeful removing of an endangered species from the wild to save it . . . And I put down my head on the steering wheel and cried."[1]

AC-9 spent fifteen years in captivity. Difficult as it had been to conceive how a creature that belonged in the sky could adjust to life in a cage, AC-9 seemed to do just that. He produced sixteen offspring for the captive-breeding program and, with each passing year, more of his progeny—including Arizona's Condor 122—flew free. Following years of reintroductions, a growing number of AC-9's fans clamored for his return to the wild despite his value to the breeding program and ongoing concerns regarding the dangers of lead. Finally, on May 1, 2002, Jan—now in her seventies and with a grandson by her side—joined a gathering of condor enthusiasts to watch AC-9 return to the sky. When he extended his great wings and was borne aloft by the wind at California's Arundell Cliffs, a euphoric Jan felt a sense of completion. At long last, she had fulfilled what had once seemed like an impossible promise to the bird whose capture had marked the end of an era—and almost the end of a species. AC-9 was free. Free again, at last.

But although AC-9 reigned supreme in the hearts and minds of fans the world over, condors in California treated the returnee like a lowly interloper. AC-9 soon frequented his old haunts, but it was several years before he regained his former stature. By 2004, though, the twenty-four-year-old AC-9 attracted a young mate that had been reintroduced to central California's Big Sur area. The pair selected a remote cave in the Sespe Condor Sanctuary and, soon after, began incubating their egg.

Unbelievably, on Easter Sunday, April 11, 2004, AC-9's chick emerged from its egg and greeted the wilderness from which its father had been removed seventeen years earlier.[2]

On Thanksgiving Day, 2004, AC-9's son—Condor 122—and Condor 119 finally fledged their first chick from the Battleship cave. Unlike the controlled plummet made by Arizona's first fledgling, this second youngster, Condor 350, extended his wings, glided out of his cave in a dramatic first flight, and landed safely at the base of the cliff. Two days earlier, AC-9's daughter, Condor 149, had fledged her first chick from a cave in the Vermilion Cliffs. Arizona now had three fledgling condors.

I was not there to witness the 2004 fledglings' first flights, and my heartache at missing these events resurfaced with subsequent condor-recovery milestones in the years to come. I had left my position as The Peregrine Fund's condor field manager at the end of May. Years later, I would learn that some of the injustices that led to my eventual departure had names like gender bias, gender pay gaps, and discrimination. Research would show that women, particularly in the STEM education fields (science, technology, engineering, and math), often were held to a higher standard than their male peers and received less recognition and remuneration for their work.[3] At the time, I just knew that my Idaho supervisor's unwillingness to recognize my position, my worth, or my contributions was demoralizing, disheartening, and demeaning.

The condor population in Arizona had doubled under my tenure—forty-five condors now flew free in Arizona—and fatalities had declined dramatically. While coordinating and supervising Arizona's field activities from 2000 to 2004, I had helped release thirty-one condors to the wild, and had designed and implemented strategies that helped newly released condors survive. These strategies had been adopted program-wide. Over the years, I had trained and supervised thirty-nine field assistants, some of whom became lifelong friends. I had managed the reams of daily data we collected, worked on scientific manuscripts and program updates, and written semimonthly "Notes from the Field" that had garnered a nationwide following and had become the most popular page on The Peregrine Fund's website. I had contributed to recovery program meetings and been a member of the team's behavior modification and nest management committees. Most important, I had devoted nearly four years to doing everything in my power to keep my charges alive and thriving.

Although the Big Sur, California, and the Baja California, Mexico, condor reintroduction projects offered me leadership positions once I left The Peregrine Fund, I was consumed by my desire to return to the refuge of the mountain country I loved. After a stint as a backcountry biologist researching small mammals in Grand Canyon National Park, I worked briefly as a wildlife biologist with an environmental consulting firm before becoming the wildlife program director for a Wyoming conservation organization. I was moving closer to the Montana mountains that spelled home for me, but for years I wondered if I ever would recover from the heartbreak of leaving the condors that had captured my heart and left such an incandescent imprint on my life.

————

Distracting me from the report I was writing, a colleague handed me some photos to identify. I shuffled through them trying to remain objective as I examined the feathered specimens they depicted. A black and chestnut body and a red eye flanked by a spray of yellow feathers identified the first bird remains as those of an Eared Grebe. The puff of yellow and olive with a jaunty black cap was a Wilson's Warbler. A falcon beak and blue-gray back were clearly those of a Merlin. I struggled for a moment with the nondescript brown back and cream belly of the fourth bird, until I spied a hint of black and yellow protruding from the bird's mostly hidden face. A Horned Lark. All victims of collisions with wind turbines.

In the previous months, I had surveyed an area slated to become a wind farm to determine whether bird activity levels there were low enough to minimize such collisions. By conducting these preconstruction surveys and overseeing searches for birds and bats killed by turbines in existing wind farms, I was getting a first-hand education into the importance of siting wind turbines in areas with low bird and bat activity. Poorly sited wind farms could have catastrophic impacts on wildlife, posing collision hazards and destroying and fragmenting habitat. I sought to minimize such impacts while working with renewable energy promoters.

Returning to my report, I started summarizing the results of the fall bird surveys I had conducted on a windswept Wyoming ridgetop. When my phone rang minutes later, I had a momentary thought that the New Year—it was January 2, 2007—was not off to a very productive start. As the voice spoke quietly, regretfully, in my ear, all thoughts of my current work fled, and I was transported to the dramatic canyon country that had once served as my office.

Moments later, I closed my flip phone and sat for a while trying to compartmentalize, trying to suppress the upwelling of emotion, trying to maintain my professionalism and return to work. It was an exercise in futility. Turning my chair at last, I met the stricken face of my supervisor, who shared an office with me and had become a friend.

"One of your condors?" she asked quietly.

"119," I responded abruptly. Captivated by my condor stories, Diane had found my "Notes from the Field" online and read them avidly. I didn't need to explain.

"I'm going home for lunch. It may be a long one," I said, as tears threatened. Diane hugged me. "Go. Take all the time you need."

As I drove to my rural home, I stared blankly at the stark white landscape that mirrored my state of mind, and I tried not to feel, not to think, not to be drawn into the abyss that threatened to overwhelm me. I made it home. I walked into my house. I wrapped my arms around my border collie. And I started to sob.

Hours later, I began writing about 119, what she had meant to me, and how monumental her loss was for me and for Grand Canyon National Park, where she had been a fixture for so many years.[4] I had admired her independent nature and adventurous spirit. Released to the wild as a two-year-old in May 1997, she had traveled farther afield than any other condor in Arizona when she'd explored the Utah-Wyoming border in 1998. She had overcome countless bouts of lead poisoning; returning her to the wild after the first episode was one of my life's unforgettable moments. She had laid the reintroduction program's first documented egg and had fledged her first chick after years of dedication.[5] I had mourned each of her failures and celebrated her eventual success. As Arizona's oldest condor, she had long seemed like its matriarch. In my last few years with her, I'd been struck by her maternal aura and the wisdom that seemed to radiate from her eyes.

Condor 119 graced Arizona's skies for nearly ten years. She died from a massive dose of lead on December 29, 2006. Condors can live sixty to seventy years. Condor 119 was eleven.

Grand Canyon personnel shared my heartbreak. Having spent more time in the canyon than any other condor, 119 was beloved, and her death was widely mourned. As a tribute, the park commissioned a condor plush toy that wore a wing tag labeled "19."

After 119's death, I distanced myself from news about the reintroduction program. There had been so many losses. And each loss of a condor I had known, I had held, I had fought for brought a measure of heartbreak. Our wild wanderer, Condor 176—perhaps my favorite condor—had disappeared in 2004. Condor 235, who had enchanted me with her comical landings and who, as a juvenile, had commandeered carcasses by lying on them with extended wings, died of lead poisoning in 2005. As did stellar young male Condor 249. Condor 196, another wide-flying, independent female, who'd once sprinted down a beach with a water jug held aloft in her bill, disappeared in 2006. Condor 227, who had charmed me with his riotous play and his long-distance flights in search of condor companions, died of lead poisoning in 2007. As did brave and beautiful 248, who had been the first of her release cohort to launch herself skyward. And so it went.

Condor 127, who raised 305—the first condor chick to fledge in Arizona's recorded history—fledged two more chicks with her mate, Condor 123, before dying of lead poisoning, at the age of fourteen, in 2009.[6]

And beloved 210, who captured the field crew's hearts while tormenting us with her challenging behavior, died of lead poisoning, at the age of thirteen, in 2013. In addition to having liver lead levels of more than 4,000 μg/dL, 210 had ten lead fragments in her digestive tract. Her last meal was a deer.[7] Prior to her death, 210 was a model adult that fledged her first chick in 2007. After her mate died of lead poisoning, she paired with Condor 122, who had lost his own mate, 119, and the two fledged a chick in 2009.

Some critics of the recovery program have claimed that the condor is a relict of the Pleistocene to explain the bird's current endangerment and their belief in the futility of trying to save the species. But the opportunistic California Condor survived into historic times, and the threats that prevent the species from thriving today are all human-caused. They include lead toxicosis, power line collisions and electrocutions, trash ingestion, intentional killing, West Nile virus, and highly pathogenic avian influenza (two diseases introduced by human activity).[8] I received a heartbreaking reminder that direct persecution was an ongoing concern when I learned that Condor 122 was shot in 2015. To the shooter, 122 may have been a large, ugly target. But to those of us who watched over him in the wild, fought to save him from lead poisoning, delighted in his pairings with Condor 119 and 210, celebrated the fledging of his two chicks, and took pride in his legacy as the son of AC-9, the shooting was a senseless, despicable act.

Although the litany of condor losses overwhelmed me, it obscured the recovery program's enormous successes as the twenty-first century proceeded.[9] My condor memories remained achingly vibrant, but time hurtled along and whenever I made an accounting of the species' status, I was thrilled by the progress being made by those who still labored to help the bird recover its place in North America's skies.

On September 25, 2019, Condor 1,000 fledged from its nest cave in Utah's Zion National Park. As I looked at photos of the chick and its parents against the backdrop of Zion's striking cliffs, I was a little stunned to discover that condor studbook numbers had reached four digits. From a population low point of 22 condors in 1982, there were now more than 500 California Condors, with over 300 of those flying free.[10]

California had finally fledged its first condor chick in November 2004, a year after Arizona's Condor 305 leaped into the record books. For a time, the danger posed to nestlings by the proliferation of trash in our environment threatened the recovery program's success.[11] Microtrash ingestion has been one of the most serious obstacles to condor reproduction, causing more nestling deaths (73 percent) than any other mortality factor.[12] Of ten nestlings hatched in California between 2001 and 2005, only one survived to fledging. Seven of eight wild chicks examined during that period had consumed microtrash. Six chicks had microtrash in their guts, and four of these nestlings died (the other two were returned to the wild after more than a year of recovery).[13]

To address this threat, biologists in California repeatedly cleared trash from nest caves and areas condors frequented.[14] For a time, bone fragments were provided at feeding areas, though this measure didn't reduce the transfer of trash from parents to nestlings.[15] Perhaps most important, biologists encouraged adult condors to forage more naturally. Because of the lead poisoning threat, biologists in California had provided carcasses at well-established feeding stations since the first reintroductions. In Arizona, though, lead poisoning was wrongly expected to be a less serious threat because of the prevalence of National Park Service lands, where hunting was forbidden. Accordingly, condors had been encouraged to forage more naturally early on; they were offered supplemental food in different locations, so the birds learned to search for it. As a result, parent condors in Arizona spent more time on the wing, searching for food, whereas condors in California remained closer to feeding stations and had more time to play with trash.[16]

Human garbage poses a global threat to wildlife. Trash ingestion and entanglement kills animals outright and impacts reproduction, growth, and health. With an estimated 5.25 trillion particles of trash, weighing approximately 270,000 tons, in our oceans, our waste causes particular harm to seabirds and other marine organisms.[17] Even seemingly innocuous trash can be dangerous. Balloons were the most likely ingested trash item to kill seabirds in a recent study.[18] Balloons are also a hazard in terrestrial environments, occurring in higher densities in some protected desert areas than western diamond-backed rattlesnakes.[19] We might be able to ignore a dropped bottlecap or cluster of deflated balloons while out walking, but it is harder to ignore graphic images of the stomach contents of dead condor or albatross nestlings, reflecting as they do our impact on the natural world.

Fortunately, the trash-consumption problem that long undermined condor recovery proved to be more tractable than other trash dilemmas, and scores of condor chicks have now fledged in the wild. Remarkably to those of us who agonized over the fates of the first wild nestlings, wild-fledged chicks are now producing their own chicks. And with each passing year, the California Condor recovery program notches new milestones. In 2011, condors in the wild outnumbered those in captivity for the first time.[20] And in 2018, twelve condor pairs in California nested over a larger geographic area than had ever been documented before.

One of the chicks that fledged in 2018 was the first to do so in California's Santa Barbara County since 1982. The chick's thirty-eight-year-old father, AC-4, had been captured in 1985. After spending *thirty* years in captivity and siring thirty young, AC-4 was released back to the wild in 2015.[21] Within two months, he returned to areas he had frequented decades earlier.

————

Although wild condors are now fledging their own young, biologists continue to bolster condor numbers through reintroductions. Condors are released to the wild in California and Arizona, and in Baja California, Mexico. A new reintroduction site that began releasing condors in northern California's Redwood National Forest in 2022 will allow condors to recolonize southern Oregon.[22] In the foreseeable future, the California Condor may reclaim the entirety of its West Coast range, from British Columbia to Baja California, Mexico.

But despite the monumental gains achieved through reintroductions, these successes mask the existential threat that lead ammunition poses to condors. Following a detailed analysis of the recovery program in 2012, scientists found that intensive management efforts—captive breeding and reintroductions; monitoring, capturing, and treating condors for lead poisoning; and providing supplemental food—concealed a lack of recovery for this critically endangered species. After determining that the bird "is not on a trajectory to a self-sustaining wild population," the scientists underscored that "without reduced lead poisoning, the California Condor will require extraordinary management efforts in perpetuity to avoid again declining to extinction in the wild."[23] Previous analyses had come to similarly disheartening conclusions.[24] A severe lead-poisoning event in 2013, in which twenty-one condors in California required treatment, provided a stark example of how quickly the condor's upward population trajectory might reverse itself without intensive oversight.[25]

Encouragingly, though, whereas many endangered species, including the Hawaiian Crow, face a host of complex and daunting threats that defy easy solutions, "if the lead exposure hazard is removed and thus lead deaths are halted or severely reduced, California Condors could once again achieve a sustainable wild population."[26] Replacing lead ammunition with nontoxic alternatives would also benefit human consumers. As a result, eliminating the threat that the condor has revealed to us in its unwitting role as a canary in the proverbial coalmine seems eminently feasible.

And yet, more than a decade after William Cornatzer revealed that venison packages for the hungry contained unacceptable levels of lead, most hunters are still using lead ammunition and are still resistant to suggestions that it would be better for their health to use nontoxic bullets. To this day, journalistic reporting about lead-poisoned condors is invariably followed by comments from readers that call the issue a hoax that is not supported by science and is perpetrated by anti-hunting environmentalists.

Denial of scientific evidence is an oft-used stratagem by those who are resistant to (or might lose financially from) proposed regulations and changes spurred by environmental and public health concerns. Disinformation campaigns and efforts to sow doubt about scientific findings were hallmarks of the battles over DDT, ozone-layer thinning, acid rain, the health impacts of tobacco smoking, and climate change.[27] All these issues, which were derided and disputed in their time, have proven to be legitimate concerns whose perils ultimately were widely recognized and accepted.

Nevertheless, public acceptance was delayed—fifty years in the case of linking cancer to smoking—by the now-familiar playbook of stakeholders and their minions sowing doubt about legitimate science, distracting the public's attention with alternative explanations, attacking the messengers, attributing false motives to them, and accusing them of the very obfuscation that they themselves employ. Gun rights advocates have adopted these strategies, disputing and rejecting extensive scientific evidence about the perils of lead-based ammunition, touting nontoxic alternatives as inadequate or expensive, and calling proposed lead-ammunition bans antihunting and anti-Second Amendment.[28]

Gun rights advocacy groups, such as the National Rifle Association (NRA) and the National Shooting Sports Foundation (NSSF)—the firearm industry's trade association—have attempted to sow doubt about the science regarding the impact of lead ammunition (which they now refer to as "traditional" ammunition) on condors and on human health. Some of their efforts are on display on the Hunt for Truth website, which was developed by organizations that oppose efforts to limit the use of lead ammunition. The coalition has claimed that "there is substantial evidence that the groups behind a [then-proposed California] ban on the use of traditional ammunition have based their claims on faulty scientific studies." Likewise, they have claimed erroneously that those who oppose the use of lead ammunition are "in many cases motivated by an underlying anti-hunting agenda."[29] To bolster their argument that lead ammunition does *not* pose a threat to condors and human health, the group provided a list of scientific studies that examine the impact of *other* lead sources on wildlife and livestock. Their argument was akin to claiming that leaded gasoline doesn't harm people because lead paint harms our dogs.[30]

In addition to sowing doubt about the dangers of lead ammunition among their constituents, gun lobbyists have had considerable success in obstructing national and statewide efforts to regulate its use. As a result, no federal agency to date has taken a comprehensive, hard look at the impacts of lead ammunition on wildlife and human health.[31] After more than a hundred organizations unsuccessfully petitioned the EPA to regulate lead ammunition under the Toxic Substances Control Act (TSCA) in 2010 and 2012, the NRA and other groups pushed legislation to permanently block the agency from doing so under TSCA and also to prevent any other federal agency from regulating lead ammunition.[32] And, after the departing Obama administration enacted an order calling for a five-year phaseout of the use of lead ammunition and fishing tackle on national wildlife refuges, the NRA, the NSSF, and other gun-rights advocacy groups successfully petitioned the Trump administration to revoke the regulation.

Nevertheless, encouraging progress has been made in protecting condors and people from the dangers of lead ammunition. In 2007, the California legislature passed the Ridley-Tree Condor Preservation Act, which banned the use of lead ammunition for hunting big game and coyotes in the thirteen counties comprising the condor's California range. Signed into law by Republican governor Arnold Schwarzenegger, the bill's passage followed years of effort by wildlife advocates and a consensus statement from scientists regarding the dangers that lead ammunition poses to condors. The California Fish and Game Commission supplemented the bill with additional lead ammunition restrictions in the condor's range.[33]

These early lead bans failed to materially reduce lead poisoning in California's condors.[34] Indeed, the state experienced one of its worst lead-poisoning events in 2013. The NRA and its associates were quick to point to lead poisoning incidents as "proof" that condors were *not* ingesting lead ammunition, since such ammunition had been banned.[35] However, regional bans were undermined by limited enforcement, the ongoing use of lead bullets in ranch management activities (for example, to protect livestock from predators or dispatch injured animals), the continued use of lead for hunting certain animals, and the prevalence of poaching.[36] And, as scientists pointed out, if only half of one percent of the carcasses that were available to condors contained lead, there was an 85–98 percent chance that a condor would feed on a lead-tainted carcass over a ten-year period.[37] Clearly, banning lead ammunition for certain types of hunting in a portion of the state provided insufficient protection for California's iconic bird.

Bolstered by an ever-growing body of research confirming the threat lead ammunition posed to condors and a host of other species, wildlife advocates pressed for a more comprehensive ban. Leading scientists, doctors, and public health experts weighed in with a 2013 consensus statement endorsing the "overwhelming scientific evidence" for the toxic effects of lead-based ammunition on human and wildlife health.[38]

Persuaded by the science and unconvinced by the dire prognostications of gun advocates that a comprehensive lead ban would lead to drastic declines in the hunting license purchases that help fund wildlife conservation, the California legislature banned the use of lead ammunition for all hunting in California in 2013.[39] Signed into law by Democratic governor Jerry Brown, the ban was phased in over five years—taking full effect on July 1, 2019—to provide California's hunters with ample time to navigate the transition.

Meanwhile, since condors in Arizona and Utah are considered a nonessential, experimental population, per the 10(j) designation in the Endangered Species Act, efforts to mitigate the lead poisoning threat in the Southwest have focused on encouraging local hunters to *voluntarily* switch to nonlead ammunition when hunting in the condors' range, and providing them with free, nonlead ammunition.[40] Those involved in these efforts point to the success of a program in which approximately 87 percent of hunters on the Kaibab now participate (about 66 percent use nonlead ammunition; others remove gut piles from the field).[41] Detractors counter that success should be measured not by hunter participation levels but by the number of lead-poisoned condors. Indeed, even if only 1 percent of available carcasses contain lead, a condor's chance of feeding on at least one contaminated carcass each year would still be high enough to lead to the extinction of the species if not for intervention and treatment by biologists.[42]

For now, it seems likely that condors will be self-sustaining in the Southwest only when scientific acceptance of the dangers of lead ammunition prevails over those who block the regulation of lead ammunition and sow doubt about its dangers. Until then, we ignore the warnings the California Condor has broadcast for decades at our collective peril. For even though hunters comprise only about 5 percent of the US population, many more consume hunter-killed meat. And while those who don't ingest game killed with lead bullets may consider themselves safe, the extent to which they may be subject to increased aggression, violence, extremism, and other unintended consequences because so many still do remains an open question.

———

Settling myself more comfortably into my chair, I trained my eyes on the nest cavity and watched avidly as Condor 190 readjusted her wings then carefully lowered her body to shelter her two-week-old nestling. A tiny, wobbly head appeared for a second before retreating into feathered comfort. Out of view, an Olive-sided Flycatcher persistently voiced its trademark call, *Quick, three beers*. A shaft of sunlight illuminated part of the large tree cavity where 190 lay on a bed of compacted woody debris. After a long interlude, 190 suddenly startled, then stood up. She vocalized softly—raspy, breathy exhalations—and I had my best view yet of her downy white nestling. Flapping its unwieldy little wings, it stumbled forward on unsteady legs. Condor 190 gently touched it with her bill then gingerly covered the nestling with her body. But the chick squirmed beneath her and 190 stood up again. Moments later, she hopped to the cavity entrance, scanned her surroundings, then glided away from the nest tree.

"190 leaves cavity," I jotted on my notepad. Then laughed at myself. Old habits die hard I thought wryly. I had no need to record the condor's behavior. I was over a thousand miles away, watching the Ventana Wildlife Society's Big Sur webcam at my office desk in Montana.[43] Riveted to my computer screen, I listened to the chick vocalize surprisingly loudly. It plucked at one of the adults' molted feathers, then nibbled on another. Followed by a bit of wood. Young as it was, the nestling was already exploring the world with its diminutive bill.

I heard a thump as twenty-three-year-old Condor 167 landed at the entrance to the cavity and walked over to his awaiting nestling. Putting its bill in 167's partially open beak, the chick received a meal from its attentive parent. Along with countless others, I was witnessing wildlife drama once reserved for biologists who braved often-perilous conditions to observe condors in California's remote wildlands. Thanks to a camera installed high in a redwood cavity, those who had followed the condor's captivating saga for decades and newcomers to the condor's story could now witness a once-near-mythical bird's daily life.

Picking up my phone, I called my sister in Vermont. "The nestling's finally in view," I gushed. Seconds later, after accessing the website, Lisa exclaimed, "It's tiny!"

Not only was I watching birds I loved, but I could also share the experience with family and friends. Lisa and I watched together as Condor 167 preened his chick then enveloped it in warm feathers. As many as five hundred people watched the Big Sur webcam each day.[44] And just as I had become invested in the futures of those condors that I watched from a camp chair in Salt Creek or a blind atop the Vermilion Cliffs, so too might those who followed this nestling's transformation from wobbly chick to gray-clad adolescent to feathered juvenile feel connected to the condor they watched grow up. And, perhaps, if all went as we hoped, in six months' time we would witness this youngster spread its wings and launch itself into a bright new world, where lead ammunition was less prevalent than it had been since settlers first colonized California.

Webcams now give anyone with a computer and an internet connection an opportunity to experience the same kinds of moments with birds that I most enjoyed as a wildlife biologist. Whether you watch parent condors gently preen their nestling, a Great Horned Owl bring prey to its hungry brood, or a downy puffin rush to greet its parent in a burrow, you can experience innumerable, usually unseen wildlife activities that go on around us all the time. Moved by the playfulness of a nestling condor or the vulnerability of an eagle taking its first flight, you might even be inspired to embark on your own feather trails. If you do, I hope you'll go where the birds are—in the wild. For it is only there that you can experience the transcendent grace of a flying condor, or hear the clarion call of an island crow announcing the dawn in a primeval forest, or see a falcon stooping with preternatural speed at an intruder over a mountain lake. When the journeys such sightings entail aren't possible, you can find the wild outside your front door. You might come to know the local birds in a nearby park. Or be inspired by the brilliance of a warbler singing from a blossom-filled tree.

While the wild outside our doors has unquestionably diminished during each of our lifetimes, for now most of the birds, including the California Condor, are still here. In my mind's eye, I see the grand silhouette of a condor drifting over Grand Canyon's sunset-touched mesas. Bold black-and-white wings trace a timeless story, stitching a world of mammoths and giant ground sloths to today's tourists and trophy elk. The great birds still enthrall all who bear witness to the poetry of their flights, inspiring new generations of conservationists. I once watched an evening crowd of tourists at Grand Canyon burst into spontaneous applause as condors wheeled and danced across the sky overhead. Winging over vast, ever-changing landscapes that have witnessed their unmistakable shadows for tens of thousands of years, condors prevail as messengers of hope for all the good that humanity can achieve when it harnesses its collective might to protect and conserve rather than to exploit and destroy.

For there is always light,
If only we're brave enough to see it,
If only we're brave enough to be it.

—Amanda Gorman

Epilogue

Conservation biologist Thomas Lovejoy, who coined the concept "biological diversity" after witnessing the Amazon's ecological riches in 1965, said, "If you take care of birds, you take care of most of the environmental problems in the world."[1] Birds are visible, vocal sentinels that alert us to a warming climate, disappearing and deteriorating habitat, declining insect populations, hidden and not-so-hidden poisons, toxic air and water, excessive persecution, resource scarcities, invasions of foreign organisms, the proliferation of harmful trash, and even outbreaks of new diseases.[2] We do well to listen and respond to the tales birds tell us through their declining numbers or their failure to thrive, since the state of the birds so often warns us of perils that threaten other organisms, including us.

Threats to birds range from the relatively straightforward—DDT threatening peregrines and lead poisoning endangering condors—to the complex. Challenges such as climate change, for example, might alter a bird's habitat, subject the species to physiological stresses, or decouple the timing between its breeding or migratory activities and peak food availability.[3] In the case of Hawaiian honeycreepers, a changing climate has increased the birds' susceptibility to introduced, mosquito-borne diseases.

The actions required to recover a species invariably depend on the nature of the threats it faces. The Peregrine Falcon became a poster bird for the ecological devastation caused by pesticides. When DDT and other toxins were banned, peregrine recovery followed. As the falcon revealed, dramatically curtailing or eliminating the principal threat to a species is fundamental to effecting the organism's recovery. Alternatively, the failure to eliminate primary threats typically undermines the recovery of a species.[4] The California

Condor is likely to remain imperiled—and dependent on ongoing conservation measures—as long as lead ammunition is used in its range. Nevertheless, returning condors to a compromised environment was instructive in clarifying the nature of the threat the birds faced, as well as teaching us that what threatened condors was also harmful to other scavengers and human hunters.

Species such as the Hawaiian Crow that face multiple complex threats will inevitably be more difficult to recover than those that face a single predominant threat. Yet reintroducing the ʻAlalā to its still-compromised realm helped tease apart the threads that led to its endangerment, revealing how the changes we inflicted on its habitat not only introduced and facilitated the spread of new predators and diseases but also tilted the playing field to benefit its principal native predator. Regardless of the threats they face, recovering imperiled birds takes tremendous work, widespread public support, sufficient funding, and a regulatory framework that precipitates and sustains recovery efforts.

The 1973 Endangered Species Act (ESA) has long served as a critical safety net for the rarest species in the United States. When President Richard Nixon signed the act into law—after it was passed unanimously by the Senate and with a 355–4 vote in the House of Representatives—he said, "Nothing is more priceless and more worthy of preservation than the rich array of animal life with which our country has been blessed. It is a many-faceted treasure, of value to scholars, scientists, and nature lovers alike, and it forms a vital part of the heritage we all share as Americans." He said, too, that the lives of future generations would be richer and America would be more beautiful in years ahead thanks to the passage of the act.[5]

Our strongest conservation law—enacted to prevent species extinctions and to recover imperiled species—the ESA has helped 99 percent of the listed species avoid extinction (saving well over 200 species). It has also increased or stabilized population numbers for about 85 percent of the bird species it has protected in the continental United States since 1967. Sixty-one percent of the Pacific Island birds protected by the ESA have increased or stabilized since listing.[6] As I watched peregrines stream across the sky, listened to ʻAlalā murmur at dusk, and laughed at the antics of condors, I was conscious always that such experiences might never have been possible without the protections the ESA bestowed on these species. Beloved birds such as the Bald Eagle, Whooping Crane, Brown Pelican, Nēnē, and Kirtland's Warbler likewise owe their turnarounds to the foresight of those who championed this powerful law conserving our biological heritage.

Although the ESA has been subjected to repeated efforts to weaken or eliminate it since its inception, the law remains broadly popular, with roughly four out of five Americans supporting it and only one in ten opposing it. Support for the law has remained stable over the last two decades.[7] Nevertheless, despite its enduring strength and its remarkable success at protecting our rarest wildlife, the ESA is not designed to forestall widespread bird population declines. Indeed, while populations of threatened and endangered birds placed under ESA protection between 1967 and 1970 *increased* by an average of 624 percent, sensitive bird species that are not protected by the ESA have undergone significant declines.[8]

In the last fifty years, North America has lost almost three billion—or nearly a third—of its birds, from almost every type of habitat. Grassland birds have undergone the steepest declines—more than 700 million adult birds, comprising thirty-one species, have disappeared—followed by boreal forest and western forest birds. Birds inhabiting Arctic tundra, eastern forests, arid lands, and coasts also have declined precipitously. Disturbingly, habitat generalists—birds not tied to particular habitats—have disappeared, as well. Notable exceptions to these losses include waterfowl and raptors, two groups that have received targeted conservation efforts—including habitat restoration, the banning of lead shot, protection from persecution, and the elimination of certain pesticides—and subsequently experienced population increases.[9]

Perhaps the most distressing feature of North America's catastrophic bird losses is the severe population declines of birds that were long considered "common." More than 90 percent of the total cumulative loss of our avifauna can be attributed to twelve bird families, including sparrows, wood warblers, blackbirds, larks, finches, flycatchers, thrushes, and aerial insectivores such as swallows, nightjars, and swifts. The loss of formerly abundant, common birds is likely to degrade and disrupt the functionality of natural habitats. While habitat loss (particularly from industrial farming, logging, and urbanization) is the leading cause of bird declines, our birdlife has experienced death by a thousand cuts, including invasive species, direct persecution, declining food resources, outdoor cats, pesticides and toxins, collisions, and climate change.

It is easy to feel disheartened by global wildlife declines, disappearing species, our exponentially growing and consuming population, and the warming climate. Climate change poses an existential threat not only to birds but to all life on our planet. Nevertheless, while addressing climate change is among the most compelling issues we face, it behooves us not to ignore the many other threats that endanger our birds. Today's stressors are often complex, intertwined, and widespread, particularly for migratory birds. But history has shown that legislation and targeted conservation actions work.[10] And given the current magnitude of bird declines, improving habitats and reducing threats to birds through widespread, individual actions is more critical than ever to helping prevent the collapse of our global birdlife.

Each of us can help birds in ways that may seem insignificant when conducted in isolation but which can change the downward trajectory for birds (and by extension, other organisms) when conducted globally by an army of us. We can keep our cats indoors; put decals on our windows to prevent bird collisions; avoid using pesticides; provide food, shelter, water, and nest boxes for birds around our homes; grow native plants and weed out exotics; avoid cutting trees and brush or mowing fields during the bird breeding season; reduce or eliminate outdoor lighting; drive less and walk or bike more; eat foods whose production is less detrimental to the environment; use nonlead ammunition; support our national wildlife refuges, national parks, and open spaces; and contribute to organizations that work to protect birds and their habitats. We can also be mindful of birds and avoid disturbing them when they are nesting, or adding to their stress when they are vulnerable.

The small changes we make in our lives to help the chickadees that grace our feeders in winter, the meadowlarks that fill our spring mornings with song, the hummingbirds that brighten our gardens, and the thrushes that serenade us in deep woods are repaid to us a thousandfold. Birds disperse seeds, shaping the habitats around us. They pollinate our crops, control our pests, clean up our landscapes, and provide significant infusions into our local and national economies. But their intangible gifts are worth infinitely more. Birds mark the passing of our seasons, awe us with their migrations, dazzle us with their diversity, intrigue us with their behavior, satisfy our competitive natures, evoke our protective instincts, impress us with their resilience, and stir us with their fragility. Their vibrant presence and visible activities affect us in ways that are often profound and transformative. With each brilliant feather that catches our eye, each song that captures our ear, each movement that sparks our curiosity, birds repay us just by going about their daily lives. If we harness our collective goodwill to help the birds that enrich *our* lives, we might be able to save *theirs*. And by saving the birds, we might just save ourselves—and the rest of life on our glorious blue-green planet, as well.

Binoculars hanging from my neck, a small notebook in my pocket, I hike up the rock-strewn trail, following my two border collies. To my right, craggy cliffs loom skyward, White-throated Swifts darting in and out of invisible crevices. To my left, the trail drops away into a steep-sided canyon where a raging chute of whitewater shuttles winter's snowfall down to a rapidly swelling river. I am in mountain country. Home. And I'm on my way to spend a bit of time with dippers.

The American Dipper was conservationist and nature writer John Muir's favorite bird. Muir described the bird, which he knew as the Water Ouzel, as a "singularly joyous and lovable little fellow . . . clad in a plain waterproof suit of bluish gray, with a tinge of chocolate on the head and shoulders." North America's only truly aquatic songbird, the dipper inhabits western mountain streams. It can forage like a duck, gleaning insects off the surface of calm pools, or dive like a nano-penguin into turbulent waters to catch small fish or pry aquatic insect larvae from subsurface rocks. Singing year-round, even in the depths of winter when ice and snow partially cover its watery world, no bird cheered Muir more in his wanderings through California's Sierra Nevada mountains. He dedicated a chapter to the "mountain stream's own darling, the hummingbird of blooming waters," in his book *The Mountains of California*.

Muir's enchanting descriptions dance through my mind while I walk along contemplating my own dipper days, when I studied the bird for my master of science (MS) degree. I remember tying a rope to my waist in a half-hearted attempt to keep myself safe, as I inched along a cliff ledge bordering the water chute above which I'm now walking to measure stream dimensions. I remember navigating these wild waters, clambering over and under downfall—a stalwart field assistant by my side—as we hiked endless miles in our waders searching for dipper nests. I remember countless happy hours sitting quietly on stream banks watching dipper parents bring food to their young. I remember slipping on slick rocks and plunging into icy waters up to my neck one February, when running to extricate a dipper from a net that I'd strung across a stream so I could capture and mark the birds with colored leg bands. I remember the wonder of finding a banded female dipper, which had raised a brood under a busy highway bridge in the valley earlier in the season, nesting again miles upstream, on a quiet wilderness cliff.[11]

I sit down by the water's edge facing a cliff where I've watched nesting dippers on and off for nearly twenty-five years. My border collies lie behind me, alert but quiet, as I raise my binoculars and look at the mossy nest the dippers have built on a cliff ledge on the other side of the roiling stream. I spot the male resting just downstream. He bobs up and down—once, twice—his eyelids flashing white feathers as he blinks. Moments later the female flies in from upstream, a pine needle in her beak. Unfazed by my presence, she disappears into the entrance hole of her large, domelike nest to build and shape the inner cup that will soon hold her eggs.

As I watch, captivated, content, I realize that aside from the obvious—the colors, the songs, the endlessly intriguing behaviors—birds have given me something more: stories. As I followed the feather trails that wove the tapestry of my life, birds gave me moments and experiences that shaped me and defined me. They gave me stories that enchanted me and stories that saddened me. They told me stories of what ailed them and stories of what made them thrive. They gave me stories to treasure and stories I could share with others. Everyone who watches birds has stories—about the bird that sparked their interest, about the bird they misidentified, about the bird that made them aware of something they'd never noticed before. The world contains a treasure trove of over 10,000 bird species about which we can learn and whose stories we can share. And just as pebbles cast across a clear pool create expanding ripples that spread outward from their source, so too may our stories—and our passion for birds—touch others, opening their eyes to the feathered world, and inspiring them to care—and conserve.

I stand slowly, joints a little stiff from sitting so long in the cold. My border collies bound forward joyously, eager to resume their hike. Eyes and ears alert to birds, I head on up the trail.

Acknowledgments

While my incandescent experiences working with Peregrine Falcons, Hawaiian Crows, and California Condors will forever loom large in my psyche, my contributions were infinitesimally small compared to the many who dedicated years, careers, and even lives to recovering these imperiled birds. We are all indebted to these conservation heroes for their tireless and selfless work.

I am grateful to Chelsea Green Publishing for its enthusiastic support of *Feather Trails*, its talented publication and marketing teams, and its professionalism. I feel fortunate to have been able to entrust this book to Matthew Derr, a passionate birder and insightful editor. Matthew and project manager Natalie Wallace steered *Feather Trails* toward publication with grace, good cheer, and unequivocal support. Thank you to copy editor Diane Durrett and proofreader Angela Boyle for their attention to detail and for smoothing out some of my rougher sentences. Melissa Jacobson did a beautiful job designing the cover. Thanks to Kirsten Drew, Sean Maher, Darrell Koerner, and Rosie Baldwin for launching the book into the world. I'm also grateful to Eliza Haun and Amalia Herren-Lage for their contributions along the way.

Thank you to biologist and friend Kira Cassidy, for her wonderful bird silhouettes and for designing my *Words for Birds* logo, which was used as a heading for the Preface and Epilogue.

Heartfelt thanks to those who answered my questions and verified my information on peregrines and peregrine recovery: Dan Berger, Barb Franklin, Grainger Hunt, and Bryan Kimsey. Thanks especially to Brian Mutch for answering my questions and inspiring me as a field biologist. It was a pleasure and privilege to engage with the abovementioned peregrine heroes. Thank you, too, to US Fish and Wildlife Service (USFWS) historian Mark Madison for information about the peregrine's federal protection, and Karissa Kovner (EPA) for information about international pesticide regulations.

ACKNOWLEDGMENTS

I am particularly indebted to the late Tom Cade, who played an instrumental role in recovering Peregrine Falcons and founded The Peregrine Fund. Not only was he unfailingly supportive and appreciative of my efforts with California Condors, but he also reviewed the peregrine section of *Feather Trails*. I relished our many conversations about peregrines, condors, and Hawaiian Crows.

I thank John Marzluff for his support in my early career as a biologist, for answering questions, and, above all, for making my work with the 'Alalā possible. I am profoundly grateful to Jay Nelson (USFWS), who provided me with invaluable information about Hawaiian Crows and their reintroductions in the 1990s. After the San Diego Zoo Wildlife Alliance inexplicably refused to answer my questions or allow me to speak to any of their captive-breeding personnel, Jay's openness, transparency, and willingness to look up program details for me were refreshing and helpful. Jay made me grateful, as ever, to the many unsung heroes in the US Fish and Wildlife Service and other government agencies who do their wildlife work with professionalism and dedication, despite scarce resources and often insurmountable odds.

Biologists Paul Banko (US Geological Survey) and Donna Ball (USFWS) were also generous with their time and expertise, answered many questions, and responded to a slew of emails. I'm grateful for their help and for their outsized roles in conserving Hawaiian Crows. Thanks, too, to biologist Glenn Klingler for allowing me to repeat his words about the 'Alalā's extinction in the wild. Steve Holmer and Brad Keitt (American Bird Conservancy) weighed in about conservation funding in Hawaii. I am glad to have worked with Peter Harrity and enjoyed his family's gracious hospitality during my time in Hawai'i. Peter kindly answered many questions, and I enjoyed reminiscing with him about our 'Alalā work. I am indebted to Sheila Conant—who witnessed the extinction of seven bird species in Hawaii and shares my love of border collies—for reviewing my 'Alalā section at a difficult time.

I am grateful to Rick Watson for hiring me as The Peregrine Fund's South America biologist and for recruiting me for my first condor work. Shawn Farry forever earned my gratitude and admiration when he introduced me to condors and first enlightened me about their behavior and management. I thank Mike Clark (Los Angeles Zoo) for generously sharing his unparalleled expertise in raising and caring for condors. Thanks also to Joe Burnett (Ventana Wildlife Society), Chris

Parish (The Peregrine Fund), and the late Mike Wallace for answering my questions and for their decades of dedication to condors. Steve Kirkland, Arianna Punzalan (both with USFWS), and Tim Hauck (The Peregrine Fund) also fielded some of my condor questions. I'm grateful to Allan Mee for information about California's nestlings and for great condor conversations. Paleontological experts Jim Mead and Steve Emslie generously discussed Grand Canyon condors in the Pleistocene and beyond, and shared resources with me. Thanks to medical toxicologists Michael Kosnett and Steven Gilbert for speaking to me about the effects of lead on wildlife and people, and to Jeff Miller (Center for Biological Diversity) for sharing his expertise and clarifying my understanding of lead regulation and policy. Any errors or deficiencies in *Feather Trails* that escaped correction are my sole responsibility.

Jan Hamber spent years watching wild condors, beginning in the late 1970s, and inspired and mentored countless biologists—particularly women. She was always available to answer my questions (and was still going into her Santa Barbara office in her nineties). I'm grateful for her friendship, encyclopedic knowledge, and lifelong dedication to condors.

Many of the condor field technicians with whom I worked earned my profound admiration and gratitude, and became dear friends. I particularly thank Jill Adams, Brandon Breen, Chris Crowe, Marta Curti, Kevin Fairhurst, and Kris Lightner, all of whom supported and encouraged me as I wrote *Feather Trails*. I cherish our memories and long-standing friendships. Fellow writers Brandon and Marta also gave me writing support and encouragement, as did friend and former Idaho coworker Dianne Van Dien. I'm especially grateful to biologist Susan Adams, with whom I coedited a forthcoming anthology of field stories by women biologists while I was working on *Feather Trails*, for her support, friendship, and all we've learned together about writing.

Among my dearest friends since our university days, Lois Morrison and Catherine Murphy have given me unparalleled support, in ways large and small, over these many years. Other dear friends who have offered support, love, and encouragement along the way include Ginny Hazen, Jen Marangelo, Beth Telford, Kristin Hall, Alea Doolittle, Micki Long, and Kim Szcodronski. I am particularly grateful to Diane Thomas, who started out as a supervisor and quickly became a friend. She shared in my condor stories, provided moral support, and cast a skilled editorial eye over much of the early manuscript.

Thanks to my half sisters, Kimberley Berlin and Dana Osborn, for their love, enthusiasm about my writing, and many kindnesses to me. I am grateful, too, for the love and support of my brother-in-law Ky Koitzsch (go Eagles!) and my uncles, Julian and Andrew Heath.

A very special thank you to my British mum—Elisabeth Osborn—who supported my every endeavor (baffling as they often were to her) with unconditional love, admiration, and pride. She would have loved to know that I had written this book (Condor 210 was her favorite bird). I hope that I may be a small echo of her singular compassion and grace.

Above all, I thank my beloved sisters and best friends, Lisa Osborn Koitzsch and Natasha Osborn. They shared in all my ups and downs as I lived and then wrote about my field work. Their enthusiastic encouragement, unparalleled support, and unwavering belief in me helped to make this book possible.

Disclaimers

Feather Trails is by no means a comprehensive portrayal of the countless individuals and entities that have worked to safeguard the bird species about which I've written. I have only been able to highlight some of those who were involved, including those who held a special interest for me.

To protect people's privacy, I sometimes did not include surnames.

Over time, scientific research contributes new knowledge, websites are revised, previously available online materials are deleted, and names of organizations, birds, and places change. I have tried to keep up with these changes and provide the most up-to-date information possible, but given the scope of *Feather Trails*, some items may well have been missed.

Notes

Preface

1. R. W. Fyfe, S. A. Temple, and T. J. Cade, "The 1975 North American Peregrine Falcon Survey," *Canadian Field-Naturalist* 90 (1976): 228–73; R. B. Berry, "History and Extinction of the Appalachian Peregrine," in *Return of the Peregrine: A North American Saga of Tenacity and Teamwork*, eds. T. Cade and W. Burnham (The Peregrine Fund, 2003), 35–55; J. H. Enderson, "Early Peregrine Experiences and Management: Rocky Mountain West," in *Return of the Peregrine*, 57–71.

Chapter 1. Beginnings

1. C. M. White et al., "Peregrine Falcon (*Falco peregrinus*)," in Birds of North America online, eds. A. Poole and F. Gill (Cornell Lab of Ornithology, 2002), https://doi.org/10.2173/bna.660.
2. White et al., "Peregrine Falcon (*Falco peregrinus*)."
3. J. J. Hickey, "Eastern Population of the Duck Hawk," *Auk* 59, no. 2 (1942): 176–204, https://doi.org/10.2307/4079550; R. M. Bond, "The Peregrine Population of Western North America," *Condor* 48, no. 3 (1946): 101–16, https://doi.org/10.2307/1364462; D. Ratcliffe, *The Peregrine Falcon*, 2nd ed. (T & A D Poyser, Ltd, 1993); White et al., "Peregrine Falcon (*Falco peregrinus*)."
4. T. Cade, "Life History Traits of the Peregrine in Relation to Recovery," in *Return of the Peregrine*, 2–11.
5. L. F. Kiff, "Commentary: Changes in the Status of the Peregrine in North America: An Overview," in *Peregrine Falcon Populations: Their Management and Recovery*, eds. T. J. Cade et al. (The Peregrine Fund, 1988), 123–39; White et al., "Peregrine Falcon (*Falco peregrinus*)."
6. C. Savage, *Peregrine Falcons* (San Francisco: Sierra Club Books, 1992).
7. Ratcliffe, *The Peregrine Falcon*, 2nd ed.
8. Savage, *Peregrine Falcons*; Peregrine Falcons were not initially protected by the 1918 Migratory Bird Treaty Act (16 U.S.C. 703-712; 40 Stat. 755)—the signature federal act that today protects all North America's native migratory birds. The species was first listed as an endangered species in 1970 (35 Federal Register 8495; June 2, 1970) under the authority of the 1969 Endangered Species Conservation Act (Public Law 91-135,835 Stat. 275). Peregrines received federal protection under the Migratory Bird Treaty Act (MBTA) on March 10, 1972 (23 U.S.T. 260; T.I.A.S. 7302), when thirty-two families of birds—including eagles, hawks, owls, and corvids—were added to a 1936 amendment of the MBTA (the Convention between the United States of America and the United Mexican States for the Protection of Migratory Birds and Game Mammals; 50 Stat. 1311; TS 912).
9. Ratcliffe, *The Peregrine Falcon*, 2nd ed. According to Ratcliffe (p. 65), "Under the war-time emergency defence regulations, the then Secretary of State for Air made on 1 July 1940 the Destruction of Peregrine Falcons Order, 1940 [No. 1164], which expired in February 1946, after the war had ended. This Order made it lawful for any properly authorised person to take or destroy Peregrines or their eggs in certain parts of Britain." A copy of this order is provided in R. B. Treleaven, *Peregrine: The Private Life of the Peregrine Falcon* (Headland Publications, 1977), 147.

10. C. M. White, P. D. Olsen, and L. F. Kiff, "Family Falconidae (Falcons and Caracaras)," in *Handbook of the Birds of the World. Vol. 2. New World Vultures to Guineafowl*, eds. J. del Hoyo, A. Elliott, and J. Sargatal (Lynx Edicions, 1994), 216–78.

11. Wikipedia, "Pigeon Keeping," http://en.wikipedia.org/wiki/Pigeon_keeping; A. Pettie, "Mike Tyson: Pugilist and Pigeon Fancier, Discover Channel, Review," *Telegraph*, March 11, 2011, http://www.telegraph.co.uk/culture/tvandradio/8376790/Mike-Tyson-pugilist-and-pigeon -fancier-Discovery-Channel-review.html.

12. Occasional illegal persecution of peregrines and other raptors by pigeon fanciers continues to this day, as do the fanciers' requests to relax laws protecting peregrines and other raptors. See, for example, P. Stirling-Aird, *Peregrine Falcon* (Firefly Books, 2012). In 2007, a group of pigeon fanciers (or "roller pigeon" enthusiasts) was charged with killing 1,000 to 2,000 raptors annually in the US. See Bruce Barcott, "Secret Agent Man: Inside the High-Stakes Life of Undercover Wildlife Agents," *Backpacker*, December 2, 2021, https://www.backpacker.com/stories/adventures/danger /secret-agent-man.

13. Britain's Protection of Birds Act (1954) included total protection for peregrines and their eggs.

14. Ratcliffe, *The Peregrine Falcon*, 2nd ed.; Savage, *Peregrine Falcons*.

15. J. J. Hickey, "Some Recollections about Eastern North America's Peregrine Falcon Population Crash," in *Peregrine Falcon Populations*, 9–16; Ratcliffe, *The Peregrine Falcon*, 2nd ed.; D. Ratcliffe, "Discovering the Causes of Peregrine Decline," in *Return of the Peregrine*, 23–33.

16. Ratcliffe, *The Peregrine Falcon*, 2nd ed.

17. Ratcliffe, *The Peregrine Falcon*, 2nd ed.

18. Ratcliffe, *The Peregrine Falcon*, 2nd ed.

19. Hickey, "Some Recollections," 9–16.

20. Hickey, "Some Recollections," 9–16; S. A. Temple, "In Memoriam—Joseph Hickey 1907–1993" (University of Wisconsin, 1993), http://images.library.wisc.edu/EcoNatRes/EFacs/PassPigeon /ppv55no04/reference/econatres.pp55n04.stemple.pdf.

21. D. Berger, "The 1964 Survey of Peregrine Eyries in the Eastern United States," in *Return of the Peregrine*, 40–41.

22. T. J. Cade et al., "Commentary: The Role of Organochlorine Pesticides in Peregrine Population Changes," in *Peregrine Falcon Populations*, 463–68; Hickey, "Some Recollections," 9–16; Berger, "The 1964 Survey of Peregrine Eyries," 40–41. Although Berger and Sindelar documented the dramatic disappearance of the eastern Peregrine Falcon, it is unlikely that it was entirely extirpated by 1964. The survey team did not check *all* known aeries and may have missed some early nest failures since part of their survey occurred later in the nesting season. Indeed, W. Spofford and J. Enderson documented an adult peregrine at Lake Willoughby in 1970.

23. T. J. Cade et al., "Preface," *Peregrine Falcon Populations*, vii–viii; Savage, *Peregrine Falcons*.

24. T. Cade and W. Burnham, "Introduction: The Madison Peregrine Conference and the Struggle to Ban DDT," in *Return of the Peregrine*, 13–21.

Chapter 2. Hacking

1. B. Heinrich, *Ravens in Winter* (Summit Books, 1989).

2. S. K. Sherrod et al., *Hacking: A Method for Releasing Peregrine Falcons and Other Birds of Prey* (The Peregrine Fund, 1982).

3. Sherrod et al., *Hacking*.

4. White et al., "Peregrine Falcon (*Falco peregrinus*)."

5. Among raptors, the degree of the size difference between males and females (sexual dimorphism) is related to the speed, agility, and availability of the prey upon which the birds feed. Carrion-, snail-, and insect-eating raptors are less sexually dimorphic in terms of size than are reptile-eaters, which in turn are less dimorphic than mammal-, fish-, and bird-eaters. Raptors that

specialize in eating birds exhibit the most sexual size dimorphism of all; I. Newton, *Population Ecology of Raptors* (Buteo Books, 1979); E. J. Temeles, "Sexual Size Dimorphism of Bird-Eating Hawks: The Effect of Prey Vulnerability," *American Naturalist* 125, no. 4 (1985): 485–99.

6. C. M. White and T. Cade, "Cliff-Nesting Raptors and Ravens along the Colville River in Arctic Alaska," *Living Bird* 10 (1971): 107–50; N. F. R. Snyder and J. W. Wiley, "Sexual Size Dimorphism in Hawks and Owls of North America," *Ornithological Monograph* 20 (1976); Newton, *Population Ecology of Raptors*; Temeles, "Sexual Size Dimorphism."

7. Newton, *Population Ecology of Raptors*; M. Andersson and R. A. Norberg, "Evolution of Reversed Sexual Dimorphism and Role Partitioning among Predatory Birds, with a Size Scaling of Flight Performance," *Biological Journal of the Linnean Society* 15 (1981): 105–30.

8. M. P. Jones, K. E. Pierce Jr., and D. Ward, "Avian Vision: A Review of Form and Function with Special Consideration to Birds of Prey," *Journal of Exotic Pet Medicine* 16, no. 2 (2007): 69–87, https://doi.org/10.1053/j.jepm.2007.03.012.

9. J. A. Waldvogel, "The Bird's Eye View," *American Scientist* 78, no. 4 (1990): 342–53; Jones, Pierce, and Ward, "Avian Vision."

10. Rods and cones are specialized cells that detect light, converting it into signals that trigger biological processes related to vision. Rods function primarily in low light conditions and allow black-and-white vision, whereas cones function mainly during the day and allow color vision.

11. H. C. Howland, S. Merola, and J. R. Basarab, "The Allometry and Scaling of the Size of Vertebrate Eyes," *Vision Research* 44, no. 17 (2004): 2043–65, https://doi.org/10.1016/j.visres.2004.03.023; M. Ruggeri et al., "Retinal Structure of Birds of Prey Revealed by Ultra-High Resolution Spectral-Domain Optical Coherence Tomography," *Investigative Ophthalmology and Visual Science* 51, no. 11 (2010): 5789–95, http://doi.org/10.1167/iovs.10-5633; T. Birkhead, *Bird Sense: What It's Like to Be a Bird* (Bloomsbury, 2012).

12. Treleaven, *Peregrine*, 68–70.

13. Other birds that have two foveae include hummingbirds, kingfishers, swallows, and shrikes. Birkhead, *Bird Sense*.

14. V. A. Tucker, "The Deep Fovea, Sideways Vision and Spiral Flight Paths in Raptors," *Journal of Experimental Biology* 203, no. 24 (2000): 3745–54, https://doi.org/10.1242/jeb.203.24.3745; Birkhead, *Bird Sense*.

15. V. A. Tucker et al., "Curved Flight Paths and Sideways Vision in Peregrine Falcons (*Falco peregrinus*)," *Journal of Experimental Biology* 203, no. 24 (2000): 3755–63, https://doi.org/10.1242/jeb.203.24.3755.

16. Cade and Burnham, eds., *Return of the Peregrine*.

17. T. Cade, "Starting The Peregrine Fund at Cornell University and Eastern Reintroduction," in *Return of the Peregrine*, 73–103.

18. Cade, "Starting The Peregrine Fund," 73–103.

19. Cade, "Starting The Peregrine Fund," 73–103.

20. Sherrod et al., *Hacking*; L. W. Oliphant and W. J. P. Thompson, "The Use of Falconry Techniques in the Reintroduction of the Peregrine," in *Peregrine Falcon Populations*, 611–17; B. W. Walton, "Restoration of the Peregrine Population in California," in *Return of the Peregrine*, 155–71.

21. Sherrod et al., *Hacking*; The Santa Cruz Predatory Bird Research Group, "Peregrine Falcon Release Methods." This webpage is no longer available on the SCPBR website but the information and a link to the archived page is available at: Wikipedia, "Hack (Flaconry)," last updated September 7, 2023, https://en.wikipedia.org/wiki/Hack_(falconry).

22. S. K. Sherrod, *Behavior of Fledgling Peregrines* (The Peregrine Fund, 1983); W. R. Spofford and D. Amadon, "Live Prey to Young Raptors—Incidental or Adaptive?" *Journal of Raptor Research* 27, no. 4 (1993): 180–84.

Chapter 3. Release

1. Sherrod et al., *Hacking*.
2. A. D. Mills et al., "The Behavior of the Japanese or Domestic Quail *Coturnix japonica*," *Neuroscience and Biobehavioral Reviews* 21, no. 3 (1997): 261–81, https://doi.org/10.1016 /s0149-7634(96)00028-0.
3. Sherrod et al., *Hacking*.
4. Newton, *Population Ecology of Raptors*; Sherrod et al., *Hacking*; Sherrod, *Behavior of Fledgling Peregrines*.
5. Personal communication with Bryan Kimsey, June 16, 2014.
6. Sherrod et al., *Hacking*.
7. R. Oakleaf and G. R. Craig, "Peregrine Restoration from a State Biologist's Perspective," in *Return of the Peregrine*, 297–311.

Chapter 4. Sky Birds

1. Peregrine vocalizations are described and explained in White et al., "Peregrine Falcon (*Falco peregrinus*)."
2. White et al., "Peregrine Falcon (*Falco peregrinus*)." Also references therein.
3. R. Oakleaf, A. Orabona, and A. Lyon-Holloran, "Conservation of birds in Wyoming," in D. W. Faulkner, *Birds of Wyoming* (Roberts and Company, 2010), 13–17.
4. Oakleaf and Craig, "Peregrine Restoration," 297–311.
5. C. M. White, "Peregrine Falcon," in *Handbook of the Birds of the World. Vol. 2. New World Vultures to Guineafowl*, 274–75; personal communication with Grainger Hunt, May 14, 2014.
6. Sherrod et al., *Hacking*; Sherrod, *Behavior of Fledgling Peregrines*.
7. L. Brown, *British Birds of Prey* (Harper Collins, 1976).
8. Cade, "Life History Traits of the Peregrine in Relation to Recovery," 2–11; P. Capainolo and C. A. Butler, *How Fast Can a Falcon Dive? Fascinating Answers to Questions about Birds of Prey* (Rutgers University Press, 2010).
9. Capainolo and Butler, *How Fast Can a Falcon Dive?*
10. Guinness Book of World Records (online), https://www.guinnessworldrecords.com/world -records/fastest-speed-for-a-race-horse.
11. C. M. White and R. W. Nelson, "Hunting Ranges and Strategies in a Tundra Breeding Peregrine and Gyrfalcon Observed from a Helicopter," *Journal of Raptor Research* 25, no. 3 (1991): 49–62.
12. K. Franklin, "Vertical Flight," *Journal of the North American Falconers' Association* 38 (1999): 68–72.
13. Cade, "Life History Traits of the Peregrine in Relation to Recovery," 2–11.
14. A. P. Payne, "The Harderian Gland: A Tercentennial Review," *Journal of Anatomy* 185, no. 1 (1994): 1–49; I. R. Schwab and D. Maggs, "The Falcon's Stoop," *British Journal of Ophthalmology* 88, no. 1 (2004): 4.
15. Capainolo and Butler, *How Fast Can a Falcon Dive?*

Chapter 5. Osiris

1. Sherrod et al., *Hacking*; Sherrod, *Behavior of Fledgling Peregrines*.
2. Sherrod, *Behavior of Fledgling Peregrines*.
3. Stirling-Aird, *Peregrine Falcon*.
4. N. Snyder and H. Snyder, *Raptors: North American Birds of Prey* (Voyageur Press, 1991); White et al., "Peregrine Falcon (*Falco peregrinus*)."
5. Sherrod et al., *Hacking*.
6. Sherrod, *Behavior of Fledgling Peregrines*.
7. Sherrod et al., *Hacking*.

8. According to Tom Cade (personal communication), there ultimately were a few cases of early departing birds (within two weeks of release) that subsequently were found in the reintroduced population.

9. Stirling-Aird, *Peregrine Falcon*.

10. For example, Treleaven, *Peregrine*, 68–70.

11. White, Olsen, and Kiff, "Family Falconidae (Falcons and Caracaras)," 216–47.

12. White and Nelson, "Hunting Ranges and Strategies."

13. White, Olsen, and Kiff, "Family Falconidae (Falcons and Caracaras)," 216–47; Capainolo and Butler, *How Fast Can a Falcon Dive?*

Chapter 6. A World Full of Poisons

1. D. Ratcliffe, "Broken Eggs in Peregrine Eyries," *British Birds* 51, no. 1 (1958): 23–26; Hickey, "Some Recollections," 9–16.

2. Hickey, "Some Recollections," 9–16; D. Ratcliffe, "Discovering the Causes of Peregrine Decline," in *Return of the Peregrine*.

3. According to Dennis Ratcliffe, although DDT did not come into widespread agricultural use in Britain until 1948 or 1949, military stockpiles of the powerful insecticide flooded the market after World War II ended in 1945 and quickly found wide-scale use in domestic, horticultural, and veterinary applications. Incredibly, as early as March 1946, *The Racing Pigeon* magazine contained an advertisement for a DDT formulation designed to control homing pigeon ectoparasites, so some peregrines may even have ingested DDT directly while consuming their prey. Indeed, Ratcliffe later tested pigeon leg bands that he found in peregrine aeries after DDT had come into widespread use and found traces of the chemical and its breakdown product (DDE) on two of them, suggesting some peregrines had eaten pigeons treated externally with DDT. (Use of DDT on pigeons ultimately was stopped because of suspicions about harmful effects.) DDT enjoyed such widespread use in Britain by 1947 that it had a high potential for contaminating a significant proportion of that nation's peregrine population. Ratcliffe, *The Peregrine Falcon*, 2nd ed, 333.

4. D. Ratcliffe, "Decrease in Eggshell Weight in Certain Birds of Prey," *Nature* 215 (1967): 208–10; Ratcliffe, *The Peregrine Falcon*, 2nd ed.; Hickey, "Some Recollections," 9–16; R. W. Risebrough, "Pesticides and Bird Populations," in *Current Ornithology, Vol 3*, ed. R. F. Johnston (Plenum Press, 1986).

5. The Bald Eagle was first given federal protection by the Bald Eagle Protection Act of 1940. This act subsequently was amended in 1962 to include Golden Eagles since nonexperts often were unable to distinguish Golden Eagles and immature Bald Eagles, leading to many inadvertent Bald Eagle deaths. In its current form, the law, which has been amended multiple times, prohibits the "take"; possession; sale; purchase; barter; offer to sell, purchase, or barter; transport; or export or import of any Bald or Golden Eagle, alive or dead, including any part, nest, or egg, unless allowed by permit (16 U.S.C. 668(a); 50 CFR 22). "Take" includes the pursuit, shooting, shooting at, poisoning, wounding, killing, capturing, trapping, collecting, molesting, or disturbing of any eagles (16 U.S.C. 668c; 50 CFR 22.3).

6. J. J. Hickey and D. Anderson, "Chlorinated Hydrocarbons and Eggshell Changes in Raptorial and Fish-Eating Birds," *Science* 162, no. 3850 (1968): 271–73, https://doi.org/10.1126/science .162.3850.271; Hickey, "Some Recollections," 9–16.

7. S. Jensen, "Report of a New Chemical Hazard," *New Scientist* 32 (1966): 612; Hickey, "Some Recollections," 9–16.

8. Hickey and Anderson, "Chlorinated Hydrocarbons"; Hickey, "Some Recollections," 9–16.

9. T. J. Cade et al., "DDE Residues and Eggshell Changes in Alaskan Falcons and Hawks," *Science* 172, no. 3986 (1971): 955–57.

10. L. Holm et al., "Embryonic Exposure to o,p'-DDT Causes Eggshell Thinning and Altered Shell Gland Carbonic Anhydrase Expression in the Domestic Hen," *Environmental Toxicology and*

Chemistry 25, no. 10 (2006): 2787–93, https://doi.org/10.1897/05-619R.1. DDE appears to cause eggshell thinning by disrupting normal functions of the shell gland that makes the hard calcareous shell of a bird's egg in the oviduct; D. B. Peakall, "DDE-Induced Eggshell Thinning: An Environmental Detective Story," *Environmental Reviews* 1, no. 1 (1993): 13–20, https://doi .org/10.1139/a93-002.

11. R. G. Heath, J. W. Spann, and J. F. Kreitzer, "Marked DDE Impairment of Mallard Reproduction in Controlled Studies," *Nature* 224 (1969): 47–48, https://doi.org/10.1038/224047a0.

12. J. L. Lincer, "DDE-Induced Eggshell-Thinning in the American Kestrel: A Comparison of the Field Situation and Laboratory Results," *Journal of Applied Ecology* 12, no. 3 (1975): 781–93, https://doi.org/10.2307/2402090.

13. Ratcliffe, *The Peregrine Falcon*, 2nd ed; D. B. Peakall, L. M. Reynolds, and M. C. French, "DDE in Eggs of the Peregrine Falcon," *Bird Study* 23, no. 3 (1976): 181–86, https://doi.org/10.1080 /00063657609476499; D. Peakall, "Settling the Matter Beyond Doubt," in *Return of the Peregrine*, 28–29.

14. R. W. Risebrough and D. B. Peakall, "The Relative Importance of the Several Organochlorines in the Decline of Peregrine Falcon Populations," in *Peregrine Falcon Populations*, 449–62; Ratcliffe, *The Peregrine Falcon*, 2nd ed.; White et al., "Peregrine Falcon (*Falco peregrinus*)."

15. T. R. Dunlap, *DDT* (Princeton University Press, 1981), 7.

16. R. Carson, *Silent Spring* (Houghton Mifflin Company, 1962), 112.

17. D. S. Greenberg, "Pesticides: White House Advisory Body Issues Report Recommending Steps to Reduce Hazard to Public," *Science* 140, no. 3569 (1963): 878–79, https://doi.org/10.1126 /science.140.3569.878; Dunlap, *DDT*.

18. Dunlap, *DDT*; C. F. Wurster, *DDT Wars: Rescuing Our National Bird, Preventing Cancer, and Creating the Environmental Defense Fund* (Oxford University Press, 2015).

19. Dunlap, *DDT*; C. F. Wurster, "The Environmental Defense Fund and the Banning of DDT," in *Return of the Peregrine*, 19; Wurster, *DDT Wars*.

20. Dunlap, *DDT*; Wurster, *DDT Wars*.

21. Wurster, *DDT Wars*.

22. Although the US is a signatory to the Stockholm Convention, it has not become a party to it since Congress hasn't authorized the implementation of all its provisions. By failing to ratify the treaty, the US lost a key opportunity to be a leading voice in the international regulation of persistent organic pollutants.

23. Environmental Protection Agency, Pacific Southwest, Region 9: Superfund—Palos Verdes Shelf, http://yosemite.epa.gov/r9/sfund/r9sfdocw.nsf/3dec8ba3252368428825742600743733/e61d 5255780dd68288257007005e9422!OpenDocument; Risebrough, "Pesticides and Bird Populations," 397–427; L. Wires, *The Double-crested Cormorant: Plight of a Feathered Pariah* (Yale University Press, 2013).

24. Environmental Protection Agency, Pacific Southwest, Region 9: Superfund—Palos Verdes Shelf, http://yosemite.epa.gov/r9/sfund/r9sfdocw.nsf/3dec8ba3252368428825742600743733/e61d5 255780dd68288257007005e9422!OpenDocument.

25. J. Moir, "New Hurdle for California Condors May Be DDT from Years Ago," *New York Times*, November 15, 2010, http://www.nytimes.com/2010/11/16/science/16condors.html; L. J. Burnett et al., "Eggshell Thinning and Depressed Hatching Success of California Condors Reintroduced to Central California," *Condor* 115, no. 3 (2013): 477–91, https://doi.org /10.1525/cond.2013.110150.

26. Burnett et al, "Eggshell Thinning."

27. B. Bienkowski, "Songbirds Dying from DDT in Michigan Yards; Superfund Site Blamed," *Environmental Health News*, July 28, 2014, https://www.ehn.org/songbirds-dying-from-ddt-in -michigan-yards-superfund-site-blamed-2645660837.html.

28. High levels of flame retardants known as polybrominated biphenyls (PBBs) have been found along with DDT in St. Louis, Michigan, backyards. The Michigan Chemical Company, which manufactured PBBs as well as DDT, is infamous for causing a large-scale chemical disaster in early 1973, when it accidentally switched bags of PBBs with those of a cattle-feed supplement that the company also produced. By the time the mix-up was discovered (in April 1974), the PBBs had ended up in mills that supplied feed for dairy cows and had entered the food chain through milk and other products. As a result, 500 contaminated Michigan farms were quarantined and approximately 30,000 cattle, 4,500 swine, 1,500 sheep, and 1.5 million chickens were destroyed, along with associated produce and animal feed. Meanwhile, thousands of people were exposed to a persistent chemical that causes cancer in laboratory animals and has been linked to endocrine and reproductive problems. Michigan Department of Community Health, "PBBs (Polybrominated Biphenyls) in Michigan—Frequently Asked Questions—2011 update," https://www.michigan.gov/documents/mdch_PBB_FAQ_92051_7.pdf.

29. Bienkowski, "Songbirds Dying from DDT."

30. M. I. Goldstein et al., "Monocrotophos-Induced Mass Mortality of Swainson's Hawks in Argentina, 1995–1996," *Ecotoxicology* 8, no. 3 (1999): 201–14; M. J. Bechard et al., "Swainson's Hawk (*Buteo swainsoni*)," version 2.0, in Birds of North America online, ed. A. Poole (Cornell Lab of Ornithology, 2010), https://doi.org/10.2173/bna.265.

31. D. Pimentel et al., "Environmental and Economic Costs of Pesticide Use," *BioScience* 42, no. 10 (1992): 750–60, https://doi.org/10.2307/1311994.

32. "The State of the Birds 2014 Report," North American Bird Conservation Initiative, US Committee, US Department of Interior, 16 pages.

33. P. Mineau and M. Whiteside, "Lethal Risk to Birds from Insecticide Use in the United States—A Spatial and Temporal Analysis," *Environmental Toxicology and Chemistry* 25, no. 5 (2006): 1214–22, https://doi.org/10.1897/05-035R.1; P. Mineau and M. Whiteside, "Pesticide Acute Toxicity Is a Better Correlate of U.S. Grassland Bird Declines than Agricultural Intensification," *PLoS ONE* 8, no. 2 (2013): e57457, https://doi.org/10.1371/journal.pone.0057457.

34. J. M. Bonmatin et al., "Environmental Fate and Exposure; Neonicotinoids and Fipronil," *Environmental Science and Pollution Research* 22, no. 1 (2015): 35–67, https://doi.org/10.1007/s11356-014-3332-7.

35. P. Mineau and C. Palmer, "The Impact of the Nation's Most Widely Used Insecticides on Birds," *American Bird Conservancy*, 2013; C. Palmer, "Note to EPA: Modernize the Tests You Run on Pesticides," *Bird Calls* (the newsletter of the American Bird Conservancy) 17, no. 2 (2013): 3, https://abcbirds.org/wp-content/uploads/2015/04/bc13jul.pdf.

36. L. U. Chensheng, K. M. Warchol, and R. A. Callahan, "*In situ* Replication of Honey Bee Colony Collapse Disorder," *Bulletin of Insectology* 65, no. 1 (2012): 99–106; L. U. Chensheng, K. M. Warchol, and R. A. Callahan, "Sub-lethal Exposure to Neonicotinoids Impaired Honey Bees Winterization before Proceeding to Colony Collapse Disorder," *Bulletin of Insectology* 67, no. 1 (2014): 125–30; T. C. Van Dijk, M. A. van Staalduinen, and J. P. van der Sluijs, "Macro-Invertebrate Decline in Surface Water Polluted with Imidacloprid," *PLoS ONE* 8, no. 5 (2013): e62374, https://doi.org/10.1371/journal.pone.0062374; L. W. Pisa et al., "Effects of Neonicotinoids and Fipronil on Non-Target Invertebrates," *Environmental Science and Pollution Research* 22, no. 1 (2015): 68–102, https://doi.org/10.1007/s11356-014-3471-x.

37. D. Goulson, "Review: An Overview of the Environmental Risks Posed by Neonicotinoid Insecticides," *Journal of Applied Ecology* 50, no. 4 (2013): 977–87, https://doi.org/10.1111/1365-2664.12111.

38. T. Brown, S. Kegley, and L. Archer, "Gardeners Beware: Bee-Toxic Pesticides Found in 'Bee-Friendly' Plants Sold at Garden Centers Nationwide," 2013, Friends of the Earth, Washington, DC; Pisa et al, "Effects of Neonicotinoids."

39. In subsequent years, after significant public concern and activism, numerous large nurseries, landscaping companies, and retailers, such as Home Depot, Lowe's, and Whole Foods, committed to eliminating neonicotinoid pesticides from their plants, but the extent to which they have done so is uncertain." Lowe's Commits to Decisive Action to Protect Bees and Other Pollinators," *Friends of the Earth* (2015), https://foe.org/news/2015-04-lowes-commits-to-protect-bees-and-pollinators.

40. Goulson, "Review: An Overview of the Environmental Risks"; Mineau and Palmer, "The Impact of the Nation's Most Widely Used Insecticides"; C. Palmer, "Birds, Bees, and Aquatic Life Threatened by Gross Underestimate of Toxicity of World's Most Widely Used Pesticide," American Bird Conservancy press release, March 20, 2013, https://abcbirds.org/article/birds-bees-and-aquatic-life-threatened-by-gross-underestimate-of-toxicity-of-worlds-most-widely-used-pesticide-2.

41. M. L. Eng, B. J. M. Stutchbury, and C. A. Morrissey, "A Neonicotinoid Insecticide Reduces Fueling and Delays Migration in Songbirds," *Science* 365, no. 6458 (2019): 1177–80, https://doi.org/10.1126/science.aaw9419.

42. B. Jarvis, "The Insect Apocalypse Is Here—What Does It Mean for the Rest of Life on Earth," *New York Times*, November 27, 2018, https://www.nytimes.com/2018/11/27/magazine/insect-apocalypse.html; M. DiBartolomeis et al., "An Assessment of Acute Insecticide Toxicity Loading (AITL) of Chemical Pesticides Used on Agricultural Land in the United States," *PLoS ONE* 14, no. 8 (2019): e0220029, https://doi.org/10.1371/journal.pone.0220029.

43. A. Van Dam, "Wait, Why Are There So Few Dead Bugs on My Windshield These Days?" *Washington Post*, October 21, 2022, https://www.washingtonpost.com/business/2022/10/21/dead-bugs-on-windshields.

44. C. A. Hallmann et al, "Declines in Insectivorous Birds Are Associated with High Neonicotinoid Concentrations," *Nature* 511 (2014): 341–43, https://doi.org/10.1038/nature13531.

45. Beyond Pesticides, "Poisoned Waterways—The Same Pesticide That Is Killing Bees Is Destroying Life in the Nation's Streams, Rivers, and Lakes," Pesticides and You (Spring 2017), https://www.beyondpesticides.org/assets/media/documents/bp-37.1-PoisonedWaterways-cited3.pdf.

46. Y. Li, R. Miao, and M. Khanna, "Neonicotinoids and Decline in Bird Biodiversity in the United States," *Nature Sustainability* 3, no. 12 (2020): 1027–35, https://doi.org/10.1038/s41893-020-0582-x.

47. M. Chagnon et al., "Risks of Large-Scale Use of Systemic Insecticides to Ecosystem Functioning and Services," *Environmental Science and Pollution Research* 22, no. 1 (2015): 119–34, https://doi.org/10.1007/s11356-014-3277-x; D. Gibbons, C. Morrissey, and P. Mineau, "A Review of the Direct and Indirect Effects of Neonicotinoids and Fipronil on Vertebrate Wildlife," *Environmental Science and Pollution Research* 22, no. 1 (2015): 103–18, https://doi.org/10.1007/s11356-014-3180-5; Pisa et al., "Effects of Neonicotinoids"; T. J. Wood and D. Goulson, "The Environmental Risks of Neonicotinoid Pesticides: A Review of the Evidence Post 2013," *Environmental Science and Pollution Research* 24, no. 21 (2017): 17285–325, https://doi.org/10.1007/s11356-017-9240-x.

48. E. Stokstad, "European Union Expands Ban of Three Neonicotinoid Pesticides," *Science*, April 27, 2018, https://www.science.org/content/article/european-union-expands-ban-three-neonicotinoid-pesticides.

49. Originally slated to complete its review of these pesticides between 2016 and 2019, the EPA now estimates its review will be completed in 2024. All pesticides sold or distributed in the United States must first be registered with the EPA, unless they qualify for an exemption. The registration process involves an initial evaluation of the pesticide, its proposed usage, and its labeling to ensure that the chemical won't adversely affect humans or the environment, but conservationists point to many flaws in this process. See the following source for an update on the EPA's risk assessment for five neonicotinoids for which decisions originally were pending in 2019: EPA, "Schedule for

Review of Neonicotinoid Pesticides," last updated October 19, 2023, https://www.epa.gov/pollinator-protection/schedule-review-neonicotinoid-pesticides.

50. This Act has been reintroduced multiple times since 2013, including in the House as H.R. 1337 in February 2019, after which it was referred to the Subcommittee on Biotechnology, Horticulture, and Research. No further action appears to have been taken with this iteration of the bill. It was again reintroduced as H.R. 4277 in June 2023 and referred to the Committee on Agriculture and Natural Resources.

51. Other cities include Seattle, Washington, and Portland, Oregon. New Jersey and Maine have passed laws restricting neonicotinoid use.

52. H. Connor, "No Refuge—How America's National Wildlife Refuges Are Needlessly Sprayed with Nearly Half a Million Pounds of Pesticides Each Year," Center for Biological Diversity, May 2018, https://www.biologicaldiversity.org/campaigns/pesticides_reduction/pdfs/No-Refuge.pdf.

53. A. Price, "Trump's Interior Department Reverses Ban of Pesticides in Wildlife Refuges," Sierra magazine, (August 9, 2018), https://www.sierraclub.org/sierra/trump-s-interior-department-reverses-ban-pesticides-wildlife-refuges; Harvard Law School Environmental and Energy Law Program, "Neonicotinoid and GMOs Ban in National Wildlife Refuges," Regulatory Tracker (October 19, 2019), https://eelp.law.harvard.edu/2019/10/neonicotinoid-ban-in-national-wildlife-refuges.

54. T. J. Cade, "The Gyrfalcon and Falconry," Living Bird 7 (1968): 237–40.

Chapter 7. Delisted

1. White et al., "Peregrine Falcon, (Falco peregrinus)."
2. S. Healy and W. A. Calder, "Rufous Hummingbird (Selasphorus rufus)," in Birds of North America online, ed. A. Poole (Cornell Lab of Ornithology, 2006), https://doi.org/10.2173/bna.53.
3. White et al., "Peregrine Falcon, (Falco peregrinus)."
4. White et al., "Peregrine Falcon, (Falco peregrinus)."
5. The Center for Conservation Biology, "FalconTrak: Investigating Peregrine Falcon Movements through Satellite Tracking," http://www.ccbbirds.org/what-we-do/research/species-of-concern/species-of-concern-projects/falcontrak.
6. White et al., "Peregrine Falcon, (Falco peregrinus)."
7. The Peregrine Fund, Operation Report, 1991. World Center for Birds of Prey (1991). One of those pairs included the adults that visited our hack site. When Dale Mutch made an aerial search for the transmittered quail that the female had carried off, he discovered that she had fed on some of the quail en route to her aerie, dropping the transmitters along the way. She had also taken one quail—along with its attached transmitter—to her nest, high on a cliff, 12 to 15 miles northwest of our hack site; personal communication with Brian Mutch, March 24, 2015.
8. The American Peregrine Falcon, which is featured in Feather Trails, is one of three Peregrine Falcon subspecies in North America. The other two are the Arctic Peregrine Falcon (Falco peregrinus tundrius), which breeds in the High Arctic tundra, and the Peale's Peregrine Falcon (Falco peregrinus pealei), which breeds in coastal southern Alaska and British Colombia.
9. W. Burnham and T. Cade, "Conclusions," in Return of the Peregrine, 349–58.
10. Cade, "Life History Traits of the Peregrine in Relation to Recovery," 2–11.
11. 64 Federal Register 46542, August 25, 1999; T. Cade and W. Burnham, "Introduction: The Madison Peregrine Conference and the Struggle to Ban DDT," in Return of the Peregrine, 13–21.
12. Risebrough, "Pesticides and Bird Populations," 397–427.
13. M. L. Morrison, "Bird Populations as Indicators of Environmental Change," in Current Ornithology, ed. R.F. Johnston, vol. 3, (Boston: Springer, 1986): 429–51.
14. 64 Federal Register, 46543, August 25, 1999. This action was taken in response to a petition submitted to the Department of Interior in 1969 by a group of raptor biologists and falconers

who'd met four years after the 1965 Madison Conference to assess further declines of peregrines in North America. A third peregrine subspecies—the Peale's Peregrine Falcon—was not listed because it showed only traces of DDT and was reproducing at near normal levels. The Peale's Peregrine subsequently was protected by a proposed rule in 1983 that designated all free-flying Peregrine Falcons in the 48 conterminous United States as endangered under the similarity-of-appearance provision of section 4(e) of the Endangered Species Act. This provision protects species that are biologically threatened or endangered by also protecting similar-looking non-endangered species or subspecies to prevent the erroneous targeting of threatened/endangered organisms.

15. Burnham and Cade, "Conclusions," 349–58.
16. 64 Federal Register 46542, August 25, 1999.
17. 64 Federal Register 46542, August 25, 1999.
18. Oakleaf and Craig, "Peregrine Restoration from a State Biologist's Perspective," 297–311.
19. Burnham and Cade, "Conclusions," 349–58.

Chapter 8. Sheep Heads, Deer Guts, and Feedlots

1. B. Heinrich, "A Birdbrain Nevermore," *Natural History* 102, no. 10 (1993): 51–56; B. Heinrich, "An Experimental Investigation of Insight in Common Ravens (*Corvus corax*)," *Auk* 112, no. 4 (1995): 994–1003, https://doi.org/10.2307/4089030. Later research suggested that rather than demonstrating insight—organizing a series of behaviors to achieve a visualized solution to a problem—string-pulling corvids might just rapidly associate an action with a reward. Even partial success—the movement of the string and food toward the puller might precipitate a repeat of the action—until the ultimate reward is obtained. Nevertheless, ever-more-involved and complicated experiments have demonstrated insight in corvids.
2. B. Heinrich, "Winter Foraging at Carcasses by Three Sympatric Corvids, with Emphasis on Recruitment by the Raven, *Corvus corax*," *Behavioral Ecology and Sociobiology* 23, no. 3 (1988): 141–56, https://doi.org/10.1007/BF00300349; B. Heinrich and J. M. Marzluff, "Do Common Ravens Yell Because They Want to Attract Others?" *Behavioral Ecology and Sociobiology* 28, no. 1 (1991): 13–21, https://www.jstor.org/stable/4600510.
3. J. M. Marzluff and J. O. McKinley, "Exploitation of Carcasses and Nesting Behavior of Common Ravens in the Snake River Birds of Prey Area," *Snake River Birds of Prey National Conservation Area Research and Monitoring Annual Report*, 1993.
4. T. J. Cade, "Using Science and Technology to Reestablish Species Lost in Nature," in *Biodiversity*, ed. E. O. Wilson (National Academy Press, 1988): 279–88.
5. K. D. Whitmore and J. M. Marzluff, "Hand-Rearing Corvids for Reintroduction: Importance of Feeding Regime, Nestling Growth, and Dominance," *Journal of Wildlife Management* 62, no. 4 (1998): 1460–79, https://doi.org/10.2307/3802013.
6. Whitmore and Marzluff, "Hand-Rearing Corvids for Reintroduction."

Chapter 9. Through the Mist

1. I have used the Americanized spelling, "Hawaii," for the state and the Hawaiian spelling, "Hawai'i," for the Big Island to differentiate for the reader.
2. J. F. Rock, "The Indigenous Trees of the Hawaiian Islands," *Hawaiian Gazette*, 1913. Cited in J. G. Giffin, J. M. Scott, and S. Mountainspring, "Habitat Selection and Management of the Hawaiian Crow," *Journal of Wildlife Management* 51, no. 2 (1987): 485–94, https://doi.org/10.2307/3801038; The Nature Conservancy, (no date), "Last Stand—The Vanishing Hawaiian Rainforest," The Nature Conservancy of Hawai'i, https://www.nature.org/media/hawaii/last_stand_web_lo.pdf.
3. R. Frost, "The Road Not Taken," in *Mountain Interval* (Henry Holt and Company, 1920).

4. R. L. Pyle and P. Pyle, "Kalij Pheasant," in *The Birds of the Hawaiian Islands: Occurrence, History, Distribution, and Status* (B. P. Bishop Museum, 2009), http://hbs.bishopmuseum.org/birds/rlp-monograph/pdfs/02-Galliformes-Procellariiformes/KAPH.pdf.

5. W. D. Duckworth et al., *Scientific Bases for the Preservation of the Hawaiian Crow* (National Academy Press, 1992).

6. R. H. MacArthur and E. O. Wilson, *The Theory of Island Biogeography* (Princeton University Press, 1967).

7. J. M. Marzluff and T. Angell, *In the Company of Crows and Ravens* (Yale University Press, 2005).

8. P. V. Kirch, "When Did the Polynesians Settle Hawai'i? A Review of 150 Years of Scholarly Inquiry and a Tentative Answer," *Hawaiian Archaeology* 12 (2011): 3–26.

9. S. L. Olson and H. F. James, "Fossil Birds from the Hawaiian Islands: Evidence for Wholesale Extinction by Man before Western Contact," *Science*, New Series 217, no. 4560 (1982): 633–35, https://doi.org/10.1126/science.217.4560.633.

10. Marzluff and Angell, *In the Company of Crows and Ravens*.

11. R. C. Fleischer and C. E. McIntosh, "Molecular Systematics and Biogeography of the Hawaiian Avifauna," *Studies in Avian Biology* 22 (2001): 51–60; P. C. Banko, "'Alalā," in *Conservation Biology of Hawaiian Forest Birds: Implications for Island Avifauna*, eds. T. K. Pratt et al. (Yale University Press, 2009), 473–86; A 2012 molecular study determined that the Hawaiian Crow's closest relative is the Rook (*Corvus frugilegus*), a Eurasian species; K. A. Jønsson, P. H. Fabre, and M. Irestedt, "Brains, Tools, Innovation and Biogeography in Crows and Ravens," *BMC Evolutionary Biology* 12, no. 1 (2012): 1–12, https://doi.org/10.1186/1471-2148-12-72.

12. P. C. Banko, D. L. Ball, and W. E. Banko, "Hawaiian Crow," in *The Birds of North America*, no. 648, eds. A. Poole and F. Gill (The Birds of North America, Inc., 2002).

13. Marzluff and Angell, *In the Company of Crows and Ravens*; Banko, Ball, and Banko, "Hawaiian Crow."

14. H. F. Sakai, C. J. Ralph, and C. D. Jenkins, "Foraging Ecology of the Hawaiian Crow, an Endangered Generalist," *Condor* 88, no. 2 (1986): 11–219, https://doi.org/10.2307/1368918.

15. R. C. L. Perkins, "Vertebrata, Pt. 4," in *Fauna Hawai'iensis; Being the Land-Fauna of the Hawaiian Islands*, ed. D. Sharp (Cambridge University Press, 1903), 365–466; cited in R. L. Pyle and P. Pyle, "Hawaiian Crow," in *The Birds of the Hawaiian Islands: Occurrence, History, Distribution, and Status* (B. P. Bishop Museum, 2009), http://hbs.bishopmuseum.org/birds/rlp-monograph/pdfs/06-PTER-TIMA/HCRO.pdf; Giffin, Scott, and Mountainspring, "Habitat Selection and Management of the Hawaiian Crow."

16. J. M. Scott, C. B. Kepler, and J. L. Sincock, "Distribution and Abundance of Hawai'i's Endemic Land Birds: Conservation and Management Strategies," in *Hawai'i's Terrestrial Ecosystems, Preservation and Management*, eds. C. P. Stone and J. M. Scott (University of Hawaii, 1985), 75–104; P. Harrity, *'Alala Restoration Project Report* (The Peregrine Fund, 1993); citations in Pyle and Pyle, "Hawaiian Crow."

17. Banko, Ball, and Banko, "Hawaiian Crow"; Marzluff and Angell, *In the Company of Crows and Ravens*.

18. Marzluff and Angell, *In the Company of Crows and Ravens*; Banko, "'Alalā," 473–86; Giffin, Scott, and Mountainspring, "Habitat Selection and Management of the Hawaiian Crow."

19. Citations in Pyle and Pyle, "Hawaiian Crow"; Giffin, Scott, and Mountainspring, "Habitat Selection and Management of the Hawaiian Crow"; Banko, Ball, and Banko, "Hawaiian Crow."

20. Marzluff and Angell, *In the Company of Crows and Ravens*, 259.

21. Banko, Ball, and Banko, "Hawaiian Crow."

22. Prerelease and release details are from Harrity, *'Alala Restoration Project Report*.

23. Harrity, *'Alala Restoration Project Report*; Definitions of 'Alalā names come from M. K. Pukui and S. H. Elbert, *Hawaiian Dictionary* (University of Hawaii Press, 1971).

24. The five crows were originally given Americanized versions of their Hawaiian names: Hiwa Hiwa, Hoapili, Lokahi, Kehau, and Malama. I believe Lokahi and Kehau were known by these

given names throughout their lives. However, I have chosen to use the current, correct Hawaiian spellings for the birds' names (with the exception of Hoapili, which is technically two words in Hawaiian: hoa pili).

25. J. M. Marzluff et al., "Captive Rearing and Hacking Alalā: A Reintroduction Plan for the Hawaiian Crow" (The Peregrine Fund, 1993). Author's collections.

26. B. Heinrich, *Mind of the Raven: Investigations and Adventures with Wolf-Birds* (HarperCollins, 1999).

27. Harrity, *'Alala Restoration Project Report*.

28. Banko, Ball, and Banko, "Hawaiian Crow"; Banko, "'Alalā," 473–86.

29. For Hawaiian Crow molt information, see Banko, Ball, and Banko, "Hawaiian Crow."

Chapter 10. Night Whispers

1. For a succinct and helpful overview of wildlife radiotelemetry, see Dylan Kesler (ed.), "Radio Telemetry," The Migratory Connectivity Project, last updated 2020, http://www.migratoryconnectivityproject.org/vhf-radios.

2. Banko, Ball, and Banko, "Hawaiian Crow."

3. J. Altmann, "Observational Study of Behaviour: Sampling Methods," *Behaviour* 49 (1974): 227–67.

4. Since I was unable to retrieve my field notebook from either The Peregrine Fund or the US Fish and Wildlife Service, I had to rely on photocopies of approximately three weeks' worth of my notes that were made before I left Hawaii. These did not include my first few days of work. As a result, the scan and activities I describe here are representative of the 'Alalā's activities during my first few weeks (I also based my descriptions on a personal diary that I kept) but are not an exact transcription of the behaviors that transpired in my first few days. Many of the activities occurred exactly as described but at a slightly later date.

5. Altmann, "Observational Study of Behaviour."

6. Giffin, Scott, and Mountainspring, "Habitat Selection and Management of the Hawaiian Crow."

7. Banko, Ball, and Banko, "Hawaiian Crow"; M. J. Walters, *Seeking the Sacred Raven: Politics and Extinction on a Hawaiian Island* (Island Press, 2006).

8. L. S. Hamilton, J. O. Juvik, and F. N. Scatena, "The Puerto Rico Tropical Cloud Forest Symposium: Introduction and Workshop Synthesis," in *Tropical Montane Cloud Forests* (Ecological Studies 110), eds. L. S. Hamilton, J. O. Juvik, F. N. Scatena (Springer Verlag, 1995), 1–23; C. Doumenge et al., "Tropical Montane Cloud Forests: Conservation Status and Management Issues," *Tropical Montane Cloud Forests*, 24–37; L. L. Loope and T. W. Giambelluca, "Vulnerability of Island Tropical Montane Cloud Forests to Climate Change, with Special Reference to East Maui, Hawaii," *Climatic Change* 39 (1998): 503–17, https://doi.org/10.1023/A:1005372118420; L. A. Bruijnzeel, *Tropical Montane Cloud Forests* (Cambridge University Press, 2010).

9. M. A. McDonald and P. R. Weissich, *Na Lei Makamae: The Treasured Lei* (University of Hawaii Press, 2003).

10. J. F. C. Rock, *The Indigenous Trees of the Hawaiian Islands* (Honolulu, TH, 1913).

11. S. Culliney et al., "Seed Dispersal by a Captive Corvid: The Role of the 'Alalā (*Corvus hawaiiensis*) in Shaping Hawai'i's Plant Communities," *Ecological Applications* 22, no. 6 (2012): 1718–32, https://doi.org/10.1890/11-1613.1.

12. Heinrich, *Mind of the Raven*, 199.

Chapter 11. Malaria

1. R. E. Warner, "The Role of Introduced Diseases in the Extinction of the Endemic Hawaiian Avifauna," *Condor* 70, no. 2 (1968): 101–20, https://doi.org/10.2307/1365954; Marzluff and Angell, *In the Company of Crows and Ravens*. It was once suspected that reservoirs of *Plasmodium* might also have been present in Hawaii seasonally (prior to the introduction of passerines, game

birds, and domestic fowl) because the parasites could have been carried to the islands in the bodies of the thousands of migratory birds that arrived each year from Canada, Alaska, and the Lower 48 states. Nevertheless, more recent research has shown that the *Plasmodium* parasite originated in the Old World rather than in the New World; J. S. Beadell et al., "Global Phylogeographic Limits of Hawaii's Avian Malaria," *Proceedings of the Royal Society B: Biological Sciences* 273, no. 1604 (2006): 2935–44, https://doi.org/10.1098/rspb.2006.3671.

2. D. A. LaPointe, C. T. Atkinson, and M. D. Samuel, "Ecology and Conservation Biology of Avian Malaria," *Annals of the New York Academy of Sciences* 1249, no. 1 (2012): 211–26 https://doi .org/10.1111/j.1749-6632.2011.06431.x.

3. Warner, "The Role of Introduced Diseases."

4. Warner, "The Role of Introduced Diseases."

5. Warner, "The Role of Introduced Diseases."

6. Warner, "The Role of Introduced Diseases."

7. At the time of these studies and during my time in Hawaii, this species was known as the Japanese White-eye.

8. Warner, "The Role of Introduced Diseases." Researchers studying Hawaiian honeycreepers in the late 1970s noted that the native Hawaiian honeycreepers they observed slept with their heads tucked into their backs and their legs tucked up. They speculate that as long as Warner's 1968 observations were correct, those birds that slept with their vulnerable parts exposed likely succumbed to avian malaria, whereas natural selection favored those native birds that slept in a more protected fashion, resulting in a mosquito/parasite-induced change in the sleeping behavior of native birds; C. van Riper III et al., "The Epizootiology and Ecological Significance of Malaria in Hawaiian Land Birds," *Ecological Monographs* 56, no. 4 (1986): 327–44, https://doi.org /10.2307/1942550.

9. L. A. Freed et al., "Increase in Avian Malaria at Upper Elevation in Hawai'i," *Condor* 107, no. 4 (2005): 753–64, https://doi.org/10.1650/7820.1.

10. Banko, Ball, and Banko, "Hawaiian Crow"; C. T. Atkinson and D. A. LaPointe, "Introduced Avian Diseases, Climate Change, and the Future of Hawaiian Honeycreepers," *Journal of Avian Medicine and Surgery* 23, no. 1 (2009): 53–63, https://doi.org/10.1647/2008-059.1.

11. US Fish and Wildlife Service, US Department of Commerce, and US Census Bureau, "2016 National Survey of Fishing, Hunting, and Wildlife-Associated Recreation," (US Fish and Wildlife Service, US Department of Commerce, and US Census Bureau, 2018). In 2022, 96.3 million people in the United States observed birds. Whether this dramatic increase in the number of bird-watchers represents a greater national interest in birds or was an artifact of changed research protocols, COVID-19 keeping people at home, or other factors is uncertain. US Fish and Wildlife Service, "2022 National Survey of Fishing, Hunting, and Wildlife-Associated Recreation," October 11, 2023.

12. LaPointe, Atkinson, and Samuel, "Ecology and Conservation Biology of Avian Malaria."

13. Pyle and Pyle, *The Birds of the Hawaiian Islands*.

14. M. P. Moulton, K. E. Miller, and E. A. Tillman, "Patterns of Success among Introduced Birds in the Hawaiian Islands," in *Evolution, Ecology, Conservation, and Management of Hawaiian Birds: A Vanishing Avifauna*, eds. J. M. Scott, S. Conant, and C. van Riper III, *Studies in Avian Biology* 22 (2001): 31–46.

15. G. C. Munro, *Birds of Hawaii* (Tongg Publishing Company, 1944); Banko, Ball, and Banko, "Hawaiian Crow."

16. Banko, Ball, and Banko, "Hawaiian Crow."

17. J. G. Massey, T. K. Graczyk, and M. R. Cranfield, "Characteristics of Naturally Acquired *Plasmodium relictum capistranoae* Infections in Native Hawaiian Crows (*Corvus hawaiiensis*) in Hawaii," *Journal of Parasitology* 82 (1996): 182–85.

18. LaPointe, Atkinson, and Samuel, "Ecology and Conservation Biology of Avian Malaria."

19. C. P. Stone, "Alien Animals in Hawai'i's Native Ecosystems: Toward Controlling the Adverse Effects of Introduced Vertebrates," in *Hawai'i's Terrestrial Ecosystems*, 251–97; C. P. Stone and S. J. Anderson, "Introduced Animals in Hawaii's Natural Areas," *Proceedings of the Thirteenth Vertebrate Pest Conference* (1988), paper 28; G. W. Cox, *Alien Species in North America and Hawaii: Impacts on Natural Ecosystems* (Island Press, 1999).

20. Stone and Anderson, "Introduced Animals in Hawaii's Natural Areas"; D. A. LaPointe, C. T. Atkinson, and S. I. Jarvi, "Managing Disease," in *Conservation Biology of Hawaiian Forest Birds*, 405–24; LaPointe, Atkinson, and Samuel, "Ecology and Conservation Biology of Avian Malaria."

21. Stone and Anderson, "Introduced Animals in Hawaii's Natural Areas."

22. Banko, Ball, and Banko, "Hawaiian Crow"; LaPointe, Atkinson, and Samuel, "Ecology and Conservation Biology of Avian Malaria."

23. Banko, Ball, and Banko, "Hawaiian Crow"; Marzluff and Angell, *In the Company of Crows and Ravens*; Atkinson and LaPointe, "Introduced Avian Diseases, Climate Change"; LaPointe, Atkinson, and Samuel, "Ecology and Conservation Biology of Avian Malaria."

24. T. L. Benning et al., "Interactions of Climate Change with Biological Invasions and Land Use in the Hawaiian Islands: Modeling the Fate of Endemic Birds Using a Geographic Information System," *Proceedings of the National Academy of Sciences* 99, no. 22 (2002): 14246–49, https://doi.org/10.1073/pnas.162372399; Atkinson and LaPointe, "Introduced Avian Diseases, Climate Change."

25. Benning et al., "Interactions of Climate Change with Biological Invasions"; LaPointe, Atkinson, and Samuel, "Ecology and Conservation Biology of Avian Malaria."

26. Freed et al., "Increase in Avian Malaria."

27. Freed et al., "Increase in Avian Malaria."

28. Freed et al., "Increase in Avian Malaria"; Atkinson and LaPointe, "Introduced Avian Diseases, Climate Change."

29. C. T. Atkinson et al., "Pathogenicity of Avian Malaria in Experimentally-Infected Hawaii Amakihi," *Journal of Wildlife Diseases* 36, no. 2 (2000): 197–204, https://doi.org/10.7589/0090-3558-36.2.197; C. T. Atkinson et al., "Wildlife Disease and Conservation in Hawaii: Pathogenicity of Avian Malaria (*Plasmodium relictum*) in Experimentally Infected Iiwi (*Vestiaria coccinea*)," *Parasitology* 111, Supplement S1 (1995): S59–S69, https://doi.org/10.1017/s003118200007582x.

30. B. L. Woodworth et al., "Host Population Persistence in the Face of Introduced Vector-Borne Diseases: Hawaii Amakihi and Avian Malaria," *Proceedings of the National Academy of Sciences of the United States of America* 102, no. 5 (2005): 1531–36, https://doi.org/10.1073/pnas.0409454102; Atkinson and LaPointe, "Introduced Avian Diseases, Climate Change."

31. Freed et al., "Increase in Avian Malaria."

32. Van Riper III et al., "The Epizootiology and Ecological Significance of Malaria."

33. Atkinson and LaPointe, "Introduced Avian Diseases, Climate Change"; LaPointe, Atkinson, and Samuel, "Ecology and Conservation Biology of Avian Malaria."

34. C. van Riper III, S. G. van Riper, and W. R. Hansen, "Epizootiology and Effect of Avian Pox on Hawaiian Forest Birds," *Auk* 119, no. 4 (2002): 929–42, https://doi.org/10.1093/auk/119.4.929; C. T. Atkinson et al., "Prevalence of Pox-like Lesions and Malaria in Forest Bird Communities on Leeward Mauna Loa Volcano, Hawaii," *Condor* 107, no. 3 (2005): 537–46, https://doi.org/10.1093/condor/107.3.537; Atkinson and LaPointe, "Introduced Avian Diseases, Climate Change."

35. See, for example, photographs of Bald Eagles and Laysan Albatrosses in W. Hansen, "Avian Pox," in *Field Manual of Wildlife Diseases: General Field Procedures and Diseases of Birds*, eds. M. Friend and J. C. Franson (US Geological Survey, 1999), 163–69, https://pubs.usgs.gov/itr/1999/field_manual_of_wildlife_diseases.pdf.

36. Warner, "The Role of Introduced Diseases"; Massey, Graczyk, and Cranfield, "Characteristics of Naturally Acquired *Plasmodium relictum capistranoae* Infections."

37. Van Riper III, van Riper, and Hansen, "Epizootiology and Effect of Avian Pox"; Hansen, "Avian Pox."

38. I. A. E. Atkinson, "A Reassessment of Factors, Particularly *Rattus rattus* L., that Influenced the Decline of Endemic Forest Birds in the Hawaiian Islands," *Pacific Science* 31, no. 2 (1977): 109–33; Van Riper III et al., "The Epizootiology and Ecological Significance"; Van Riper III, van Riper, and Hansen, "Epizootiology and Effect of Avian Pox"; LaPointe, Atkinson, and Samuel, "Ecology and Conservation Biology of Avian Malaria."

39. Atkinson et al., "Prevalence of Pox-like Lesions and Malaria"; S. I. Jarvi et al., "Diversity, Origins and Virulence of *Avipoxviruses* in Hawaiian Forest Birds," *Conservation Genetics* 9 (2008): 339–48, https://doi.org/10.1007/s10592-007-9346-7; LaPointe, Atkinson, and Samuel, "Ecology and Conservation Biology of Avian Malaria"; M. D. Samuel et al., "The Epidemiology of Avian Pox and Interaction with Avian Malaria in Hawaiian Forest Birds," *Ecological Monographs* 88, no. 4 (2018): 621–37, https://doi.org/10.1002/ecm.1311.

40. Jarvi et al., "Diversity, Origins and Virulence of *Avipoxviruses*."

41. E. A. VanderWerf, A. Cowell, and J. L. Rohrer, "Distribution, Abundance, and Conservation of Oʻahu ʻElepaio in the Southern Leeward Koʻolau Range," *Elepaio* 57 (1997): 99–105. Cited in LaPointe, Atkinson, and Samuel, "Ecology and Conservation Biology of Avian Malaria"; Van Riper III, van Riper, and Hansen, "Epizootiology and Effect of Avian Pox."

42. Van Riper III, van Riper, and Hansen, "Epizootiology and Effect of Avian Pox"; Atkinson et al., "Prevalence of Pox-like Lesions and Malaria."

43. Giffin, Scott, and Mountainspring, "Habitat Selection and Management of the Hawaiian Crow"; C. D. Jenkins et al., "Disease-Related Aspects of Conserving the Endangered Hawaiian Crow," in *Disease Related Aspects of Conserving Threatened Bird Populations*, ed. J. E. Cooper (International Council for Bird Preservation, 1989); Van Riper III, van Riper, and Hansen, "Epizootiology and Effect of Avian Pox."

44. Personal communication with Peter Harrity, June 6, 2017.

45. M. D. Samuel et al., "The Dynamics, Transmission, and Population Impacts of Avian Malaria in Native Hawaiian Birds: A Modeling Approach," *Ecological Applications* 21, no. 8 (2011): 2960–73, https://doi.org/10.1890/10-1311.1; J. A. Ahumada et al., "Modeling the Epidemiology of Avian Malaria and Pox in Hawaii," in *Conservation Biology of Hawaiian Forest Birds*, 331–55; LaPointe, Atkinson, and Samuel, "Ecology and Conservation Biology of Avian Malaria."

46. Atkinson et al., "Prevalence of Pox-like Lesions and Malaria."

47. Personal communication with Dr. Sheila Conant, September 22, 2023.

48. I. Iturbe-Ormaetxe, T. Walker, and S. L. O'Neill, "*Wolbachia* and the Biological Control of Mosquito-Borne Disease," *EMBO Reports* 12, no. 6 (2011): 508–18, https://doi.org/10.1038/embor.2011.84; T. Harvey-Samuel et al., "*Culex quinquefasciatus*: Status as a Threat to Island Avifauna and Options for Genetic Control," *CABI Agriculture and Bioscience* 2, no. 1 (2021): 1–21, https://doi.org/10.1186/s43170-021-00030-1. *Wolbachia* is a common, naturally occurring bacteria that infects the cells of approximately three-quarters of the Earth's insect species. In Hawaii, *Wolbachia* is found in many native and non-native insects, including mosquitoes. A multi-agency partnership called "Birds, Not Mosquitoes" plans to launch this conservation tool in 2023, https://www.birdsnotmosquitoes.org.

49. W. Liao et al., "Mitigating Future Avian Malaria Threats to Hawaiian Forest Birds from Climate Change," *PloS ONE* 12, no. 1 (2017): e0168880, https://doi.org/10.1371/journal.pone.0168880.

50. M. D. Samuel et al., "Facilitated Adaptation for Conservation: Can Gene Editing Save Hawaii's Endangered Birds from Climate Driven Avian Malaria?" *Biological Conservation* 241 (2020): 10839, https://doi.org/10.1016/j.biocon.2019.108390.

51. M. Specter, "The DNA Revolution," *National Geographic* (August 2016), 30–59; Liao et al., "Mitigating Future Avian Malaria Threats"; B. J. Novak, T. Maloney, and R. Phelan, "Advancing a New Toolkit for Conservation: From Science to Policy," *The CRISPR Journal* 1, no. 1 (2018): 11–15, https://doi.org/10.1089/crispr.2017.0019.
52. Marzluff and Angell, *In the Company of Crows and Ravens*.

Chapter 12. Forest Intruders

1. Molecular research has shown that there may be two subspecies of hoary bats in Hawaii, which some scientists have recommended elevating to separate species. A. B. Baird et al., "Molecular Systematic Revision of Tree Bats (Lasiurini): Doubling the Native Mammals of the Hawaiian Islands," *Journal of Mammalogy* 96, no. 6 (2015): 1255–74; Nature Conservancy, "Last Stand."
2. Giffin, Scott, and Mountainspring, "Habitat Selection and Management of the Hawaiian Crow"; Banko, Ball, and Banko, "Hawaiian Crow"; Banko, "Alalā."
3. D. Pimentel, "Nonnative Species in the World," in *Biological Invasions: Economic and Environmental Costs of Alien Plant, Animal, and Microbe Species*, ed. D. Pimentel (CRC Press, 2011), 1–7.
4. J. C. Greenway Jr., *Extinct and Vanishing Birds of the World* (American Commission for International Wildlife Protection, 1958), referenced in Van Riper III et al., "The Epizootiology and Ecological Significance"; National Research Council, *The Scientific Bases for the Preservation of the Hawaiian Crow* (National Academy Press, 1992).
5. Hawaii Department of Land and Natural Resources, "Native Birds of Hawai'i," 2023, https://dlnr.hawaii.gov/wildlife/birds.
6. C. Farmer, "Battling Ants in Hawai'i—Small Invaders, Big Problem," *Bird Conservation*, Spring 2015, 7–10, https://abcbirds.org/ISSUU/bird-conservation-spring-2015.
7. F. Courchamp, J. L. Chapuis, and M. Pascal, "Mammal Invaders on Islands: Impact, Control and Control Impact," *Biological Reviews* 78, no. 3 (2003): 347–83, https://doi.org/10.1017/s1464793102006061.
8. G. D. Lindsey et al., "Small Mammals as Predators and Competitors," in *Conservation Biology of Hawaiian Forest Birds*, 274–93; P. M. Vitousek et al., "Human Domination of Earth's Ecosystems," *Science* 277, no. 5325 (1997): 494–99, https://doi.org/10.1126/science.277.5325.494.
9. Lindsey et al., "Small Mammals as Predators and Competitors."
10. BirdLife International, "Invasive Alien Species Have Been Implicated in Nearly Half of Recent Bird Extinctions," BirdLife State of the World's Birds, 2008, http://datazone.birdlife.org/sowb/casestudy/invasive-alien-species-have-been-implicated-in-nearly-half-of-recent-bird-extinctions-; Lindsey et al., "Small Mammals as Predators and Competitors." Also references therein.
11. J. S. Athens, "*Rattus exulans* and the Catastrophic Disappearance of Hawai'i's Native Lowland Forest," *Biological Invasions* 11, no. 7 (2009): 1489–1501, https://doi.org/10.1007/s10530-008-9402-3; Lindsey et al., "Small Mammals as Predators and Competitors."
12. Lindsey et al., "Small Mammals as Predators and Competitors."
13. Lindsey et al., "Small Mammals as Predators and Competitors."
14. Atkinson, "A Reassessment of Factors"; Lindsey et al., "Small Mammals as Predators and Competitors."
15. D. R. Towns, I. A. E. Atkinson, and C. H. Daugherty, "Have the Harmful Effects of Introduced Rats on Islands Been Exaggerated?" *Biological Invasions* 8 (2006): 863–91, https://doi.org/10.1007/s10530-005-0421-z; IPBES, "IBPES Invasive Alien Species Assessment: Summary for Policymakers" (2023), IPBES Secretariat, Bonn, Germany.
16. P. H. Baldwin, "The Life History of the Laysan Rail," *Condor* 49, no. 1 (1947): 14–21, https://doi.org/10.2307/1364423; J. M. Scott et al., "Forest Bird Communities of the Hawaiian Islands: Their Dynamics, Ecology and Conservation," *Studies in Avian Biology* no. 9 (Cooper

Ornithological Society, 1986); Stone and Anderson. "Introduced Animals in Hawaii's Natural Areas"; BirdLife International, "Laysan Rail: *Zaporria palmeri*," IUCN Red List of Threatened Species, October 1, 2016, https://dx.doi.org/10.2305/IUCN.UK.2016-3.RLTS.T22692672 A93363618.en; BirdLife International, "Kauai Oo: *Moho braccatus*," IUCN Red List of Threatened Species, October 1, 2016. https://dx.doi.org/10.2305/IUCN.UK.2016-3.RLTS .T22704323A93963628.en.

17. Stone and Anderson, "Introduced Animals in Hawaii's Natural Areas"; BirdLife International, "Ou: *Psittirostra psittacea*," IUCN Red List of Threatened Species, 2018, https://dx.doi.org /10.2305/IUCN.UK.2016-3.RLTS.T22720734A126791352.en.

18. P. Q. Tomich, *Mammals in Hawaii*, 2nd ed. (Bishop Museum Press, 1986); Stone and Anderson, "Introduced Animals in Hawaii's Natural Areas."

19. The species has also been classified as *Herpestes javanicus*. Based on more recent molecular evidence, it is now *Urva auropunctata*. M. Patou et al., "Molecular Phylogeny of the Herpestidae (Mammalia, Carnivora) with a Special Emphasis on the Asian Herpestes," *Molecular Phylogenetics and Evolution* 53, no. 1 (2009): 69–80.

20. A. Barun et al., "Impact of the Introduced Small Indian Mongoose (*Herpestes auropunctatus*) on Abundance and Activity Time of the Introduced Ship Rat (*Rattus rattus*) and the Small Mammal Community on Adriatic Islands, Croatia," *NeoBiota* 11 (2011): 51–61, https://doi.org/10.3897 /neobiota.11.1819; M. W. Fall et al., "Rodents and Other Vertebrate Invaders in the United States," in *Biological Invasions*, 381–410; W. C. Pitt, R. T. Sugihara, and A. R. Berentsen, "Effect of Travel Distance, Home Range, and Bait on the Management of Small Indian Mongooses, *Herpestes auropunctatus*," *Biological Invasions* 17, (2015): 1743–59, https://doi.org/10.1007 /s10530-014-0831-x.

21. Banko, Ball, and Banko, "Hawaiian Crow."

22. S. R. Loss, T. Will, and P. P. Marra, "The Impact of Free-Ranging Domestic Cats on Wildlife of the United States," *Nature Communications* 4, no. 1 (2013): 1–8, https://doi.org/10.1038 /ncomms2380.

23. US Fish and Wildlife Service, "Threats to Birds" (2016), https://www.fws.gov/birds/bird -enthusiasts/threats-to-birds.php.

24. F. M. Medina et al., "A Global Review of the Impacts of Invasive Cats on Island Endangered Vertebrates," *Global Change Biology* 17, no. 11 (2011): 3503–10, https://doi.org/10.1111/j.1365 -2486.2011.02464.x; P. P. Marra and C. Santella, *Cat Wars: The Devastating Consequences of a Cuddly Killer* (Princeton University Press, 2016).

25. Courchamp, Chapuis, and Pascal, "Mammal Invaders on Islands."

26. B. M. Fitzgerald and C. R. Veitch, "The Cats of Herekopare Island, New Zealand; Their History, Ecology and Affects [sic] on Birdlife," *New Zealand Journal of Zoology* 12, no. 3 (1985): 319–30, https://doi.org/10.1080/03014223.1985.10428285.

27. BirdLife International, "Invasive Alien Species Have Been Implicated."

28. Tomich, *Mammals in Hawaii*; Stone and Anderson, "Introduced Animals in Hawaii's Natural Areas."

29. C. Farmer and G. Sizemore, "For Rare Hawaiian Birds, Cats Are Unwelcome Neighbors," American Bird Conservancy, (2016), https://abcbirds.org/for-rare-hawaiian-birds-cats -unwelcome.

30. Medina et al., "A Global Review of the Impacts of Invasive Cats."

31. Marra and Santella, *Cat Wars*.

32. S. K. Johnson and P. T. J. Johnson, "Toxoplasmosis: Recent Advances in Understanding the Link Between Infection and Host Behavior," *Annual Review of Animal Biosciences* 9 (2021): 249–64, https://doi.org/10.1146/annurev-animal-081720-111125.

33. Marra and Santella, *Cat Wars*; Johnson and Johnson, "Toxoplasmosis."

34. Marra and Santella, *Cat Wars*.

35. This is the number of monk seals killed as of November 2022. The first documented death was in 2001. National Oceanic and Atmospheric Administration (NOAA), 2020. "The Toll of Toxoplasmosis: Protozoal Disease Has Now Claimed the Lives of 12 Monk Seals and Left Another Fighting to Survive," US Department of Commerce, NOAA, February 20, 2020: https://www.fisheries.noaa.gov/feature-story/toll-toxoplasmosis-protozoal-disease-has-now -claimed-lives-12-monk-seals-and-left.

36. T. M. Work et al., "Mortality Patterns in Endangered Hawaiian Geese (Nene; *Branta sandvicensis*)," *Journal of Wildlife Diseases* 51, no. 3 (2015): 688–95, https://doi.org/10.7589/2014-11-256; Marra and Santella, *Cat Wars*; Johnson and Johnson, "Toxoplasmosis."

37. T. M. Work et al., "Fatal Toxoplasmosis in Free-Ranging Endangered 'Alala from Hawaii," *Journal of Wildlife Diseases* 36, no. 2 (2000): 205–12, https://doi.org/10.7589/0090-3558-36.2.205.

38. Marra and Santella, *Cat Wars*.

39. Marzluff and Angell, *In the Company of Crows and Ravens*.

40. J. Marzluff and T. Angell, *Gifts of the Crow: How Perception, Emotion, and Thought Allow Smart Birds to Behave Like Humans* (Simon and Schuster, 2013).

41. D. Pimentel, "Environmental and Economic Costs Associated with Alien Invasive Species in the United States," in *Biological Invasions*, 411–30.

42. IPBES, "IBPES Invasive Alien Species Assessment: Summary for Policymakers" (2023), IBPES secretariat, Bonn, Germany.

Chapter 13. Extinct

1. M. Ficken, "Avian Play," *Auk* 94, no. 3 (1977): 573–82, https://doi.org/10.1093/auk/94.3.573.

2. B. Heinrich and T. Bugnyar, "Just How Smart Are Ravens?" *Scientific American* 296, no. 4 (2007): 64–71.

3. Ficken, "Avian Play"; C. S. Savage, *Bird Brains: The Intelligence of Crows, Ravens, Magpies, and Jays* (Sierra Club Books, 1995); Marzluff and Angell, *In the Company of Crows and Ravens*.

4. G. R. Hunt, "Manufacture and Use of Hook-Tools by New Caledonian Crows," *Nature* 379, no. 6562 (1996): 249–51, https://doi.org/10.1038/379249a0; Marzluff and Angell, *In the Company of Crows and Ravens*.

5. Marzluff and Angell, *In the Company of Crows and Ravens*; A. H. Taylor et al., "Complex Cognition and Behavioural Innovation in New Caledonian Crows," *Proceedings of the Royal Society of London B: Biological Sciences* 277, no. 1694 (2010): 2637–43, https://doi.org/10.1098/rspb.2010.0285.

6. For a video of "007" solving the eight-step puzzle to acquire a piece of food, see I. Saul, "This Crow Is the Smartest Bird You've Ever Seen," *Huffington Post*, February 7, 2014, http://www .huffingtonpost.com/2014/02/06/crow-smartest-bird_n_4738171.html.

7. C. Rutz et al., "Discovery of Species-wide Tool Use in the Hawaiian Crow," *Nature* 537, no. 7620 (2016): 403–7, https://doi.org/10.1038/nature19103; S. Kaplan, "These Genius Crows Almost Went Extinct before Scientists Discovered They Can Use Tools," *Washington Post*, September 14, 2016. *Note:* I could not verify this *Washington Post* quote with the 'Alalā's captive-breeding program manager because San Diego Zoo Global (since renamed San Diego Wildlife Alliance) would not allow me to speak with him.

8. C. Rutz and J. J. H. St Clair, "The Evolutionary Origins and Ecological Context of Tool Use in New Caledonian Crows," *Behavioural Processes* 89, no. 2 (2012): 153–65, https://doi.org/10.1016/j .beproc.2011.11.005; Rutz et al., "Discovery of Species-wide Tool Use."

9. The following video shows captive 'Alalā erecting their head feathers and provides information about raising the species in captivity: San Diego Zoo, "Alala Conservation," YouTube video, 2:40, June 14, 2011, https://www.youtube.com/watch?v=3TXY9pnEg7M&feature=related.

10. Mist nets are long nylon- or polyester-mesh nets that biologists use to trap birds and bats. The volleyball-like nets are suspended between two poles and are almost invisible at a distance. When

birds fly into the nets, they become tangled in the mesh, where they are removed for banding or examination by biologists.

11. Banko, "'Alalā," 473–86.

12. Walters, *Seeking the Sacred Raven*, 241.

13. Walters, *Seeking the Sacred Raven*.

14. Marzluff and Angell, *In the Company of Crows and Ravens*.

15. D. H. Janzen and P. S. Martin, "Neotropical Anachronisms: The Fruits the Gomphotheres Ate," *Science* 215, no. 4528 (1982): 19–27, https://www.jstor.org/stable/1688516; P. R. Guimarães Jr., M. Galetti, and P. Jordano, "Seed Dispersal Anachronisms: Rethinking the Fruits Extinct Megafauna Ate," *PloS ONE* 3, no. 3 (2008): e1745, https://doi.org/10.1371/journal.pone .0001745; Culliney et al., "Seed Dispersal by a Captive Corvid."

16. Culliney et al., "Seed Dispersal by a Captive Corvid."

17. D. Pauly, "Anecdotes and the Shifting Baseline Syndrome of Fisheries," *Trends in Ecology & Evolution* 10, no. 10: 430. Specifically, Pauly wrote of fisheries scientists who accepted as a baseline the size and species composition of fish stocks that they encountered at the beginning of their careers, and used these metrics to evaluate future changes. Others have applied the concept of shifting baseline syndrome more broadly.

18. Walters, *Seeking the Sacred Raven*.

Chapter 14. Captive

1. Banko, Ball, and Banko, "Hawaiian Crow."

2. M. C. Perry, C. S. Bond, and E. J. R. Lohnes, "Winston Edgar Banko," Washington Biologists' Field Club, USGS Patuxent Wildlife Research Center, 2007. *Note:* This remembrance is no longer available on the USGS PWRC website. A copy is available in the author's collections.

3. Walters, *Seeking the Sacred Raven*.

4. Walters, *Seeking the Sacred Raven*.

5. Banko, "'Alalā," 473–86; Pyle and Pyle, "Hawaiian Crow."

6. Personal communication with Peter Harrity, June 6, 2017.

7. For a video about the Keauhou Bird Conservation Center, see Kamehameha Schools, "Native Bird Project — Huaka'i 'Āina Ho'oilina (3/10)," YouTube video, 4:59, November 25, 2008, https:// www.youtube.com/watch?v=0kKFWQ-SbyA.

8. K. Castanera, "Nine Hawaiian Crow Chicks Hatch at Hawaii Island Endangered Bird Sanctuary, Bringing Total World Population to 114," *Hawai'i Magazine* (August 28, 2014), http://www .hawaiimagazine.com/blogs/hawaii_today/2014/8/18/nine_endangered_alala_hatch_at _keauhou_bird_conservation_center_on_Hawaii_island.

9. Banko, "'Alalā," 473–86.

10. US Fish and Wildlife Service "Revised Recovery Plan for the 'Alalā (*Corvus hawaiiensis*)," (2009).

11. Banko, "'Alalā," 473–86; P. E. A. Hoeck et al., "Effects of Inbreeding and Parental Incubation on Captive Breeding Success in Hawaiian Crows," *Biological Conservation* 184 (2015): 357–64, https://doi.org/10.1016/j.biocon.2015.02.011; P. W. Hedrick et al., "The Influence of Captive Breeding Management on Founder Representation and Inbreeding in the 'Alalā, the Hawaiian Crow," *Conservation Genetics* 17, no. 2 (2016): 369–78, https://doi.org/10.1007 /s10592-015-0788-z.

12. Walters, *Seeking the Sacred Raven*.

13. Banko, Ball, and Banko, "Hawaiian Crow"; personal communication with Donna Ball, US Fish and Wildlife Service, May 23, 2017; US Fish and Wildlife Service, "Revised Recovery Plan for the 'Alalā."

14. Personal communication with Peter Harrity, June 6, 2017; C. Kuehler et al., "Reintroduction of Hand-Reared Alala *Corvus hawaiiensis* in Hawaii," *Oryx* 29, no. 04 (1995): 261–66, https://doi .org/10.1017/S0030605300021256.

15. Walters, *Seeking the Sacred Raven.*
16. Personal communication with Donna Ball, US Fish and Wildlife Service, May 23, 2017; personal communication with Peter Harrity, June 6, 2017.
17. Personal communication with Donna Ball, US Fish and Wildlife Service, May 23, 2017.
18. Personal communication with Donna Ball, US Fish and Wildlife Service, October 12, 2017.
19. Personal communication with Donna Ball, US Fish and Wildlife Service, May 23, 2017; personal communication with Peter Harrity, June 6, 2017.
20. Personal communication with Donna Ball, US Fish and Wildlife Service, May 23, 2017.
21. Personal communication with Jay Nelson, US Fish and Wildlife Service, November 17, 2017.
22. Biologists attempted one final release, in January 1999. Three of these birds were juveniles. The fourth (a female named Puanani) had been released in 1997 and recaptured in 1998. After two birds from the 1999 release cohort died or disappeared, the two remaining ʻAlalā (including Puanani) were permanently returned to captivity.
23. Work et al., "Fatal Toxoplasmosis"; Banko, Ball, and Banko, "Hawaiian Crow"; Banko, "'Alalā," 473–86; personal communication with Jay Nelson, US Fish and Wildlife Service, November 13, 2017.
24. Banko, "'Alalā," 473–86; personal communication with Paul Banko, US Geological Survey, March 27, 2017.
25. B. B. Druker, "Rearing and Environmental Factors Influencing Aberrant, Affiliative, Agonistic, and Nest-Building Behaviors in the ʻAlala (*Corvus hawaiiensis*)," Master of Science thesis, 2000, McGill University, Montreal, Canada. Available from ProQuest Information and Learning (UMI No. 0-612-75303-4).
26. Personal communication with Donna Ball, US Fish and Wildlife Service, May 23, 2017.
27. S. Bebus, "Hawaii Bird Program: Open House," San Diego Zoonooz, San Diego Zoo Global (2009). This blog is no longer available on the San Diego Wildlife Alliance website.
28. L. Komarcyzk, "Corvid Cupid (Parts 1 and 2)," Zoonooz, San Diego Zoo Global (2010). This blog is no longer available on the San Diego Wildlife Alliance website.
29. For a short video on captive-rearing the ʻAlalā, see San Diego Zoo, "Alala Conservation," YouTube video, 2:40, June 14, 2011, https://www.youtube.com/watch?v=3TXY9pnEg7M. For a general video about ʻAlalā recovery, see Hawaiian Skies, "Flight of the Alala," YouTube video, 7:49, March 21, 2012, https://www.youtube.com/watch?v=OR2VgTMj61A.
30. Personal communication with Jay Nelson, US Fish and Wildlife Service, November 13, 2017.
31. Pyle and Pyle, "Hawaiian Crow."

Chapter 15. Second Chances

1. Walters, *Seeking the Sacred Raven.*
2. US Fish and Wildlife Service, "Revised Recovery Plan for the ʻAlalā."
3. Banko, "'Alalā," 473–86.
4. Duckworth et al., *Scientific Bases for the Preservation of the Hawaiian Crow.*
5. US Fish and Wildlife Service, "Revised Recovery Plan for the ʻAlalā ."
6. P. Tummons, "Hawaiʻi's Forest Birds: Is Recovery Possible? FWS Biologists Discuss the Prospects," *Environment Hawaii* 14, no. 9 (March 2004). This source consisted of an interview with J. Burgett (who made this comment) and other key personnel in the US Fish and Wildlife Service's Honolulu office.
7. Banko, "'Alalā," 473–86.
8. US Fish and Wildlife Service, "Revised Recovery Plan for the ʻAlalā."
9. S. A. Temple, "The Problem of Avian Extinctions," in *Current Ornithology*, ed. R.F. Johnston, vol. 3 (Boston: Springer, 1986): 453–85, https://doi.org/10.1007/978-1-4615-6784-4_11.
10. Banko, "'Alalā," 473–86; US Fish and Wildlife Service, "Revised Recovery Plan for the ʻAlalā."

11. US Fish and Wildlife Service, "Revised Recovery Plan for the 'Alalā ."

12. US Fish and Wildlife Service, "Revised Recovery Plan for the 'Alalā"; 'Alalā Project, "Release," Hawaii.gov (2017), http://dlnr.hawaii.gov/alalaproject/release.

13. Stone and Anderson, "Introduced Animals in Hawaii's Natural Areas"; American Bird Conservancy, "New Tools to Help Conserve Hawaiian Birds," Bird Conservation, American Bird Conservancy (Spring 2015): 28–29; A. B. Shiels et al., "Effectiveness of Snap and A24-Automated Traps and Broadcast Anticoagulant Bait in Suppressing Commensal Rodents in Hawaii," *USDA Wildlife Services—Staff Publications* no. 2295, *Human–Wildlife Interactions* 13, no. 2 (2019): 226–37.

14. J. D. Eisemann and C. E. Swift, "Ecological and Human Health Hazards from Broadcast Application of 0.005% Diphacinone Rodenticide Baits in Native Hawaiian Ecosystems," in *Proceedings of Vertebrate Pest Conference* 22 (2006): 413–33; E. B. Spurr et al., "Aerial-Broadcast [sic] Application of Diphacinone Bait for Rodent Control in Hawaii: Efficacy and Non-Target Species Risk Assessment," Technical Report HCSU-071, Hawai'i Cooperative Studies Unit, University of Hawai'i, Hilo, Hawaii (2015).

15. P. A. Dunlevy and E. W. Campbell III, "Assessment of Hazards to Non-Native Mongooses (*Herpestes auropunctatus*) and Feral Cats (*Felis catus*) from the Broadcast Application of Rodenticide Bait in Native Hawaiian Forests," USDA National Wildlife Research Center Staff Publications, (2002): Paper 480, http://digitalcommons.unl.edu/icwdm_usdanwrc/480; Eisemann and Swift, "Ecological and Human Health Hazards," 413–33; A. B. Shiels et al., "Large-scale Aerial Baiting to Suppress Invasive Rats in Hawaii: Efficacy of Diphacinone and Associated Risks," *Proceedings of the Vertebrate Pest Conference* 29 (2020): Paper No. 42, 5 pp.

16. E. A. VanderWerf and D. G. Smith, "Effects of Alien Rodent Control on Demography of the O'ahu 'Elepaio, an Endangered Hawaiian Forest Bird," *Pacific Conservation Biology* 3, no. 2 (2002): 73–81, https://doi.org/10.1071/PC020073; P. C. Banko et al., "Increased Nesting Success of Hawaii Elepaio in Response to the Removal of Invasive Black Rats," *Condor* 121, no. 2 (2019): 1–12, https://doi.org/10.1093/condor/duz003.

17. A. M. Kilpatrick, "Facilitating the Evolution of Resistance to Avian Malaria in Hawaiian Birds," *Biological Conservation* 128 no. 4 (2006): 475–85, https://doi.org/10.1016/.biocon.2005.10.014.

18. M. Hoffman et al., "The Impact of Conservation on the Status of the World's Vertebrates," *Science* 330, no. 6010 (2010): 1503–9, https://doi.org/10.1126/science.1194442.

19. Hoffman et al., "The Impact of Conservation"; D. Luther et al., "Conservation Action Implementation, Funding and Population Trends of Birds Listed Under the Endangered Species Act," *Biological Conservation* 197 (2016): 229–34.

20. Culliney et al., "Seed Dispersal by a Captive Corvid."

21. R. J. Camp et al., "Population Trends of Forest Birds at Hakalau Forest National Wildlife Refuge, Hawai'i," *Condor* 112, no. 2 (2010): 196–212, https://doi.org/10.1525/cond.2010.080113.

22. Fall et al., "Rodents and Other Vertebrate Invaders," 381–410; Pimentel, "Environmental and Economic Costs Associated with Alien Invasive Species in the United States," 411–30.

23. For President Clinton's Executive Order 13112, see https://www.invasivespeciesinfo.gov/laws/execorder-13112.shtml#sec3. For President Obama's Executive Order "Safeguarding the Nation from the Impacts of Invasive Species," see https://obamawhitehouse.archives.gov/the-press-office/2016/12/05/executive-order-safeguarding-nation-impacts-invasive-species.

24. Fall et al., "Rodents and Other Vertebrate Invaders," 381–410. For more information on The Hawaii Invasive Species Council, see http://dlnr.hawaii.gov/hisc.

25. Luther et al., "Conservation Action Implementation, Funding and Population Trends." This study corroborated similar findings in earlier analyses: J. K. Miller et al., "The Endangered Species Act: Dollars and Sense?" *Bioscience* 52 (2002): 163–68, https://doi.org/10.1641/0006 -3568(2002)052[0163:TESADA]2.0.CO;2; K. E Gibbs and D. J. Currie, "Protecting

Endangered Species: Do the Main Legislative Tools Work?" *PLoS ONE* 7, no. 5 (2012): e35730, https://doi.org/10.1371/journal.pone.0035730.

26. M. Restani and J. M. Marzluff, "Funding Extinction? Biological Needs and Political Realities in the Allocation of Resources to Endangered Species Recovery," *BioScience* 52, no. 2 (2002): 169–77, https://doi.org/10.1641/0006-3568(2002)052[0169:FEBNAP]2.0.CO;2; Luther et al., "Conservation Action Implementation, Funding and Population Trends," *Biological Conservation*.

27. S. Holmer, *Endangered Species Act: A Record of Success* (American Bird Conservancy, 2016).

28. D. L. Leonard Jr., "Recovery Expenditures for Birds Listed under the US Endangered Species Act: The Disparity between Mainland and Hawaiian Taxa," *Biological Conservation* 141, no. 8 (2008): 2054–61.

29. S. Conant and D. Leonard, "Book Reviews—*Seeking the Sacred Raven: Politics and Extinction on a Hawaiian Island*, by Jerome Mark Walters," *Condor* 110, no. 1 (2008): 188.

30. Leonard, "Recovery Expenditures for Birds," *Biological Conservation*. Reasons for the inadequate funding of endangered Hawaiian birds include a lack of awareness about Hawaii's avifauna, particularly by mainland US residents; a lack of familiarity among Hawaiians with their native birds because these are often restricted to inaccessible and remote high-elevation forests; a lack of political attention to the recovery needs of Hawaii's birds and a dearth of lawsuits compelling conservation of the state's avifauna; Hawaii's small population, which impacts the amount of conservation funds the state receives and minimizes the level of funds generated from initiatives such as tax check-offs and the sale of "wildlife" license plates; the small geographic range of imperiled Hawaiian birds; and the conflicting desires of Hawaii's Division of Forestry and Wildlife to provide public hunting opportunities (in the form of feral ungulates) and to conserve endangered species, with which introduced ungulates are incompatible.

31. Holmer, *Endangered Species Act*.

32. Conant and Leonard, "Book Reviews—*Seeking the Sacred Raven*." Sheila Conant saw living individuals of seven endemic Hawaiian forest bird species that are now extinct.

33. American Bird Conservancy, "Threatened Birds Recovering Thanks to Endangered Species Act Protection," press release, June 25, 2016.

34. US Department of Interior, "Biden-Harris Administration Announces Nearly $16 Million Through the President's Investing in America Agenda to Prevent the Imminent Extinction of Hawaiian Forest Birds," press release, June 27, 2023.

35. US Fish and Wildlife Service, "Revised Recovery Plan for the 'Alalā"; E. A. VanderWerf, *Hawaiian Bird Conservation Action Plan* (Pacific Rim Conservation, 2012): https://www.pacific rimconservation.org/wp-content/uploads/2013/10/Hawaiian%20Crow.pdf.

36. The following video shows the Keauhou Bird Conservation Center, provides a glimpse into captive-rearing 'Alalā, shows the hatching of the year's first chick in 2016, and previews the upcoming 2016 reintroduction: San Diego Zoo, "Saving Hawaiian Birds from Extinction," YouTube video, 5:39, September 29, 2014, https://www.youtube.com/watch?v=qeYqtubC0iY.

37. Big Island Now, "'Alalā Preparing for Life in Hawaiian Forest," BigIslandNow.com (November 21, 2016): http://bigislandnow.com/2016/11/21/alala; J. Epping, "DLNR: Two Miles of Pu'u Maka'ala NAR Fencing Destroyed," BigIslandNow.com (June 23, 2015): http://bigislandnow. com/2015/06/23/dlnr-two-miles-of-puu-makaala-nar-fencing-destroyed. The long-standing conflict between hunters, who want to retain feral ungulates on the landscape for hunting, and conservationists, who want to remove feral ungulates to restore Hawaii's forests and save its endemic species, complicates the use of fencing as a conservation tool to protect native habitat.

38. For a view of the 2016 release cohort in the acclimation aviary in the Pu'u Maka'ala Natural Area Reserve, see Big Island Video News, "'Alala Almost Ready For Release (Nov. 18, 2016)," YouTube video, 2:09, November 19, 2016, https://www.youtube.com/watch?v=QV1fUxrzcuI.

39. Big Island Now, "Project to Celebrate First Release of 'Alalā," BigIslandNow.com, (November 17, 2016): https://bigislandnow.com/2016/11/17/project-to-celebrate-first-release-of -%CA%BBalala/; I. Ashe, "Experts Work to Improve 'Alalā's Chances," *Hawai'i Tribune Herald*, (March 21, 2017), http://dlnr.hawaii.gov/alalaproject/2017/03/21/hawai%ca%bbi-tribune -herald-experts-work-to-improve-%ca%bbalalas-chances.

40. A. Lieberman, "Alala Egg That Changed the Future," Zoonooz, San Diego Zoo Global (January 8, 2013). This blog is no longer available on the San Diego Wildlife Alliance website.

41. I. Ashe, "Native Crows to Be Released into the Wild in November," *Hawai'i Tribune-Herald*, August 27, 2016, https://www.westhawaiitoday.com/2016/08/27/hawaii-news/native-crows -to-be-released-into-the-wild-in-november.

42. For a short video about the 2016 reintroduction with several views of the newly released 'Alalā, see Big Island Video News, "Hawaiian Crow Released After Going Extinct In Wild (Dec. 16, 2016)," YouTube video, 1:36, December 16, 2016, https://www.youtube.com/watch?v=K_JxhGyI0qI.

43. Ashe, "Experts Work to Improve 'Alalā's Chances."

44. US Fish and Wildlife Service, "Revised Recovery Plan for the 'Alalā."

45. US Fish and Wildlife Service, "Revised Recovery Plan for the 'Alalā."

46. 'Alalā Project, "Sights and Sounds Used to Train 'Alalā to Avoid Predators After Release," press release, September 14, 2018, https://dlnr.hawaii.gov/alalaproject/2018/09/14/press-release- sights-and-sound-used-to-train-%CA%BBalala-to-avoid-predators-after-release; A. L. Greggor et al., "Pre-Release Training, Predator Interactions and Evidence for Persistence of Anti-Predator Behavior in Reintroduced Alalā, Hawaiian Crow," *Global Ecology and Conservation* 28 (2021): e01658, https://doi.org/10.1016/j.gecco.2021.e01658.

47. Personal communication with Paul Banko, US Geological Survey, March 27, 2017.

48. L. Loope et al., "Guidance Document for Rapid 'Ōhi'a Death: Background for the 2017–2019 ROD Strategic Response Plan," University of Hawaii, College of Tropical Agriculture and Human Resources (2016); L. B. Fortini et al., "The Evolving Threat of Rapid 'Ōhi'a Death (ROD) to Hawai'i's Native Ecosystems and Rare Plant Species," *Forest Ecology and Management* 448 (2019): 376–85, https://doi.org/10.1016/j.foreco.2019.06.025; R. J. Camp et al., "Large-Scale Tree Mortality from Rapid Ohia Death Negatively Influences Avifauna in Lower Puna, Hawaii Island, USA," *Condor* 121, no. 2 (2019): 1–16, https://doi.org/10.1093/condor/duz007.

49. D. Ferry, "What's Killing Hawaii's Trees?" *Outside* (March 10, 2016): https://www.outsideonline .com/2060691/whats-killing-hawaiis-trees; "Rapid 'Ōhi'a Death," College of Tropical Agriculture and Human Resources. University of Hawaii (2017): https://cms.ctahr.hawaii.edu/rod/Home. aspx; B. C. Luiz et al., "A Framework for Establishing a Rapid 'Ōhi'a Death Resistance Program," *New Forests* (January 25, 2022): 1–24, https://doi.org/10.1007/s11056-021-09896-5.

50. S. A. Yaremych et al., "West Nile Virus and High Death Rate in American Crows," *Emerging Infectious Diseases* 10, no. 4 (2004): 709–11, https://doi.org/10.3201/eid1004.030499; D. A. Lapointe et al., "Experimental Infection of Hawaiian 'Amakihi (*Hemignathus virens*) with West Nile Virus and Competence of a Co-occurring Vector, *Culex quinquefasciatus*: Potential Impacts on Endemic Hawaiian Avifauna," *Journal of Wildlife Diseases* 45, no. 2 (2009): 257–71, https://doi.org/10.7589/0090-3558-45.2.257.

51. T. K. Pratt, "Preface," in *Conservation Biology of Hawaiian Forest Birds*, xiii–xvi.

Chapter 16. Lead Shock

1. U. Valdez and S. A. H. Osborn, "Observations on the Ecology of the Black-and-Chestnut Eagle (*Oroaetus isidori*) in a Montane Forest of Southeastern Peru," *Ornitología Neotropical* 15, no. 1 (2003): 31–40.

2. Condors can be held and processed (bled or instrumented with radio-transmitters) by two skilled condor handlers, but field crews often are composed of novices who need incremental exposure

to handling condors to keep the birds and themselves safe. As with most animal work, handling condors has evolved over time. Animal keeper Mike Clark, of the Los Angeles Zoo, has been instrumental in developing handling techniques that cause the least stress and discomfort for the condor. In particular, biologists typically now control a condor's head solely by grasping the bird's bill to minimize the bird's discomfort at having a hand around the back of its head and neck.

3. N. F. R. Snyder and N. J. Schmitt, "California Condor (*Gymnogyps californianus*)," in *The Birds of North America*, no. 610, eds. A. Poole and F. Gill (The Birds of North America, Inc., 2002).

4. There is some variability in how scientists interpret levels of lead toxicity in condors. These categories follow the convention of P. T. Redig, "An Investigation into the Effects of Lead Poisoning on Bald Eagles and Other Raptors: Final Report," Minnesota Endangered Species Program Study lOOA-100B, University of Minnesota, 1984. As cited in O. H. Pattee et al., "Lead Hazards within the Range of the California Condor," *Condor* 92, no. 4 (1990): 931–37, https://doi.org/10.2307/1368729. Also see D. M. Frey and J. Maurer, "Assessment of Lead Contamination Sources Exposing California Condors," California Department of Fish and Game (2003). In addition to µg/dL, lead values are now sometimes expressed in ng/ml. 0.01 ppm = 1 µg/dL = 10 ng/ml.

5. S. Farry, "Notes from the Field—California Condors in Grand Canyon Area. March 20, 2000 to April 12, 2000," The Peregrine Fund, 2000. Author's collections. These notes are also archived at the Santa Barbara Museum of Natural History.

6. S. Farry, "Notes from the Field—California Condors in Grand Canyon Area. April 24, 2000 to April 30, 2000," The Peregrine Fund, 2000. Author's collections. These notes are also archived at the Santa Barbara Museum of Natural History.

7. Arizona Vacation Guide—Grand Canyon South Rim: http://www.arizona-leisure.com/south-rim-grand-canyon.html.

8. N. Snyder and H. Snyder, *The California Condor: A Saga of Natural History and Conservation* (Academic Press, 2000); S. A. H. Osborn, *Condors in Canyon Country: The Return of the California Condor to the Grand Canyon Region* (Grand Canyon Association, 2007).

9. S. D. Emslie, "Age and Diet of Fossil California Condors in Grand Canyon, Arizona," *Science* 237, no. 4816 (1987): 768–70, https://www.jstor.org/stable/1699210; Osborn, *Condors in Canyon Country*.

10. K. Stager, "The Role of Olfaction in Food Location by the Turkey Vulture (*Cathartes aura*)," *Los Angeles County Museum Contributions in Science* 81 (1964): 1–63; S. A. Smith and R. A. Paselk, "Olfactory Sensitivity of the Turkey Vulture (*Cathartes aura*) to Three Carrion-Associated Odorants," *Auk* 103, no. 3 (1986): 586–92, https://doi.org/10.1093/auk/103.3.586.

11. Advisory Committee on Childhood Lead Poisoning Prevention of the Centers for Disease Control and Prevention, "Low Level Lead Exposure Harms Children: A Renewed Call for Primary Prevention," Centers for Disease Control and Prevention, 2012.

12. Following these events, sixteen condors remained in the wild. In addition to the five deaths, four newly released juveniles had been returned to captivity temporarily, after showing a lack of wariness around people and a disregard for their own safety.

Chapter 17. Hard Lessons

1. Although the last egg to be laid in the wild (prior to the capture of the last remaining wild birds in 1987) was in 1986, the last successful wild hatching was in 1984.

2. Osborn, *Condors in Canyon Country*.

3. My decisions, informed as they were by years of field experience and a master's degree in biology, were often challenged by relatively inexperienced, male field assistants. Interestingly, when my instructions were conveyed by a male colleague (a Grand Canyon biologist with less experience and an unfinished degree), they were never questioned.

4. T. S. Eliot, "The Hollow Men," 1925, https://allpoetry.com/The-Hollow-Men.

5. In 2009, the International Union of Geological Sciences revised the time boundary between the Pleistocene and the preceding Pliocene from 1.8 million years before the present (BP) to the currently accepted 2.588 million years BP. As a result, publications prior to 2009 may use either definition of the Pleistocene period.

6. R. De Saussure, "Remains of the California Condor in Arizona Caves," *Plateau* 29, no. 2 (1956): 44–45; P. Brodkorb, "Catalogue of Fossil Birds. Part 2 (Anseriformes through Galliformes)," *Bulletin of the Florida State Museum Biological Sciences* 8, no. 3 (1964): 195–335; P. W. Parmalee, "California Condor and Other Birds from Stanton Cave, Arizona," *Journal of the Arizona Academy of Sciences* 5 (1969): 204–206 (and references therein); D. D. Simons, "Interactions between California Condors and Humans in Prehistoric Far Western North America," in *Vulture Biology and Management*, eds. S. R. Wilbur and J. A. Jackson (University of California Press, 1983), 470–94; D. W. Steadman and N. G. Miller, "California Condor Associated with Spruce-Jack Pine Woodland in the Late Pleistocene of New York," *Quaternary Research* 28, no. 3 (1987): 415–26, https://doi.org/10.1016/0033-5894(87)90008-1; Emslie, "Age and Diet of Fossil California Condors"; Snyder and Snyder, *The California Condor*.

7. Simons, "Interactions between California Condors and Humans," 470–94; Snyder and Snyder, *The California Condor*.

8. D. L. Fischer, *Early Southwest Ornithologists, 1528–1900* (University of Arizona Press, 2001); Snyder and Snyder, *The California Condor*.

9. Simons, "Interactions between California Condors and Humans," 470–94; Snyder and Snyder, *The California Condor*. Also, references therein; C. P. Chamberlain et al., "Pleistocene to Recent Dietary Shifts in California Condors," *Proceedings of the National Academy of Sciences of the USA* 102, no. 46 (2005): 16707–11, https://doi.org/10.1073/pnas.0508529102.

10. M. Lewis, W. Clark, and Members of the Corps of Discovery, *The Journals of the Lewis and Clark Expedition*, ed. G. Moulton (University of Nebraska—Lincoln Libraries Electronic Text Center, 2002), https://lewisandclarkjournals.unl.edu/item/lc.jrn.1806-02-17#ln28021710.

11. Cattle were introduced to California in the 1770s (during the Spanish colonial period) and proliferated rapidly, numbering approximately 75,000 animals by 1800. Chamberlain et al., "Pleistocene to Recent Dietary Shifts in California Condors."

12. Emslie, "Age and Diet of Fossil California Condors"; Snyder and Snyder, *The California Condor*; Snyder and Schmitt, "California Condor." Paleontologist Steven D. Emslie argues that California condors died out in the inland West, following the extinction of the Pleistocene megafauna, and were restricted solely to the Pacific Coast, where they fed on whales, seals, and fish. He believes that historical sightings of condors in the inland West represented a return or reinvasion of condors to parts of their former range, which was initiated by the introduction of large herds of cattle, horses, and sheep, beginning in the 1700s.

13. C. G. Thelander and M. Crabtree, eds., *Life on the Edge: A Guide to California's Natural Resources* (Wildlife. BioSystems Books, 1994).

14. Snyder and Snyder, *The California Condor*.

15. Snyder and Snyder, *The California Condor*; N. R. R. Snyder, "Limiting Factors for Wild California Condors," in *California Condors in the 21st Century*, eds. A. Mee and L. S. Hall, Series in Ornithology 2 (Nuttall Ornithological Club and The American Ornithologists' Union, 2007), 9–33.

16. M. W. Lyon Jr., "Occurrence of California Vulture in Idaho," *Journal of the Washington Academy of Sciences* 8 (1918): 25–28. Referenced in Snyder and Snyder, *The California Condor*.

17. Snyder and Snyder, *The California Condor*; Snyder and Schmitt, "California Condor."

18. L. Kiff, "The California Condor Recovery Programme," in *Raptors at Risk. World Working Group on Birds of Prey and Owls*, eds. R. D. Chancellor and B.-U. Meyburg (Hancock House Publishers, 2000), 307–19.

19. Snyder and Snyder, *The California Condor*.

20. D. Darlington, *In Condor Country: A Portrait of a Landscape, Its Denizens, and Its Defenders* (Houghton Mifflin Company, 1987).

21. Snyder and Snyder, *The California Condor*.

22. Snyder and Snyder, *The California Condor*; Snyder and Schmitt, "California Condor." The oldest known California Condor, a captive bird known as Topa Topa (studbook #1) that was hatched in 1966 and taken into captivity in 1967, celebrated his 57th birthday at the Los Angeles Zoo in 2023.

23. Snyder and Snyder, *The California Condor*.

24. Conversation with the author in 2004, as recounted in Osborn, *Condors in Canyon Country*.

25. Kiff, "The California Condor Recovery Programme," 307–19.

26. N. F. R. Snyder and E. V. Johnson, "Photographic Censusing of the 1982–1983 California Condor Population," *Condor* 87, no. 1 (1985): l–13, https://doi.org/10.2307/1367123; Snyder and Snyder, *The California Condor*.

27. D. L. Janssen et al., "Lead Poisoning in Free-Ranging California Condors," *Journal of the American Veterinary Medical Association* 189, no. 9 (1986): 1115–17.

28. Conversation with the author in 2004, as recounted in Osborn, *Condors in Canyon Country*.

29. Condors initially were bred in captivity at the Los Angeles Zoo and the San Diego Zoo. The Peregrine Fund joined the condor captive-breeding effort in 1993, the Oregon Zoo began breeding condors in 2003, and the Chapultepec Zoo in Mexico began breeding condors in 2014.

30. Snyder and Snyder, *The California Condor*, 296.

31. Osborn, *Condors in Canyon Country*.

32. One of the men eventually received a $1,500 fine and two years of probation for the crime; J. Hulse, "2 Men Accused of Trying to Shoot California Condor," *Los Angeles Times*, March 18, 1993, http://articles.latimes.com/1993-03-18/news/mn-12341_1_california-condor.

33. In 2023, twenty-one Arizona and Utah condors died in less than two months after contracting a highly pathogenic avian influenza (HPAI) virus, highlighting the risk of novel diseases to birds and people alike. Condors in other states were unaffected.

34. A. Rea, "California Condor Captive Breeding: A Recovery Proposal," *Environment Southwest* 492 (1981): 8–12.

35. Osborn, *Condors in Canyon Country*.

36. Endangered Species Act of 1973 (ESA; 16 U.S.C. § 1531 et seq.).

37. Osborn, *Condors in Canyon Country*; Associated Press, "Condors Sighted at Mesa Verde," July 3, 1999. This article is no longer available online.

Chapter 18. Growing Up

1. R. G. Clark and R. D. Ohmart, "Spread-Winged Posture of Turkey Vultures: Single or Multiple Function?" *Condor* 87, no. 3 (1985): 350–55, https://doi.org/10.2307/1367215.

2. J. E. Heath, "Temperature Fluctuation in the Turkey Vulture," *Condor* 64, no. 3 (1962): 234–35, https://doi.org/10.2307/1365205; D. E. Hatch, "Energy Conserving and Heat Dissipating Mechanisms of the Turkey Vulture," *Auk* 87, no. 1 (1970): 111–24, https://doi.org/10.2307/4083662; O. Bahat, I. Choshniak, and D. C. Houston, "Nocturnal Variation in Body Temperature of Griffon Vultures," *Condor* 100, no. 1 (1998): 168–71, https://doi.org/10.2307/1369911.

3. D. C. Houston, "A Possible Function of Sunning Behavior by Griffon Vultures, *Gyps* spp., and Other Large Soaring Birds," *Ibis* 122, no. 3 (1980): 366–69, https://doi.org/10.1111/j.1474-919X.1980.tb00892.x.

4. Snyder and Snyder, *The California Condor*; Snyder and Schmitt, "California Condor."

5. Biologists in California suspected that one of their condors had laid an egg that failed to hatch in 2000, but had been unable to confirm it.

6. National Public Radio, "Condor Egg," Living on Earth, April 13, 2001, http://www.loe.org/shows/shows.html?programID=01-P13-00015#feature7.

7. Osborn, *Condors in Canyon Country*.

8. Heinrich, *Mind of the Raven*, 70–71.

9. This recounting actually combines two separate play episodes that I witnessed; the first involved eleven birds on the beach and the second involved the condors playing with the water jug; Osborn, *Condors in Canyon Country*.

10. Snyder and Snyder, *The California Condor*.

11. Osborn, *Condors in Canyon Country*.

12. After additional transgressions, Condor 224 was returned to the captive breeding facility in Boise, Idaho, and rereleased the following year.

13. Snyder and Snyder, *The California Condor*; Osborn, *Condors in Canyon Country*.

Chapter 19. Shot

1. Calf carcasses were acquired from a dairy farm (calves being the unwanted byproducts of getting mother cows to produce milk). Saddened as I was by the calves' demise (I was never able to face going to the dairy to pick up the dead calves; other crewmembers kindly went in my stead), I was nonetheless glad that the calves were put to valuable use as condor food.

2. When a juvenile condor first landed on the mock power pole, it inevitably lifted a foot as it felt the mild shock in its feet, then quickly took flight and landed on a more comfortable perch. After several attempted landings, the youngsters avoided the perch altogether.

3. Prior to this release, juvenile condors had to exit the pen through a side door—which many took hours to find—on release day. Before juveniles were transported to the release pen for the February 2001 release, project director Chris Parish retrofitted the pen with a new front gate that could be raised by a hidden crewmember, facilitating the condors' transition to freedom.

4. The egg that became Condor 240 was taken from 176 (and 187)'s captive parents, and hatched in an incubator on April 11, 2001. Condor 240 was then raised by human caretakers—wearing a condor arm puppet—at The Peregrine Fund's World Center for Birds of Prey captive breeding facility in Boise, Idaho. Meanwhile, his condor parents laid a replacement egg that became 248, who hatched at the same facility on May 8, 2001, and was raised by the parent condors. As an example of the similarities I noticed among siblings, 176, 187, 223, 240, and 248 were strong-flying siblings that ranged widely and were ferocious in the hand, whereas 203, 210, 224, and 243 were siblings (all puppet-raised) that were particularly unwary around people and had to be recaptured for poor behavior. Interestingly, the parent-raised siblings of this last group (235, 257, and 280) exhibited stellar behavior and were not recaptured. The majority of our "problem" condors were the offspring of only three captive pairs.

5. Heartbreakingly, Condor 252 was killed by coyotes a week later. His untimely death taught us that the sheer bands of scree that divided the cliff layers forming the lower part of the Vermilion Cliffs—an area we had thought was safe for perching condors—actually were accessible to agile canids. After 252's death (during a week of storms and poor visibility), the field crew scaled the cliffs and talus to haze inexperienced condors that tried to roost in this area, and we never again lost juveniles post-release during my tenure as field manager.

6. Mortality and necropsy reports for Condor 186. Author's collections.

7. In later years, as the condor flock grew, the numbering schemes for tags became increasingly complicated, with dots and letters being used as stand-ins for the last two digits of a studbook number.

8. S. Osborn, "Notes from the Field—May 16–31, 2002," The Peregrine Fund, 2002. Author's collections. These notes are also archived at the Santa Barbara Museum of Natural History.

9. This number includes 186 being shot by an arrow. Data courtesy of the US Fish and Wildlife Service.

10. Osborn, *Condors in Canyon Country*.

11. D. M. Fry and J. R. Maurer, "Assessment of Lead Contamination Sources Exposing California Condors," Final report to the California Department of Fish and Game, 2003.

12. J. Moir, *Return of the Condor: The Race to Save Our Largest Bird from Extinction* (Lyons Press, 2006).

13. Forest Watch, "In Memory of Condor 526."

Chapter 20. A Sisyphean Predicament

1. In 2002 and 2003, we documented 130 carcasses found by our condors. The birds likely found many more since so much of the area in which they ranged was inaccessible to us; S. A. H. Osborn and C. Olson, "Foraging on Non-proffered Carcasses by California Condors (*Gymnogyps californianus*) Reintroduced in Northern Arizona," unpublished manuscript, author's collections, 2005.

2. The six-cow strike actually occurred in August 2003 rather than in August 2002 as I have presented it here.

3. Fry and Maurer, "Assessment of Lead Contamination Sources."

4. W. G. Hunt, C. N. Parish, S. C. Farry, et al., "Movements of Introduced California Condors in Arizona in Relation to Lead Exposure," in *California Condors in the 21st Century*, 79–96.

5. Hunt et al., "Movements of Introduced California Condors," in *California Condors in the 21st Century*, 79–96.

6. I did not have a record of the exact value of 187's lead test, so I provided a representative value for a bird that was releasable after lead testing.

7. A close examination of data collected over a number of years bore this out: Condors most often had high lead levels during the fall hunting season on the Kaibab Plateau (and during fall hunting seasons in California), and condors that spent more time on the Kaibab Plateau—where they most often found hunter-killed deer—were more likely to have elevated lead levels when tested; Hunt et al., "Movements of Introduced California Condors," 79–96.

8. In 2003, we began trapping condors before the hunting season began and held them in the flight pen for the duration of the season, an unsustainable strategy that nevertheless kept our birds safe for a time.

9. G. B. Grinnell, "Lead Poisoning," *Forest Stream* 42, no. 6 (1894): 117–18; F. C. Bellrose, "Lead Poisoning as a Mortality Factor in Waterfowl Populations," *Illinois Natural History Survey Bulletin* 27 (1959): 234–88, https://doi.org/10.21900/j.inhs.v27.172. Hunters typically shoot waterfowl with shotgun shells, which are packed with small pellets—previously lead, now steel—that spray widely when fired. Bellrose conservatively estimated that approximately 1,400 lead shot pellets were deposited in marshes and lakes for every duck that was killed (pp. 249–50).

10. US Fish and Wildlife Service, "Proposed Use of Steel Shot for Hunting Waterfowl in the United States," Final Environmental Impact Statement, Department of the Interior, 1976.

11. W. L. Reichel et al., "Pesticide, PCB, and Lead Residues and Necropsy Data for Bald Eagles from 32 States—1978–81," *Environmental Monitoring and Assessment* 4 (1984): 395–403, https://doi.org/10.1007/BF00394177.

12. Snyder, "Limiting Factors for Wild California Condors," 9–33. This was an average annual rate over a four-year period. Yearly rates were 18.9%, 16.7%, 43.2%, and 27.5%.

13. Snyder, "Limiting Factors for Wild California Condors," 9–33.

14. L. Miller, "Succession in the Cathartine Dynasty," *Condor* 44, no. 5 (1942): 212–13, https://doi.org/10.2307/1364131.

15. W. D. Toone and M. P. Wallace, "The Extinction in the Wild and Reintroduction of the California Condor (*Gymnogyps californianus*)," in *Creative Conservation*, eds. P. J. S. Olney, G. M. Mace, and A. T. C. Feistner (Springer, 1994); Snyder, "Limiting Factors for Wild California Condors," 9–33; The fourth condor that was found dead had been shot.

16. Snyder, "Limiting Factors for Wild California Condors," 9–33.

17. Summarized in B. A. Rideout et al., "Patterns of Mortality in Free-Ranging California Condors (*Gymnogyps californianus*)," *Journal of Wildlife Diseases* 48, no. 1 (2012): 95–112, https://doi .org/10.7589/0090-3558-48.1.95.

18. Pattee et al., "Lead Hazards within the Range of the California Condor."

19. W. G. Hunt et al., "Bullet Fragments in Deer Remains: Implications for Lead Exposure in Avian Scavengers," *Wildlife Society Bulletin* 34, no. 1 (2006): 167–70, https://doi.org/10.2193/0091 -7648(2006)34[167:BFIDRI]2.0.CO;2; M. D. Grund et al., "Bullet Fragmentation and Lead Deposition in White-Tailed Deer and Domestic Sheep," *Human-Wildlife Interactions* 4, no. 2 (2010): 257–65, https://www.jstor.org/stable/24863845; F. Gremse et al., "Performance of Lead-Free versus Lead-Based Hunting Ammunition in Ballistic Soap," *PLoS ONE* 9, no. 7 (2014): p.e102015, https://doi.org/10.1371/journal.pone.0102015.

20. Hunt et al., "Bullet Fragments in Deer Remains."

21. S. R. Beissinger et al., "Science Links Lead Ammunition to Lead Exposure in California Condors (*Gymnogyps californianus*)—Statement of Scientific Agreement," July 10, 2007, https://www .biologicaldiversity.org/species/birds/California_condor/pdfs/Condor-Lead-Science.pdf.

22. When lead was used as an additive for gasoline, it seeped into the ground, washed into waterways, and was blown into the air as vehicle exhaust.

23. A. M. Scheuhammer and D. M. Templeton, "Use of Stable Isotope Ratios to Distinguish Sources of Lead Exposure in Wild Birds," *Ecotoxicology* 7 (1998): 37–42, https://doi.org/10.1023 /A:1008855617453.

24. M. E. Church et al., "Ammunition is the Principal Source of Lead Accumulated by California Condors Reintroduced to the Wild," *Environmental Science and Technology* 40 (2006): 6143–50.

25. N. F. R. Snyder, E. V. Johnson, and D. A. Clendenen, "Primary Molt of California Condors," *Condor* 89, no. 3 (1987): 468–85, https://doi.org/10.2307/1368637; Fry and Maurer, "Assessment of Lead Contamination Sources Exposing California Condors"; M. E. Finkelstein et al., "Feather Lead Concentrations and [207]Pb/[206]Pb Ratios Reveal Lead Exposure History of California Condors (*Gymnogyps californianus*)," *Environmental Science and Technology* 44, no. 7 (2010): 2639–47, https://doi.org/10.1021/es903176w.

26. Finkelstein et al., "Feather Lead Concentrations."

27. Finkelstein et al., "Feather Lead Concentrations."

28. Snyder, "Limiting Factors for Wild California Condors," 9–33.

29. Rideout et al., "Patterns of Mortality in Free-Ranging California Condors." There were a total of 135 condor deaths from October 1992 through December 2009. The 2012 study documenting these fatalities examined 98 carcasses but could only determine cause of death for 76 birds. A large proportion of unrecoverable birds also likely died of lead poisoning.

30. Rideout et al., "Patterns of Mortality in Free-Ranging California Condors."

31. C. N. Parish, W. R. Heinrich, and W. G. Hunt, "Lead Exposure, Diagnosis, and Treatment in California Condors Released in Arizona," in *California Condors in the 21st Century*, 97–108.

32. M. Fry et al., "Lead Intoxication Kinetics in Condors from California. Abstract," in *Ingestion of Lead from Spent Ammunition: Implications for Wildlife and Humans*, eds. R. T. Watson et al. (The Peregrine Fund, 2009), 266.

33. The Peregrine Fund, *California Condor Releases in Arizona: Notes from the Field, 2010.* Author's collections. These notes are also archived at the Santa Barbara Museum of Natural History.

34. C. P. Woods et al., "Survival and Reproduction of California Condors Released in Arizona," in *California Condors in the 21st Century,* 79–96; J. R. Walters et al., "Status of the California Condor (*Gymnogyps californianus*) and Efforts to Achieve Its Recovery," *Auk* 127, no. 4 (2010): 969–1001, https://doi.org/10.1525/auk.2010.127.4.969; M. E. Finkelstein et al., "Lead Poisoning and the Deceptive Recovery of the Critically Endangered California Condor,"

Proceedings of the National Academy of Sciences 109, no. 28 (2012): 11449–54, https://doi.org /10.1073/pnas.1203141109.

35. Finkelstein et al., "Lead Poisoning and the Deceptive Recovery."

36. W. Stansley and D. E. Roscoe, "The Uptake and Effects of Lead in Small Mammals and Frogs at a Trap and Skeet Range," *Archives of Environmental Contamination and Toxicology* 30 (1996): 220–26, https://doi.org/10.1007/BF00215801; I. J. Fisher, D. J. Pain, and V. G. Thomas, "A Review of Lead Poisoning from Ammunition Sources in Terrestrial Birds," *Biological Conservation* 131, no. 3 (2006): 421–32, https://doi.org/10.1016/j.biocon.2006.02.018; M. A. Tranel and R. O. Kimmel, "Impacts of Lead Ammunition on Wildlife, the Environment, and Human Health: A Literature Review and Implications for Minnesota," in *Ingestion of Lead from Spent Ammunition*, 318–37.

37. J. H. Schulz et al., "Spent-Shot Availability and Ingestion on Areas Managed for Mourning Doves," *Wildlife Society Bulletin* 30, no. 1 (2002): 112–20, https://www.jstor.org/stable/3784644; J. H. Schulz et al., "Acute Lead Toxicosis in Mourning Doves," *Journal of Wildlife Management* 70, no. 2 (2006): 413–21, https://doi.org/10.2193/0022-541X(2006)70[413:ALTIMD]2.0.CO;2.

38. J. L. Kramer and P. T. Redig, "Sixteen Years of Lead Poisoning in Eagles, 1980–1995: An Epizootiologic View," *Journal of Raptor Research* 31, no. 4 (1997): 327–32.

39. B. Bedrosian and D. Craighead, "Blood Lead Levels of Bald and Golden Eagles Sampled during and after Hunting Seasons in the Greater Yellowstone Ecosystem. Extended Abstract," in *Ingestion of Lead from Spent Ammunition*, 219–20; R Domenech and H. Langner, "Blood-Lead Levels of Fall Migrant Golden Eagles in West-Central Montana. Extended Abstract," in *Ingestion of Lead from Spent Ammunition*, 221–22; V. A. Slabe et al., "Demographic Implications of Lead Poisoning for Eagles across North America," *Science* 375, no. 6582 (2022): 779–82.

40. D. Craighead and B. Bedrosian, "Blood Lead Levels of Common Ravens with Access to Big-Game Offal," *Journal of Wildlife Management* 72, no. 1 (2008): 240–45, https://jstor.org/stable /25097524; D. Craighead and B. Bedrosian, "A Relationship between Blood Lead Levels of Common Ravens and the Hunting Season in the Southern Yellowstone Ecosystem," in *Ingestion of Lead from Spent Ammunition*, 202–5.

41. M. C. Smith et al., "Lead Shot Poisoning in Swans: Sources of Pellets within Whatcom County, WA, USA, and Sumas Prairie, BC, Canada. Extended Abstract," in *Ingestion of Lead from Spent Ammunition*, 274–77; personal communication with Michael C. Smith, August 2015.

42. W. E. Cornatzer, E. F. Fogarty, and E. W. Cornatzer, "Qualitative and Quantitative Detection of Lead Bullet Fragments in Random Venison Packages Donated to the Community Action Food Centers of North Dakota, 2007," in *Ingestion of Lead from Spent Ammunition*, 154–56.

43. Grund et al., "Bullet Fragmentation and Lead Deposition," *Human-Wildlife Interactions*.

44. For Wisconsin, see Wisconsin Department of Health and Family Services, Health Consultation: The Potential for Ingestion Exposure to Lead Fragments in Venison in Wisconsin, November 4, 2008, https://www.atsdr.cdc.gov/hac/pha/LeadFragmentsinVenison/Venison%20and%20 Lead%20HC%20110408.pdf.

45. D. Avery and R. T. Watson, "Distribution of Venison to Humanitarian Organizations in the USA and Canada," in *Ingestion of Lead from Spent Ammunition*, 157–60.

46. US Environmental Protection Agency, "Lead in Toys and Toy Jewelry," 2004, https://archive. epa.gov/multimedia/web/multimedia/lead_REMOVEDFEB23/toys.html#:~:text=In%20 2004%2C%20the%20threat%20of,sold%20widely%20in%20vending%20machines.

47. D. C. Bellinger et al., "Health Risks from Lead-Based Ammunition in the Environment—A Consensus Statement of Scientists," *Microbiology and Environmental Toxicology* (2013), http:// escholarship.org/uc/item/6dq3h64x.

48. US Fish and Wildlife Service, "2011 National Survey of Fishing, Hunting, and Wildlife-Associated Recreation—National Overview," (Issued August 2012), US Department of the

Interior, US Fish and Wildlife Service. Yearly numbers of hunters are variable. (The 2012 report provided numbers of adults and children that hunted.) According to the 2022 National Survey, 14.4 million people 16 years and older hunted in the US.

49. P. Johansen et al., "Lead Shot from Hunting as a Source of Lead in Human Blood," *Environmental Pollution* 142, no.1 (2006): 93–97, https://doi.org/10.1016/j.envpol.2005.09.015; L. J. S. Tsuji et al., "The Identification of Lead Ammunition as a Source of Lead Exposure in First Nations: The Use of Lead Isotope Ratios," *Science of the Total Environment* 393, nos. 2–3 (2008): 291–98, https://doi.org/10.1016/j.scitotenv.2008.01.022; S. Iqbal et al., "Hunting with Lead: Association between Blood Levels and Wild Game Consumption," *Environmental Research* 109 (2009): 952–59.

50. A. McKean, "Update: Lead in Your Meat? Are High-Velocity Bullets Tainting Your Venison?" *Outdoor Life*, December 22, 2008. This article is no longer available on the *Outdoor Life* website. However, these quotations are repeated in the following article: A. McKean, "Is Game Meat Shot with Lead Safe to Eat?" *Outdoor Life*, October 15, 2019, https://www.outdoorlife.com/is-game-meat-shot-with-lead-safe-to-eat.

51. E. K. Silbergeld, J. Schwartz, and K. Mahaffey, "Lead and Osteoporosis: Mobilization of Lead from Bone in Postmenopausal Women," *Environmental Research* 47, no. 1 (1988): 79–94, https://doi.org/10.1016/s0013-9351(88)80023-9; H. Needleman, "Lead Poisoning," *Annual Review of Medicine* 55 (2004): 209–22, https://doi.org/10.1146/annurev.med.55.091902.103653; M. J. Kosnett, "Health Effects of Low Dose Lead Exposure in Adults and Children, and Preventable Risk Posed by the Consumption of Game Meat Harvested with Lead Ammunition," in *Ingestion of Lead from Spent Ammunition*, 24–33; M.A. Pokras and M. R. Kneeland, "Understanding Lead Uptake and Effects across Species Lines: A Conservation Medicine Based Approach," in *Ingestion of Lead from Spent Ammunition*, 7–22.

52. Needleman, "Lead Poisoning"; K. M. Cecil et al., "Decreased Brain Volume in Adults with Childhood Lead Exposure," *PLoS Medicine* 5, no. 5 (2008): e112, https://doi.org/10.1371/journal.pmed.0050112; Pokras and Kneeland, "Understanding Lead Uptake and Effects," 7–22; Kosnett, "Health Effects of Low Dose Lead Exposure," in *Ingestion of Lead from Spent Ammunition*, 24–33.

53. Fry and Maurer, "Assessment of Lead Contaminations Sources Exposing California Condors."

54. S. B. Muldoon et al., "Effects of Blood Lead Levels on Cognitive Function of Older Women," *Neuroepidemiology* 15, no. 2 (1996): 62–72, https://doi.org/10.1159/000109891; B. S. Schwartz et al., "Past Adult Lead Exposure is Associated with Longitudinal Decline in Cognitive Function," *Neurology* 55, no. 8 (2000): 1144–50, https://doi.org/10.1212/wnl.55.8.1144; M. G. Weisskopf et al., "Cumulative Lead Exposure and Cognitive Performance among Elderly Men," *Epidemiology* 18, no. 1 (2007): 59–66, https://doi.org/10.1097/01.ede.0000248237.35363.29.

55. M. Lustberg and E. Silbergeld, "Blood Lead Levels and Mortality," *Archives of Internal Medicine* 162, no. 21 (2002): 2443–49, https://doi.org/10.1001/archinte.162.21.2443.

56. A. Menke et al., "Blood Lead below 0.48 μmol (10 μg/dL) and Mortality among US Adults," *Circulation* 114 (2006): 1388–94.

57. R. L. Canfield et al., "Intellectual Impairment in Children with Blood Lead Concentrations below 10 μg per Deciliter," *New England Journal of Medicine* 348, no. 16 (2003): 1517–26, https://doi.org/10.1056/NEJMoa022848; Needleman, "Lead Poisoning," (and references therein); T. A. Jusko et al., "Blood Lead Concentrations <10 μg/dL and Child Intelligence at 6 Years of Age," *Environmental Health Perspectives* 116, no. 2 (2007): 243–48, https://doi.org/10.1289/ehp.10424; D. C. Bellinger, "Neurological and Behavioral Consequences of Childhood Lead Exposure," *PLoS Medicine* 5, no. 5 (2008): e115, https://doi.org/10.1371/journal.pmed.0050115; Kosnett, "Health Effects of Low Dose Lead Exposure," 24–33.

58. S. D. Grosse et al., "Economic Gains Resulting from the Reduction in Children's Exposure to Lead in the United States," *Environmental Health Perspectives* 110, no.6 (2002): 563–69, https://doi.org/10.1289/ehp.02110563.

59. K. N. Dietrich et al., "Early Exposure to Lead and Juvenile Delinquency," *Neurotoxicology and Teratology* 23, no. 6 (2001): 511–18, https://doi.org/10.1016/s0892-0362(01)00184-2; Needleman, "Lead Poisoning," (and references therein); D. C. Bellinger, "Very Low Lead Exposures and Children's Neurodevelopment," *Current Opinion in Pediatrics* 20, no. 2 (2008): 172–77, https://doi.org/10.1097/mop.0b013e3282f4f97b.

60. P. B. Stretesky and M. J. Lynch, "The Relationship between Lead Exposure and Homicide," *Archives of Pediatrics and Adolescent Medicine* 155, no. 5 (2001): 579–82, https://doi.org/10.1001/archpedi.155.5.579; Needleman, "Lead Poisoning"; J. P. Wright et al., "Association of Prenatal and Childhood Blood Lead Concentrations with Criminal Arrests in Early Adulthood," *PLoS Medicine* 5, no. 5 (2008): e101, https://doi.org/10.1371/journal.pmed.0050101; M. P. Taylor et al., "The Relationship between Atmospheric Lead Emissions and Aggressive Crime: An Ecological Study," *Environmental Health* 15 (2016): 1–10, https://doi.org/10.1186/s12940-016-0122-3.

61. Wright et al., "Association of Prenatal and Childhood Blood Lead Concentrations."

62. R. Nevin, "How Lead Exposure Relates to Temporal Changes in IQ, Violent Crime, and Unwed Pregnancy," *Environmental Research* 83, no. 1 (2000): 1–22, https://doi.org/10.1006/enrs.1999.4045; Stretesky and Lynch, "The Relationship between Lead Exposure and Homicide"; J. W. Reyes, "Environmental Policy as Social Policy? The Impact of Childhood Lead Exposure on Crime," *BE Journal of Economic Analysis & Policy* 7, no. 1 (2007), https://doi.org/10.2202/1935-1682.1796; Taylor et al., "The Relationship between Atmospheric Lead Emissions and Aggressive Crime."

63. Stretesky and Lynch, "The Relationship between Lead Exposure and Homicide."

64. Needleman, "Lead Poisoning"; Reyes, "Environmental Policy as Social Policy?"; K. Drum, "Lead: America's Real Criminal Element. The Hidden Villain behind Violent Crime, Lower IQs, and Even the ADHD Epidemic," *Mother Jones*, Jan/Feb 2013, https://www.motherjones.com/environment/2016/02/lead-exposure-gasoline-crime-increase-children-health.

65. J. M. Braun et al., "Exposures to Environmental Toxicants and Attention Deficit Hyperactivity Disorder in U.S. Children," *Environmental Health Perspectives* 114, no. 12 (2006): 1904–9, https://doi.org/10.1289/ehp.9478; Kosnett, "Health Effects of Low Dose Lead Exposure," 24–33. One study (Braun et al., 2006) found that 290,000 cases of ADHD in US children 4 to 15 years of age are attributable to environmental lead exposure.

66. Kosnett, "Health Effects of Low Dose Lead Exposure," 24–33; Bellinger et al., "Health Risks from Lead-Based Ammunition in the Environment"; A. Bernhoft et al., "Wildlife and Human Health Risks from Lead-Based Ammunition in Europe: A Consensus Statement by Scientists" (2014), http://www.zoo.cam.ac.uk/leadammunitionstatement.

67. J. M. Arnemo et al., "Health and Environmental Risks from Lead-Based Ammunition: Science versus Socio-Politics," *EcoHealth* 13 (2016): 618–22, https://doi.org/10.1007/s10393-016-1177-x.

68. R. K. Stroud and W. G. Hunt, "Gunshot Wounds: A Source of Lead in the Environment," in *Ingestion of Lead from Spent Ammunition*, 119–25.

69. J. Knott et al., "Comparison of the Lethality of Lead and Copper Bullets in Deer Control Operations to Reduce Incidental Lead Poisoning; Field Trials in England and Scotland," *Conservation Evidence* 6 (2009): 71–78; Grund et al., "Bullet Fragmentation and Lead Deposition"; A. Trinogga et al., "Are Lead-Free Hunting Rifle Bullets as Effective at Killing Wildlife as Conventional Lead Bullets? A Comparison Based on Wound Size and Morphology," *Science of the Total Environment* 443, no. 15 (2013): 226–32, https://doi.org/10.1016/j.scitotenv.2012.10.084; V. G. Thomas, "Lead-Free Hunting Rifle Ammunition: Product Availability, Price, Effectiveness, and Role in Global Wildlife Conservation," *Ambio* 42 (2013): 737–45, https://doi.org/10.1007/s13280-012-0361-7; Gremse et al., "Performance of Lead-Free versus Lead-Based Ammunition."

70. Thomas, "Lead-Free Hunting Rifle Ammunition."

71. The Institute for Wildlife Studies, "Hunting with Non-lead" program provides hunters with information about hunting with non-lead to help them navigate this transition: https:// huntingwithnonlead.org.

72. Thomas, "Lead-Free Hunting Rifle Ammunition."

Chapter 21. No Tags

1. I first heard the phrase "prisoners of hope" used by Senator Cory Booker. Although it comes from a biblical passage—Zechariah 9:12—I use it to refer to those who remain hopeful in spite of circumstances that often seem hopeless.

2. Although reintroductions started in California in 1992, all reintroduced birds were recaptured in 1994 because of behavioral problems (approaching people and landing on human structures). Reintroductions in southern California began again in 1995, a year before reintroductions were initiated in Arizona.

3. The ailing embryo successfully hatched, becoming Condor 262. He was released to the wild at the Hopper Mountain National Wildlife Refuge in 2002. In 2019, webcam viewers around the world watched Condor 262 successfully raise a chick of his own in the wild in California: Tyler Hayden, "Meet the Condor Cam's New Stars," *Santa Barbara Independent*, June 10, 2019, https://www .independent.com/2019/06/10/meet-the-condor-cams-new-stars.

4. US Fish and Wildlife Service, "California Condor Recovery Plan, Third Revision" (1996): 62 pages.

5. US Fish and Wildlife Service, "California Condor Chick Hatched in the Wild Dies," press release (June 27, 2001). This press release is no longer available, but much of its information is contained in the following article: M. Surman, "Condor Hatched in Wild is Killed," *Los Angeles Times*, June 28, 2001, https://www.latimes.com/archives/la-xpm-2001-jun-28-me-16011-story.html.

6. Osborn, *Condors in Canyon Country*.

7. Osborn, *Condors in Canyon Country*. Personal communication with Allan Mee, November 2001.

8. Based on anatomical, morphological, behavioral, molecular, and early DNA analyses, California condors and other New World vultures (family Cathartidae) were believed to be more closely related to storks (Ciconiidae) than to Old World vultures (Accipitridae). However, more recent DNA analyses refute this close association with storks and suggest that New World vultures are more closely related to Old World vultures. Nevertheless, the taxonomic placement of New World vultures remains uncertain; S. J. Hackett et al., "A Phylogenomic Study of Birds Reveals Their Evolutionary History," *Science* 320, no. 5884 (2008): 1763–68, https://doi.org/10.1126 /science.1157704.

9. D. C. Houston, A. Mee, and M. McGrady, "Why Do Condors and Vultures Eat Junk?: The Implications for Conservation," *Journal of Raptor Research* 41, no. 3 (2007): 235–39.

10. Houston, Mee, and McGrady, "Why Do Condors and Vultures Eat Junk?"; A. Mee et al., "Junk Ingestion and Nestling Mortality in a Reintroduced Population of California Condors *Gymnogyps californianus*," *Bird Conservation International* 17, no. 2 (2007): 119–30, https://doi .org/10.1017/S095927090700069X.

11. P. J. Mundy and J. A. Ledger, "Griffon Vultures, Carnivores and Bones," *South African Journal of Science* 72 (1976): 106–10; P. R. K. Richardson, P. J. Mundy, and I. Plug, "Bone Crushing Carnivores and Their Significance to Osteodystrophy in Griffon Vulture Chicks," *Journal of Zoology* 210, no. 1 (1986): 23–43, https://doi.org/10.1111/j.1469-7998.1986.tb03618.x.

12. P. C. Benson, I. Plug, and J. C. Dobbs, "An Analysis of Bones and Other Materials Collected by Cape Vultures at the Kransberg and Blouberg Colonies, Limpopo Province, South Africa," *Ostrich* 75 (2004): 118–32, https://doi.org/10.2989/00306520409485423.

13. Houston, Mee, and McGrady, "Why Do Condors and Vultures Eat Junk?"; Mee et al., "Junk Ingestion and Nestling Mortality."

14. B. Heinrich, "Neophilia and Exploration in Juvenile Common Ravens, *Corvus corax*," *Animal Behaviour* 50, no. 3 (1995): 695–704, https://doi.org/10.1016/0003-3472(95)80130-8; Heinrich, *Mind of the Raven*.

15. Houston, Mee, and McGrady, "Why Do Condors and Vultures Eat Junk?"

16. A. Mee, J. A. Hamber, and J. Sinclair, "Low Nest Success in a Reintroduced Population of California Condors," in *California Condors in the 21st Century*, 163–84; Mee et al., "Junk Ingestion and Nestling Mortality." This nestling's cause of death was unclear. Its growth was retarded and it's digestive tract contained some trash (eleven pieces of plastic and one piece of glass). It had gone without food for at least six to eight days, in hot weather, prior to its death and likely died of dehydration. It had histological evidence of visceral gout, a condition in which decreased kidney function leads to uric acid (nitrogenous waste) accumulating in the blood and body fluids.

17. Osborn, *Condors in Canyon Country*; Mee et al., "Junk Ingestion and Nestling Mortality."

18. Snyder and Snyder, *The California Condor*.

19. K. Bastone, "The 10 Most Dangerous Hikes in America: Bright Angel Trail, AZ—Baked or Broiled?" *Backpacker*, Outside, 2019 (original version, 2008), https://www.backpacker.com/survival/bright-angel-trail-grand-canyon-america-s-most-dangerous-hikes.

20. Osborn, *Condors in Canyon Country*.

21. J. I. Mead, "The Last 30,000 Years of Faunal History within the Grand Canyon, Arizona," *Quaternary Research* 15, no. 3 (1981): 311–26, https://doi.org/10.1016/0033-5894(81)90033-8; J. I. Mead and A. M. Phillips III, "The Late Pleistocene and Holocene Fauna and Flora of Vulture Cave, Grand Canyon, Arizona," *The Southwestern Naturalist* 26, no. 3 (1981): 257–8, https://doi.org/10.2307/3670906; author correspondence with J. I. Mead, Northern Arizona University, August to October 2005.

22. P. S. Martin, B. E. Sabels, and D. Shutler, "Rampart Cave Coprolite and Ecology of the Shasta Ground Sloth," *American Journal of Science* 259, no. 2 (1961): 102–7; Mead, "The Last 30,000 Years of Faunal History"; L. Nelson, *Ice Age Mammals of the Colorado Plateau* (Northern Arizona University, 1990); M. Kropf, J. I. Mead, and R. S. Anderson, "Dung, Diet, and the Paleoenvironment of the Extinct Shrub-Ox (*Euceratherium collinum*) on the Colorado Plateau, USA," *Quaternary Research* 67, no. 1 (2007): 143–51, https://doi.org/10.1016/j.yqres.2006.10.002; author correspondence with J. I. Mead, Northern Arizona University, August to October 2005.

23. A. Long, R. M. Hansen, and P. S. Martin, "Extinction of the Shasta Ground Sloth," *Geological Society of America Bulletin* 85, no. 12 (1974): 1843–48, https://doi.org/10.1130/0016-7606(1974)85<1843:EOTSGS>2.0.CO;2; Nelson, *Ice Age Mammals of the Colorado Plateau*; S. B. Cooke et al., "Anthropogenic Extinction Dominates Holocene Declines of West Indian Mammals," *Annual Review of Ecology, Evolution, and Systematics* 48 (2017): 301–27, https://doi.org/10.1146/annurev-ecolsys-110316-022754.

24. Emslie, "Age and Diet of Fossil California Condors."

25. Emslie, "Age and Diet of Fossil California Condors"; Chamberlain et al, "Pleistocene to Recent Dietary Shifts in California Condors."

26. E. Coues, "List of the Birds of Fort Whipple, Arizona: With Which Are Incorporated All Other Species Ascertained to Inhabit the Territory; With Brief Critical and Field Notes, Descriptions of New Species, etc." *Proceedings of the Academy of Natural Sciences of Philadelphia* 18 (1866): 39–100, https://www.jstor.org/stable/4059674; A. Phillips, J. Marshall, and G. Monson, *The Birds of Arizona* (University of Arizona Press, 1964).

27. H. Brown, "The California Vulture in Arizona," *Auk* 16 (1899): 272.

28. S. N. Rhoads, "The Birds of Southeastern Texas and Southern Arizona Observed during May, June, and July 1891," *Proceedings of the Academy of Natural Sciences of Philadelphia* 44 (1892): 98–126, https://www.jstor.org/stable/4061847.

29. Phillips, Marshall, and Monson, *The Birds of Arizona*; Rea, "California Condor Captive Breeding."

30. The first condor fatality from West Nile virus occurred in California in 2005. The condor was a three-month-old nestling that had not been vaccinated. West Nile virus was also determined to be the primary cause of death of a one-and-a-half-year-old condor (Condor 450) that had been vaccinated twice. However, this bird also suffered from respiratory aspergillosis; Rideout et al., "Patterns of Mortality in Free-Ranging California Condors."

31. The South Rim's Bright Angel Trail passes through the approximately 270-million-years-old (myo) Kaibab Limestone, the 273 myo Toroweap Formation, the 275 myo Coconino Sandstone, the 280 myo Hermit Formation, the 300 myo Supai Group, the 340 myo Redwall Limestone, the 505 myo Muav Limestone, and the 515 myo Bright Angel Shale, before arriving at the Havasupai Gardens Campground on the Tonto Plateau. Rock-layer ages are constantly being refined and different sources provide different ages. A. Mathis and C. Bowman, "Telling Time at Grand Canyon National Park," National Park Service, Grand Canyon National Park, 2018, https://www.nps.gov/articles/age-of-rocks-in-grand-canyon.htm.

32. Snyder and Schmitt, "California Condor"; also, photographs of known-age condor nestlings by Mike Clark of the Los Angeles Zoo. Author's collections.

33. Snyder and Schmitt, "California Condor."

34. Snyder and Snyder, *The California Condor*.

Chapter 22. Condor in the Coal Mine

1. Jan Hamber's moving account of AC-9's capture and her involvement with wild condors in California are chronicled in a tribute video by Jeff McLoughlin: Good Eye Films, "Jan Hamber - 40 Years with the California Condor," YouTube video, 8:57, March 16, 2017, https://www.youtube.com/watch?v=Lz9qEpY0BqU.

2. AC-9 remained in the wild until June 2016 when he disappeared into a remote canyon in the Sespe Wilderness and is presumed to have died.

3. For example: C.A. Moss-Racusin et al., "Science Faculty's Subtle Gender Biases Favor Male Students," *Proceedings of the National Academy of Sciences* 109, no. 41 (2012): 16474–79, https://doi.org/10.1073/pnas.1211286109; C. Corbett and C. Hill, *Solving the Equation: The Variables for Women's Success in Engineering and Computing* (American Association of University Women, 2015).

4. S. A. H. Osborn, "Goodbye to 119," Nature Notes—Grand Canyon National Park 13, no. 1 (2007): 1–2.

5. After fledging Condor 350 in November 2004, 119 and 122 laid another egg in March 2006, but during the incubation period, 122 was diagnosed with acute lead poisoning that required many months of treatment (he returned to the wild in September 2006). Condor 119 was unable to successfully incubate and hatch their egg on her own. Arizona Game and Fish Department, "California Condor AZ Updates—April 12, 2006 and May 15, 2006," author's collections.

6. Heartbreakingly, Condor 305 died in March 2005. The recorded cause of death was poor body condition.

7. The Peregrine Fund, "Winter Results for Arizona–Utah Condor Program: Preventable Deaths Remain Focus of Recovery Effort," 2013, https://peregrinefund.org/news-release/winter-results-arizona-utah-condor-program-preventable-deaths-remain-focus-recovery; National Park Service, "SW CA Condor Update— 2013-07 (July)," 2013, https://www.nps.gov/articles/sw-ca-condor-2013-07.htm.

8. Rideout et al, "Patterns of Mortality in Free-Ranging California Condors."

9. Of 352 condors released between 1992 and 2009, 135 (or 38 percent) died; Rideout et al, "Patterns of Mortality in Free-Ranging California Condors."

10. As of December 31, 2023, there were 561 California Condors in the world, with 217 in captivity and 344 in the wild. There were 206 condors in the wild in California, 90 in Arizona and Utah, and 48 in Baja California, Mexico.

11. Mee et al., "Junk Ingestion and Nestling Mortality."

12. Rideout et al., "Patterns of Mortality in Free-Ranging California Condors."

13. Houston, Mee, and McGrady, "Why Do Condors and Vultures Eat Junk?"; Mee, Hamber, and Sinclair, "Low Nest Success in a Reintroduced Population"; Mee et al., "Junk Ingestion and Nestling Mortality," *Bird Conservation International*.

14. Houston, Mee, and McGrady, "Why Do Condors and Vultures Eat Junk?"; Rideout et al., "Patterns of Mortality in Free-Ranging California Condors."

15. Mee, Hamber, and Sinclair, "Low Nest Success in a Reintroduced Population."

16. Mee, Hamber, and Sinclair, "Low Nest Success in a Reintroduced Population"; Mee et al., "Junk Ingestion and Nestling Mortality." Also, S. Osborn and A. Mee, unpublished data and conversations in 2003 and 2004.

17. M. R. Gregory, "Environmental Implications of Plastic Debris in Marine Settings—Entanglement, Ingestion, Smothering, Hangers-On, Hitch-Hiking and Alien Invasions," *Philosophical Transactions of the Royal Society B: Biological Sciences* 364, no. 1526 (2009): 2013–25, https://doi.org/10.1098/rstb.2008.0265; M. Eriksen et al., "Plastic Pollution in the World's Oceans: More than 5 Trillion Plastic Pieces Weighing over 250,000 Tons Afloat at Sea," *PLoS ONE* 9, no. 12 (2014): e111913, https://doi.org/10.1371/journal.pone.0111913; L. Roman et al., "A Quantitative Analysis Linking Seabird Mortality and Marine Debris Ingestion," *Scientific Reports* 9, no. 1 (2019): 1–7, https://doi.org/10.1038/s41598-018-36585-9.

18. Roman et al., "A Quantitative Analysis Linking Seabird Mortality."

19. E. R. Zylstra, "Accumulation of Wind-Dispersed Trash in Desert Environments," *Journal of Arid Environments* 89 (2013): 13–15, https://doi.org/10.1016/j.jaridenv.2012.10.004.

20. K. Ridler, "California Condors Reach Key Survival Milestone in the Wild," Associated Press, February 23, 2016, https://www.cbsnews.com/sanfrancisco/news/california-condors-reach-key-survival-milestone-in-the-wild.

21. US Fish and Wildlife Service, "Home for the Holidays: California Condor AC-4 Returns to the Wild," press release, December 30, 2015, https://www.fws.gov/press-release/2015-12/home-holidays-california-condor-ac-4-returns-wild-after-30-years; B. Baker, "Wild California Condor Chick Fledges in Santa Barbara County for First Time Since 1982," Los Padres Forestwatch, December 10, 2018. AC-4's captive-reared offspring included Molloko, who hatched from the first egg laid in captivity; Arizona's Condor 116, who died of lead poisoning on a Colorado River beach; and Condor 158, who underwent surgery that saved him from lead poisoning in 2000.

22. US Fish and Wildlife Service, "Establishment of a Nonessential Experimental Population of the California Condor in the Pacific Northwest," *Federal Register* 84, no. 66 (2019): 13587–603, https://www.govinfo.gov/content/pkg/FR-2019-04-05/pdf/2019-06293.pdf.

23. Finkelstein et al., "Lead Poisoning and the Deceptive Recovery."

24. V. J. Meretsky et al., "Demography of the California Condor: Implications for Reestablishment," *Conservation Biology* 14, no. 4 (2000): 957–67, https://doi.org/10.1046/j.1523-1739.2000.99113.x; Walters et al., "Status of the California Condor."

25. J. Kim, "Record Number of Endangered California Condors Treated for Lead Poisoning," 89.3 KPCC: Environment and Science, (2013), https://www.scpr.org/news/2013/11/08/40258/record-number-of-endangered-condors-treated-for-le.

26. Finkelstein et al., "Lead Poisoning and the Deceptive Recovery," 11453.

27. N. Oreskes and E. M. Conway, *Merchants of Doubt: How a Handful of Scientists Obscured the Truth on Issues from Tobacco Smoke to Global Warming* (Bloomsbury Publishing, 2011). The politicization of measures to control Covid-19 is just the latest such denialism regarding environmental and public-health related science.

28. Arnemo et al., "Health and Environmental Risks from Lead-Based Ammunition."

29. Hunt for Truth recently revised its website. See: https://www.huntfortruth.org. Its arguments and original information remain, but in a slightly different format. (Digital screenshots of the original pages with these quotations are available in the author's collections). Also see: D. Starin, "Eat Lead! Condor Survival versus National Rifle Association," *Ecologist*, July 7, 2014, https://theecologist.org/2014/jul/07/eat-lead-condor-survival-versus-national-rifle-association; *Ecologist* editors, "Hunting for Truth: We Respond to the NRA's Retraction Demand," 2014, https://theecologist.org/2014/jul/11/hunting-truth-we-respond-nras-retraction-demand; L. Peeples, "In the Battle Over Lead Ammunition, Science Collides with Culture," *Undark Magazine*, 2017, https://undark.org/2017/01/30/lead-ammunition-bullets-hunting-copper.

30. The list of scientific studies that "bolstered" Hunt for Truth's arguments on their original website is available at: https://web.archive.org/web/20170319070811/http://www.huntfortruth.org/science/scientific-opinions/papersstudies.

31. Personal communication with Jeff Miller, Center for Biological Diversity, September 6, 2021, and August 9, 2023.

32. Rather than examining the evidence the petitioners provided regarding the dangers posed by lead ammunition, the EPA denied the 2010 petition, which requested a ban on lead ammunition, claiming that an exclusion clause prevented the agency from addressing lead ammunition under the TSCA. In response to the 2012 petition, which asked the agency to regulate lead ammunition, the EPA claimed that they didn't have the authority to do so. Center for Biological Diversity et al., "Petition to the Environmental Protection Agency to Regulate Lead Bullets and Shot under the Toxic Substances Control Act" (2012), https://www.epa.gov/sites/production/files/2015-10/documents/tsca_ammo_petition_3-13-12.pdf; R. Hawkins, "EPA Shoots Down Lead Shot Regulation: Lead Ammo's Unreasonable Risk to Human Health and the Environment, and the Special Situation of the California Condor," *Golden Gate University Environmental Law Journal* 5, no. 2 (2012): 533–66, https://digitalcommons.law.ggu.edu/cgi/viewcontent.cgi?article=1086&context=gguelj; 115th Congress, 2017–2018. H.R. 3668 Sportsmen's Heritage and Recreational Enhancement Act (or SHARE Act), Title I, Section 103: https://www.congress.gov/bill/115th-congress/house-bill/3668/text; A. McGlashen, "If Lead Ammunition Is Bad for People and the Environment, Why Do We Still Use It?" *Ensia*, April 14, 2018, https://medium.com/ensia/if-lead-ammunition-is-bad-for-people-and-the-environment-why-do-we-still-use-it-9d3baefcabb9.

33. California Department of Fish and Game, "Methods Authorized for Taking Big Game, Section 353, Title 14, CCR," Office of Administrative Law, Sacramento, California, 2008. Prior to these bans, gun advocates claimed in expert testimony to the California Fish and Game Commission that non-lead .22 caliber rimfire bullets were impossible to produce. Four months after the lead bans, these bullets were made commercially available; Center for Biological Diversity et al., "Petition to the Environmental Protection Agency to Regulate Lead Bullets and Shot."

34. Finkelstein et al., "Lead Poisoning and the Deceptive Recovery."

35. See "Summary of Current Lead Ammunition Regulations in California," Hunt for Truth Association, https://web.archive.org/web/20161221183049/http://www.huntfortruth.org/legal/state-regulations/california/. Hunt for Truth also pointed to one incident in which two condors at Pinnacles National Park ingested lead paint chips that were flaking off a fire lookout tower as evidence that condors were getting lead from sources other than lead ammunition: "California Condor," Hunt for Truth Association, https://www.huntfortruth.org/california-condor. Condor biologists recognized and acknowledged this atypical lead-paint ingestion incident. In this case, the blood lead isotope ratio of the poisoned birds matched the isotope ratio of the lead-based paint collected from the fire tower and did not match lead ammunition; Finkelstein et al., "Lead Poisoning and the Deceptive Recovery."

36. US Fish and Wildlife Service, "California Condor (*Gymnogyps californianus*) 5-year Review: Summary and Evaluation" (2013), https://ecos.fws.gov/docs/tess/species_nonpublish/2041.pdf.

37. Finkelstein et al., "Lead Poisoning and the Deceptive Recovery."

38. Bellinger et al., "Health Risks from Lead-Based Ammunition in the Environment."

39. No such declines were seen after the 2007 lead ban. Personal communication with Jeff Miller, Center for Biological Diversity, March 18, 2020.

40. Initiated by the Arizona Game and Fish Department in 2003, these efforts were expanded to southern Utah by the Utah Division of Wildlife Resources, in 2010.

41. Arizona Game and Fish Department (website), California Condor Recovery—Lead Reduction by Year Graph and Chart (Percentages are an average calculated from 2015–2019), https://www.azgfd.com/wildlife/speciesofgreatestconservneed/raptor-management/california-condor-recovery.

42. Walters et al., "Status of the California Condor"; Finkelstein et al., "Lead Poisoning and the Deceptive Recovery." According to Finkelstein et al. (2012), if restrictions were in place that resulted in only 1 percent of carcasses containing lead, the annual probability that a condor would feed on one or more contaminated carcasses would only be reduced to 31–53 percent.

43. Ventana Wildlife Society, Condor Cams: https://www.ventanaws.org/condor_cam.html.

44. Personal communication with Joe Burnett, Ventana Wildlife Society, September 10, 2021. This webcam can be viewed only when condors are nesting in the redwood cavity, but Ventana also has condor webcams at their release sites (Big Sur and San Simeon) that can be viewed year-round.

Epilogue

1. Thomas Lovejoy appears to have made this statement during National Geographic's launch of its 2018 "Year of the Bird"—an effort to celebrate and draw attention to the world of birds.

2. M. L. Morrison, "Bird Populations as Indicators of Environmental Change," *Current Ornithology*, ed. R.F. Johnston, vol. 3, chapter 10 (Boston: Springer, 1986): 429–51; P. Doherty, *Their Fate is Our Fate: How Birds Foretell Threats to Our Health and Our World* (The Experiment, 2013).

3. Doherty, *Their Fate is Our Fate.*

4. Meretsky et al., "Demography of the California Condor"; Finkelstein et al., "Lead Poisoning and the Deceptive Recovery."

5. R. Nixon, Statement upon signing the Endangered Species Act, 16 U.S.C. § 1531 (December 28, 1973), https://www.endangered.org/campaigns/wild-success-endangered-species-act-at-40/richard-nixon.

6. Center for Biological Diversity, "A Wild Success—A Systematic Review of Bird Recovery Under the Endangered Species Act" (2016), https://www.esasuccess.org/pdfs/WildSuccess.pdf.

7. J. T. Bruskotter et al., "Support for the U.S. Endangered Species Act over Time and Space: Controversial Species Do Not Weaken Public Support for Protective Legislation," *Conservation Letters* 11, no. 6 (2018): e12595, https://doi.org/10.1111/conl.12595; Ohio State University, "Vast Majority of Americans Support Endangered Species Act Despite Increasing Efforts to Curtail It: Political and Business Interests Don't Appear to Align with Public's View," *ScienceDaily*, 2018, https://www.sciencedaily.com/releases/2018/07/180719121800.htm.

8. Center for Biological Diversity, "A Wild Success—A Systematic Review of Bird Recovery Under the Endangered Species Act."

9. K. V. Rosenberg et al., "Decline of the North American Avifauna," *Science* 366, no. 6461 (2019): 120–24, https://doi.org/10.1126/science.aaw1313.

10. Rosenberg et al., "Decline of the North American Avifauna."

11. S. A. H. Osborn, "Itinerant Breeding and Mate Switching by an American Dipper," *The Wilson Bulletin* 112, no. 4 (2000): 539–41, https://www.jstor.org/stable/4164277.

Index

INDEX

About the Author

Lisa Knitzsch

Sophie A. H. Osborn is an award-winning environmental author and wildlife biologist whose work has included the study and conservation of more than a dozen bird species in the Americas and small mammal research in the backcountry of Grand Canyon National Park. She has contributed to reintroduction efforts for several endangered birds and served as the field manager for the California Condor Recovery Program in Arizona for nearly four years. Her books include *Condors in Canyon Country*, which won the 2007 National Outdoor Book Award for Nature and the Environment. Her writings have been published in *BirdWatching*, *Wyoming Wildlife*, and *Sojourns* magazines.